LINEAR ALGEBRA

WITH COMPUTER APPLICATIONS

ONALD I. ROTHENBERG

Linear Algebra
with Computer Applications

MORE THAN 80 SELF-TEACHING GUIDES TEACH PRACTICAL SKILLS FROM ACCOUNTING TO ASTRONOMY, MANAGEMENT TO MICROCOMPUTERS. LOOK FOR THEM ALL AT YOUR FAVORITE BOOKSTORE.

STGs on mathematics:

Background Math for a Computer World, 2nd ed., Ashley
Business Mathematics, Locke
Business Statistics, 2nd ed., Koosis
Finite Mathematics, Rothenberg
Geometry and Trigonometry for Calculus, Selby
Linear Algebra with Computer Applications, Rothenberg
Math Shortcuts, Locke
Math Skills for the Sciences, Pearson
Practical Algebra, Selby
Quick Algebra Review, Selby
Quick Arithmetic, Carman & Carman
Quick Calculus, Kleppner & Ramsey
Statistics, 2nd ed., Koosis
Thinking Metric, 2nd ed., Gilbert & Gilbert
Using Graphs and Tables, Selby

Linear Algebra
with Computer Applications

RONALD I. ROTHENBERG

Associate Professor of Mathematics
Queens College
City University of New York

John Wiley & Sons, Inc.
New York • Chichester • Brisbane • Toronto • Singapore

Library of Congress Cataloging in Publication Data

Rothenberg, Ronald I., 1936–
   Linear algebra with computer applications.

   (Wiley self-teaching guide)
   Includes index.
   1. Algebras, Linear.   2. Algebras, Linear—Data
processing.   I. Title.   II. Series.
QA184.R68   1983   512′.5   82–24806
ISBN: 0–471–09652–0 (pbk.)

Printed in the United States of America

83 84 10 9 8 7 6 5 4 3

Preface

Linear algebra has become a standard part of the undergraduate mathematical training of students in such diverse fields as physics, chemistry, biology, engineering (all types), business, economics, operations research, sociology, psychology, and computer science. This book presents an elementary treatment of linear algebra, and the companion subject, matrix algebra, with emphasis on computational techniques, manipulation of symbols, and the understanding of basic concepts (often through calculations, verifications, and proofs for special cases). The book contains several applications in the economics, business, and social science areas and a treatment of computer and programmable calculator techniques for doing matrix calculations. For the latter topics, listings of programs and printouts pertaining to computer and calculator runs are presented.

This book, with its self-instructional format, is designed for study on one's own or as a supplement to current linear (matrix) algebra textbooks.

A one-year course in eleventh-year high school mathematics (intermediate algebra with some trigonometry) is assumed as a prerequisite. Calculus is not assumed as a prerequisite.

The level in this book is suitable for students in their freshman or sophomore year of college. The book should also be useful to people at different levels (including graduate school) because of the emphasis on computational techniques and the plentiful supply of numerical examples.

People with personal computers should also find the book useful. In addition to the computer topics referred to previously, a discussion of available software for doing matrix calculations on personal computers is found in Section B of Chapter 9.

The book starts out with a chapter on matrix algebra (Chapter 1) and continues with a chapter dealing with linear systems, with a major emphasis on the Gauss–Jordan elimination method (Chapter 2). These two chapters provide an adequate preparation for someone who is merely interested in rudimentary matrix manipulations.

The next two chapters (Chapter 3 on determinants and Chapter 4 on geometrical properties) are both very practical chapters that deal with some fundamental ideas. They also provide important background material for the following three chapters.

Taken collectively, Chapters 5, 6, and 7 provide the material for the heart of an elementary course in linear algebra. The headings of these chapters are, respectively, *Vector Spaces, Linear Transformations,* and *Eigenvalues and Eigenvectors.* In the interest of both simplicity and practicality, most of the examples in these chapters deal with matrices.

Chapter 8 is concerned with some interesting applications, namely, Markov processes (with applications to market research and business) and the Leslie model of population growth. In Chapter 9 the emphasis is on numerical methods and on using computers and programmable calculators to do matrix calculations. Programs written in FORTRAN and BASIC are included.

In this book, as in other John Wiley Self-Teaching Guides, statements of objectives are given at the beginning of each chapter. In addition, each chapter is subdivided into many segments, called frames, which are numbered consecutively in each chapter. Each frame presents some new information and concludes with a question to be answered or a problem to be solved (sometimes just one part or several parts of a problem). The question or problem is separated from the answer or solution by a dashed line (----). Ideally, the reader should determine the answer/solution without looking below the dashed line.

Each chapter ends with a self-test, followed by a list of correct answers to the test questions. Accompanying each answer is a list of frames where topics pertaining to the question are covered.

ACKNOWLEDGMENTS

I am pleased to acknowledge the comments and suggestions of student readers Deborah M. Chaskin, Beradino D'Aquila, and Gary Katz, each of whom attended and graduated from Queens College. My thanks go to the following editorial and production people connected with the John Wiley & Sons Self-Teaching Guides Division: Catherine Dillon, Dianne Littwin, Alicia Conklin, and Maria Colligan. Also, Rachel Hockett of Cobb/Dunlop Publisher Services was very helpful.

My appreciation is extended to the two subject reviewers of the manuscript for their suggestions and encouraging words: Prof. David Carlson of the Oregon State Univ. Mathematics Dept., and Prof. Joan Dykes of the Northern Virginia Community College Mathematics Dept. Much of the computer work connected with this book was made easier because of the helpful nature of John Leong and Alan Boord, both programming analysts, and the sound administrative ability of Dr. Seymour Goodman. All were connected with the Queens College Computing Center.

Inspiration for the current project came from a fondness for linear and matrix algebra, and a belief that I might be able to write a book on this subject matter which students would find helpful.

Finally, I would like to indicate my appreciation for the understanding and encouragement of my wife, Olga, who recently received her Ph.D. degree from the CUNY Graduate Center.

DEDICATION

This book is dedicated to my former Ph.D. advisor, Professor Joe Mauk Smith (Dept. of Chemical Engineering, University of California at Davis), a man who helped me both academically and personally.

Contents

Linear Algebra
with Computer Applications

CHAPTER ONE
Elementary Matrix Algebra

Matrix algebra is a branch of *linear algebra*. Matrices (plural of matrix) are useful in studying systems of linear equations and are used for formulation and analysis of many subjects that arise in economics, business, the natural sciences, the social sciences, and computer science. In this chapter we shall study some of the fundamental rules, properties, and manipulations of matrix algebra.

OBJECTIVES

When you complete this chapter you should be able to

- Identify the entries of a matrix and know how to perform the operations of matrix addition and multiplication of a matrix by a number.
- Perform operations with column and row vectors.
- Multiply matrices.
- Apply matrix methods to practical situations, such as classification of industrial and business data.
- Find the transpose of a matrix and understand its properties.

A. INTRODUCTION TO MATRICES

1. A *matrix* is a rectangular array of numbers.* The following are matrices:

(a) $\begin{bmatrix} 3 & 2 & 1 \\ 4 & 6 & 5 \end{bmatrix}$,   (b) $\begin{bmatrix} 3 \\ -4 \end{bmatrix}$,   (c) $[7, \ 6, \ 5]$,   (d) $\begin{bmatrix} 6 & 4 & 5 \\ 2 & 1 & -3 \\ 8 & -2 & 7 \end{bmatrix}$

---

* The "numbers" referred to here are to be considered as real numbers unless otherwise stated. Often, when referring to real numbers, we shall omit the word *real*. Actually, most of the linear algebra properties considered in this book apply equally well to complex numbers. A complex number is a number of the form $c = a + ib$, where a and b are real numbers and where $i = \sqrt{-1}$.

The decision to limit most of the development in this book to the real number case is based on the fact that in the analysis of most practical situations through linear algebra, it will be sufficient to use real numbers.

The numbers in the matrix are called *entries* or *elements*. A matrix is said to have a *size* $m \times n$ (or m by n), where m denotes the number of rows (horizontal lines) and n denotes the number of columns (vertical lines).

For example, the size for the matrix cited as (a) above is 2×3. What are the sizes for the other three preceding matrices?

— — — — — — — — — —

   (b) 2×1,   (c) 1×3,   (d) 3×3.

2. We will usually denote matrices by uppercase letters (A, B, M, etc.). Lowercase letters, such as a, b, c, r, q, usually will denote numbers; in discussions on matrices it is common to use the word *scalar* for a number. In a general discussion of a matrix A, we refer to the entry in row i and column j as a_{ij}, which is read "a sub i,j." Thus, the general 2×3 matrix A is written

$$A = \begin{bmatrix} a_{11} & a_{12} & a_{13} \\ a_{21} & a_{22} & a_{23} \end{bmatrix}$$

For example, a_{12} is the entry in row 1 and column 2 and a_{23} is the entry in row 2 and column 3. Here, a_{23} is read "a sub 2, 3."

Note that the row label increases in the "top to bottom direction" (row 1, row 2, etc.) while the column label increases in the "left to right" direction (column 1, column 2, etc.). Thus, row 2 is the second horizontal line from the top and column 3 is the third vertical line from the left.

Suppose the matrix cited as (d) in frame 1 is called B. Identify all the entries of B by using the b_{ij} notation.

— — — — — — — — — —

   $b_{11} = 6$, $b_{12} = 4$, $b_{13} = 5$, $b_{21} = 2$, $b_{22} = 1$, $b_{23} = -3$, $b_{31} = 8$, $b_{32} = -2$, and $b_{33} = 7$. For example, $b_{31} = 8$ is the entry in row 3 and column 1.

3. Matrices arise in a natural way when one tries to record information that involves a two-way classification of data. Suppose an appliance store stocks a particular brand of television set in three sizes—small (12 inches), medium (17 inches), and large (21 inches)—and in the two types—black & white and color. The inventory (list of goods available in stock) at the end of May could be recorded in matrix form as follows, where the matrix is arbitrarily denoted as S:

$$S = \begin{array}{c} \\ \\ \end{array} \begin{matrix} \text{Sm.} & \text{Med.} & \text{Lg.} \\ \begin{bmatrix} 10 & 18 & 14 \\ 17 & 15 & 11 \end{bmatrix} & & \end{matrix} \begin{array}{l} \text{B \& W} \\ \text{Color} \end{array}$$

For example, $s_{11} = 10$ indicates the store has available 10 small, black & white sets in stock, while s_{23} indicates that 11 large, color sets are in stock.

Interpret the entries s_{13} and s_{22}.

— — — — — — — — — —

The s_{13} entry indicates 14 large, black & white sets in stock while the s_{22} entry indicates 15 medium, color sets in stock.

4. If a matrix has m rows and n columns, we say the size of the matrix is $m \times n$. Thus, matrices A and B have the same size if they have the same number of rows, and columns, respectively. Two matrices A and B are said to be *equal*, indicated by $A = B$, if they have the same size and if their corresponding entries are equal.

For example,

if $\quad A = \begin{bmatrix} 7 & 2 & 5 \\ 6 & 4 & -9 \end{bmatrix} \quad$ and $\quad B = \begin{bmatrix} b_{11} & 2 & 5 \\ 6 & b_{22} & -9 \end{bmatrix},$

then for A and B to be equal, we must have $b_{11} = 7$ and $b_{22} = 4$.

Consider the three matrices C, D, and E below. Indicate whether or not equality exists between any pair. Give reasons for your answers.

$C = \begin{bmatrix} 5 & -7 \\ 2 & 4 \end{bmatrix}, \quad D = \begin{bmatrix} 5 & 2 \\ -7 & 4 \end{bmatrix}, \quad E = \begin{bmatrix} 5 & -7 & 0 \\ 2 & 4 & 0 \end{bmatrix}$

— — — — — — — — — —

$C \neq D$ ($\neq$ means *not* equal), since corresponding entries are not equal. $C \neq E$, since C and E have different sizes; likewise, $D \neq E$.

5. A matrix with one row is called a *row vector*. Thus, a row vector with k entries is the same as a $1 \times k$ matrix. A matrix with one column is called a *column vector*. Thus, a column vector with k entries is the same as a $k \times 1$ matrix.

We shall often denote vectors by lowercase letters with boldface type, as in **x** or **b**. We read **x** as "vector x." Other symbols used in practice are a letter with an arrow above, as in $\vec{c}$, a letter with a caret above, as in $\hat{c}$, or a letter with a bar above, as in $\bar{c}$.

The entries of vectors are often referred to as components. In addition, a single subscript is usually used. Thus, for a k-component row vector **a**, we have the symbolism

$$\mathbf{a} = [a_1, a_2, \ldots, a_k],$$

where the components are separated by commas.

The general k-component column vector **b** is denoted by

$$\mathbf{b} = \begin{bmatrix} b_1 \\ b_2 \\ \vdots \\ b_k \end{bmatrix} \qquad \text{where the components are } b_1, b_2, \text{ etc.}$$

Here, as for row vectors, we use a single subscript for components.
Which of the matrices of frame 1 are vectors?

— — — — — — — — — — —

$\begin{bmatrix} 3 \\ -4 \end{bmatrix}$ is a two-component column vector, and $[7, 6, 5]$ is a three-component
row vector. If the latter is called **c**, then $c_1 = 7$, $c_2 = 6$, and $c_3 = 5$.

6. A zero vector is a vector for which every component is zero. We will denote
a zero vector by the boldface zero symbol, **0**. Thus,

$$\mathbf{0} = \begin{bmatrix} 0 \\ 0 \\ \vdots \\ 0 \end{bmatrix} \qquad \text{and} \qquad \mathbf{0} = [0, 0, \dots, 0]$$

are symbols for a zero column vector and row vector, respectively. The number
of components, and whether a zero vector is a row or column vector, will be
clear from the nature of the discussion. A three-component zero column vector
is

$$\mathbf{0} = \begin{bmatrix} 0 \\ 0 \\ 0 \end{bmatrix}$$

Now we consider the definition for matrix addition (or sum).

Definition 1.1 (Matrix Addition):

Given two matrices A and B of the same size. The sum of A and B, written
as $A + B$, is the matrix obtained by adding corresponding entries of A and
B. Addition (i.e., sum) is *not defined* for matrices with different sizes.

Consider the 2×3 matrices A and B:

$$A = \begin{bmatrix} 2 & 6 & 1 \\ 4 & 3 & 8 \end{bmatrix}, \qquad B = \begin{bmatrix} 6 & -3 & 4 \\ 2 & 5 & 1 \end{bmatrix}$$

The row 1, column 1, entry of $A + B$ is $2 + 6 = 8$. Determine the entire matrix $A + B$.

- - - - - - - - -

$$A + B = \begin{bmatrix} 2 + 6 & 6 + (-3) & 1 + 4 \\ 4 + 2 & 3 + 5 & 8 + 1 \end{bmatrix} = \begin{bmatrix} 8 & 3 & 5 \\ 6 & 8 & 9 \end{bmatrix}$$

Here the size of $A + B$ is also 2×3.

7. In general, if A and B are both $m \times n$ matrices, then the sum matrix $A + B$ will also be $m \times n$.

Refer to the appliance store situation of frame 3, which presents the matrix S, for television sets in stock at the end of May. Suppose the store gets a shipment of television sets on June 1, as given by the matrix J:

$$J = \begin{bmatrix} \text{Sm.} & \text{Med.} & \text{Lg.} \\ 6 & 10 & 8 \\ 8 & 12 & 7 \end{bmatrix} \begin{matrix} \text{B \& W} \\ \text{Color} \end{matrix}$$

The matrix representing the total amounts of sets in stock after the shipment arrives is given by $S + J$, where

$$S + J = \begin{bmatrix} \text{Sm.} & \text{Med.} & \text{Lg.} \\ 16 & 28 & 22 \\ 25 & 27 & 18 \end{bmatrix} \begin{matrix} \text{B \& W} \\ \text{Color} \end{matrix}$$

Thus, for example, after the shipment arrives, there are 27 medium-sized, color T.V.'s in stock.

Certain properties for matrix addition follow directly from properties for addition of numbers. These are summarized in the following theorem.

Theorem 1.1

For matrices A, B, and C, all of the same size,

(a) $A + B = B + A$; i.e., addition of matrices is *commutative*.
(b) $(A + B) + C = A + (B + C)$; i.e., addition of matrices is *associative*.

Because of the equality in (b), we define the triple matrix sum $A + B + C$ as equal to the common value given in (b).

The proof of this theorem depends merely on the commutative and associative properties for addition of real numbers and the definition of matrix addition.

The parentheses in $(A + B) + C$ means that *first* A and B are added, and then the resulting matrix is added to C. Suppose

$$A = \begin{bmatrix} 4 & -3 \\ 6 & 5 \end{bmatrix}, \qquad B = \begin{bmatrix} 8 & 2 \\ -3 & 7 \end{bmatrix}, \qquad C = \begin{bmatrix} 1 & 0 \\ 9 & 2 \end{bmatrix}$$

Then,

$$(A + B) + C = \left(\begin{bmatrix} 4 & -3 \\ 6 & 5 \end{bmatrix} + \begin{bmatrix} 8 & 2 \\ -3 & 7 \end{bmatrix} \right) + \begin{bmatrix} 1 & 0 \\ 9 & 2 \end{bmatrix}$$

$$= \begin{bmatrix} 12 & -1 \\ 3 & 12 \end{bmatrix} + \begin{bmatrix} 1 & 0 \\ 9 & 2 \end{bmatrix} = \begin{bmatrix} 13 & -1 \\ 12 & 14 \end{bmatrix}$$

Show that computation of $A + (B + C)$ results in the same final matrix. Note that the parentheses here means that *first* B and C are added and then the resulting matrix is added to A.

— — — — — — — — — —

$$A + (B + C) = \begin{bmatrix} 4 & -3 \\ 6 & 5 \end{bmatrix} + \left(\begin{bmatrix} 8 & 2 \\ -3 & 7 \end{bmatrix} + \begin{bmatrix} 1 & 0 \\ 9 & 2 \end{bmatrix} \right)$$

$$= \begin{bmatrix} 4 & -3 \\ 6 & 5 \end{bmatrix} + \begin{bmatrix} 9 & 2 \\ 6 & 9 \end{bmatrix} = \begin{bmatrix} 13 & -1 \\ 12 & 14 \end{bmatrix}$$

8.   A matrix for which all the entries are zero is called a zero matrix and is denoted by **O**. The following are zero matrices:

$$\begin{bmatrix} 0 & 0 \\ 0 & 0 \end{bmatrix}, \qquad \begin{bmatrix} 0 & 0 \\ 0 & 0 \\ 0 & 0 \end{bmatrix}, \qquad \begin{bmatrix} 0 & 0 & 0 & 0 \\ 0 & 0 & 0 & 0 \end{bmatrix}, \qquad [0]$$

The sizes are 2×2, 3×2, 2×4, and 1×1, respectively. If A is any matrix and **O** is a zero matrix of the same size, then

$$A + \mathbf{O} = \mathbf{O} + A = A$$

Because of this last property, the matrix **O** can be thought of as the additive identity for matrix addition. Note that the number 0 is the additive identity for ordinary addition of numbers since $k + 0 = 0 + k = k$ for any number k.

Definition 1.2 (Scalar Multiplication):

The *product of a scalar* (number) k *and a matrix* A, written as kA, is the matrix obtained by multiplying each entry of A by k.

Thus, if A is a general 2×3 matrix, we have

$$kA = k\begin{bmatrix} a_{11} & a_{12} & a_{13} \\ a_{21} & a_{22} & a_{23} \end{bmatrix} = \begin{bmatrix} ka_{11} & ka_{12} & ka_{13} \\ ka_{21} & ka_{22} & ka_{23} \end{bmatrix}$$

Note that A and kA will have the same size.

The *negative of a matrix* A, denoted by $-A$, is defined to be the scalar -1 times A. That is,

$$-A = (-1)A$$

Suppose $A = \begin{bmatrix} 6 & 5 & 4 \\ 1 & -2 & 3 \end{bmatrix}$. Determine (a) $3A$, and (b) $-A$.

- - - - - - - - -

(a) $3A = 3\begin{bmatrix} 6 & 5 & 4 \\ 1 & -2 & 3 \end{bmatrix} = \begin{bmatrix} 3 \cdot 6 & 3 \cdot 5 & 3 \cdot 4 \\ 3 \cdot 1 & 3 \cdot (-2) & 3 \cdot 3 \end{bmatrix} = \begin{bmatrix} 18 & 15 & 12 \\ 3 & -6 & 9 \end{bmatrix}$.

(b) $-A = (-1)\begin{bmatrix} 6 & 5 & 4 \\ 1 & -2 & 3 \end{bmatrix} = \begin{bmatrix} -6 & -5 & -4 \\ -1 & 2 & -3 \end{bmatrix}$.

9.   Next, we consider the definition for matrix subtraction.

Definition 1.3 (Matrix Subtraction):

The *difference* between matrices A and B of the same size, denoted by $A - B$, and referred to as A *minus* B, is defined by $A - B = A + (-B)$. That is, $A - B$ is equivalent to adding A and $(-B)$.

Suppose that $A = \begin{bmatrix} 6 & 5 & 4 \\ 1 & -2 & 3 \end{bmatrix}$ and $B = \begin{bmatrix} 3 & 8 & 5 \\ 2 & 1 & 1 \end{bmatrix}$. Determine (a) $-B$,

(b) $A - B$, and (c) $A - A$.

- - - - - - - - -

(a) $-B = (-1)\begin{bmatrix} 3 & 8 & 5 \\ 2 & 1 & 1 \end{bmatrix} = \begin{bmatrix} -3 & -8 & -5 \\ -2 & -1 & -1 \end{bmatrix}$.

(b) $A - B = A + (-B) = \begin{bmatrix} 6 & 5 & 4 \\ 1 & -2 & 3 \end{bmatrix} + \begin{bmatrix} -3 & -8 & -5 \\ -2 & -1 & -1 \end{bmatrix}$

$\qquad = \begin{bmatrix} 3 & -3 & -1 \\ -1 & -3 & 2 \end{bmatrix}$.

(c) $A - A = \begin{bmatrix} 6 + (-6) & 5 + (-5) & 4 + (-4) \\ 1 + (-1) & (-2) + 2 & 3 + (-3) \end{bmatrix} = \begin{bmatrix} 0 & 0 & 0 \\ 0 & 0 & 0 \end{bmatrix} = \mathbf{O}$.

10. Note that for any matrix A, we have that $A - A$ is the zero matrix of the same size. Observe that a similar result holds with respect to numbers, namely, that $k - k = 0$, for any number k. Also, observe that the difference $A - B$ can be calculated by subtracting each entry of B from the corresponding entry of A. Thus, for the previous example,

$$A - B = \begin{bmatrix} 6 & 5 & 4 \\ 1 & -2 & 3 \end{bmatrix} - \begin{bmatrix} 3 & 8 & 5 \\ 2 & 1 & 1 \end{bmatrix} = \begin{bmatrix} 3 & -3 & -1 \\ -1 & -3 & 2 \end{bmatrix}$$

Several properties involving multiplication of a matrix by a scalar are given in the next theorem.

Theorem 1.2

For matrices A and B of the same size, and for any two scalars h and k, the following rules hold:

(a) $(h + k)A = hA + kA$,
(b) $h(A + B) = hA + hB$,
(c) $hk(A) = h(kA)$.

Verify property (b) with respect to the matrices A and B of frame 9 if $h = 3$. The left side of (b) indicates that you should first form the matrix sum $A + B$ and then multiply by 3, while the right side of (b) indicates that you should first form the matrices $3A$ and $3B$ and then add $3A$ and $3B$ together.

— — — — — — — — —

Left side: $3(A + B) = 3\left(\begin{bmatrix} 6 & 5 & 4 \\ 1 & -2 & 3 \end{bmatrix} + \begin{bmatrix} 3 & 8 & 5 \\ 2 & 1 & 1 \end{bmatrix} \right)$

$$= 3 \begin{bmatrix} 9 & 13 & 9 \\ 3 & -1 & 4 \end{bmatrix} = \begin{bmatrix} 27 & 39 & 27 \\ 9 & -3 & 12 \end{bmatrix}$$

Right side: $3A + 3B = 3 \begin{bmatrix} 6 & 5 & 4 \\ 1 & -2 & 3 \end{bmatrix} + 3 \begin{bmatrix} 3 & 8 & 5 \\ 2 & 1 & 1 \end{bmatrix}$

$$= \begin{bmatrix} 18 & 15 & 12 \\ 3 & -6 & 9 \end{bmatrix} + \begin{bmatrix} 9 & 24 & 15 \\ 6 & 3 & 3 \end{bmatrix}$$

$$= \begin{bmatrix} 27 & 39 & 27 \\ 9 & -3 & 12 \end{bmatrix}$$

We end up with the same matrix, thus verifying property (b) of Theorem 1.2 for the particular matrices and the scalar 3.

ELEMENTARY MATRIX ALGEBRA 9

11. We don't want to forget that one of the reasons we study matrices is because of their practical applications. We have already focused on two practical examples dealing with using matrices for a two-way classification of data (frames 3 and 7). Now we will make use of matrices S and J listed in frames 3 and 7 in another practical application. These matrices represent the inventory at the end of May and the shipment arriving in June, respectively. Suppose we let the matrix D represent the television sets sold during the month of June. For example, suppose

$$D = \begin{bmatrix} 9 & 19 & 17 \\ 16 & 21 & 16 \end{bmatrix} \begin{matrix} \text{B \& W} \\ \text{Color} \end{matrix}$$

with columns Sm. Med. Lg.

This indicates that $9 (= d_{11})$ small, black & white and $21 (= d_{22})$ medium-sized, color T.V.'s are sold during June, for example. It should be clear that the matrix given by $S + J - D$ represents the inventory of T.V. sets at the *end* of June. (Recall that inventory refers to goods in stock.) Determine the matrix $S + J - D$ and interpret several entries in the matrix. (Note that one way of computing the entries of $S + J - D$ is by first adding corresponding entries of S and J and then subtracting the corresponding entry of D.)

— — — — — — — — — —

$$S + J - D = \begin{bmatrix} 10 & 18 & 14 \\ 17 & 15 & 11 \end{bmatrix} + \begin{bmatrix} 6 & 10 & 8 \\ 8 & 12 & 7 \end{bmatrix} - \begin{bmatrix} 9 & 19 & 17 \\ 16 & 21 & 16 \end{bmatrix}$$

$$= \begin{bmatrix} 7 & 9 & 5 \\ 9 & 6 & 2 \end{bmatrix} \begin{matrix} \text{B \& W} \\ \text{Color} \end{matrix}$$

with columns Sm. Med. Lg.

As a partial interpretation, we see there will be nine medium-sized, black & white T.V. sets and two large, color sets left in stock at the end of June.

B. MATRIX MULTIPLICATION

12. We shall now define what is meant by matrix multiplication. Let A and B be matrices in which the number of columns of A is equal to the number of rows of B. Then AB, which is read as the *product of A and B* (or *A times B*), is a matrix that has the same number of rows as A and the same number of columns as B.

Definition 1.4 (Matrix Multiplication):

Suppose A is $m \times p$ and B is $p \times n$. Then the product matrix AB is an $m \times n$ matrix. Let us temporarily rename AB as C. The computation of c_{ij}, the entry in the ith row (i.e., row i) and jth column (i.e., column j) of matrix C, is indicated as follows:

$$c_{ij} = a_{i1}b_{1j} + a_{i2}b_{2j} + \cdots + a_{ip}b_{pj}$$

There are p terms in this sum. A schematic diagram showing this multiplication appears on Figure 1.1.

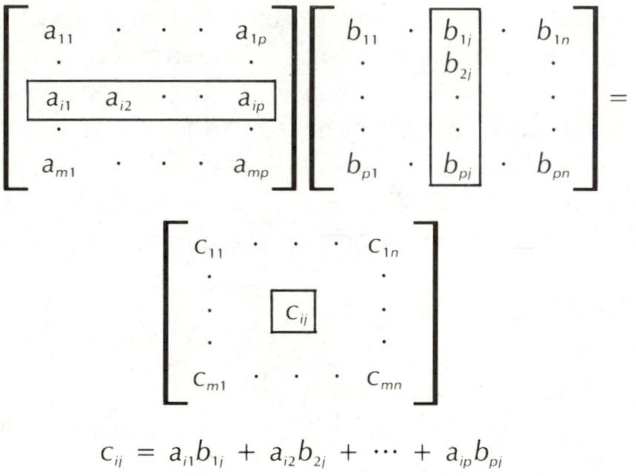

$$c_{ij} = a_{i1}b_{1j} + a_{i2}b_{2j} + \cdots + a_{ip}b_{pj}$$

FIGURE 1.1   Matrix multiplication: $C = AB$.

As indicated by the boxed-in strips in Figure 1.1, to find the entry c_{ij} we multiply the corresponding entries of row i of A and column j of B and add the resulting products.

Notes: (a) If the number of columns of A is not equal to the number of rows of B—for example, if A is $m \times p$ and B is $r \times n$, where $p \neq r$—then the product AB is *not defined*.

(b) The product BA is usually not equal to the product AB. In fact, if A is $m \times p$ and B is $p \times n$, and m is unequal to n, then AB is an $m \times n$ matrix but BA is not even defined.

The following example will illustrate matrix multipliction.

Example: Suppose the 3×2 matrix A and 2×4 matrix B are given as follows:

$$A = \begin{bmatrix} 1 & 2 \\ 3 & 4 \\ 5 & 6 \end{bmatrix}, \qquad B = \begin{bmatrix} 7 & 8 & 9 & 1 \\ 2 & 3 & 4 & 5 \end{bmatrix}$$

Determine the product matrix $C = AB$.

Solution: First, observe that C will be 3×4, since A is 3×2 and B is 2×4. Also, observe that the matrix product BA is not defined, since the number of columns of B (4) is not equal to the number of rows of A (3).

For the calculation of c_{23}, we form products between the entries in row 2 of A and column 3 of B and then add the products. Thus,

$$c_{23} = 3 \cdot 9 + 4 \cdot 4 = 27 + 16 = 43$$

For another example, c_{11} is computed from products between the entries of row 1 of A and column 1 of B. Thus,

$$c_{11} = 1 \cdot 7 + 2 \cdot 2 = 7 + 4 = 11$$

Also

$$c_{12} = 1 \cdot 8 + 2 \cdot 3 = 8 + 6 = 14.$$

Compute the rest of the entries of the matrix $AB = C$ and display the matrix AB.

— — — — — — — — —

Some of the calculations of typical entries are as follows:

$$c_{24} = 3 \cdot 1 + 4 \cdot 5 = 3 + 20 = 23$$
$$c_{31} = 5 \cdot 7 + 6 \cdot 2 = 35 + 12 = 47$$
$$c_{33} = 5 \cdot 9 + 6 \cdot 4 = 45 + 24 = 69$$

Thus, matrix $C = AB$ is the following 3×4 matrix:

$$AB = \begin{bmatrix} 11 & 14 & 17 & 11 \\ 29 & 36 & 43 & 23 \\ 47 & 58 & 69 & 35 \end{bmatrix}$$

13. Here are further notes on the characteristics of matrix multiplication.

Notes: (i) Suppose A is $m \times p$. Then both AB and BA will be defined if B is $p \times m$. In this case, AB will be $m \times m$ and BA will be $p \times p$. (ii) If A and B are both $m \times m$, then AB and BA both will be defined and will have the same size, namely, $m \times m$. Usually, for this case, AB will not equal BA.

For example, if $A = \begin{bmatrix} 1 & 2 \\ 3 & 4 \end{bmatrix}$ and $B = \begin{bmatrix} 4 & 2 \\ 3 & 1 \end{bmatrix}$, you should verify that

$$AB = \begin{bmatrix} 10 & 4 \\ 24 & 10 \end{bmatrix} \text{ and } BA = \begin{bmatrix} 10 & 16 \\ 6 & 10 \end{bmatrix}; \text{ that is, } AB \neq BA.$$

The following theorem summarizes several rules of matrix multiplication. It is assumed that the products and sums are defined.

Theorem 1.3

Matrix multiplication satisfies the following properties:

(a) $(AB)C = A(BC)$ (Associative Law)
(b) $A(B + C) = AB + AC$ (Left Distributive Law)
(c) $(B + C)A = BA + CA$ (Right Distributive Law)
(d) $k(AB) = (kA)B = A(kB)$,   where k is a scalar

Note: Because of property (a), we define the triple matrix product ABC as equal to the common value given in (a): thus, $ABC = (AB)C = A(BC)$.

Suppose matrices A, B, and C are given by

$$A = \begin{bmatrix} 1 & 2 \\ 0 & 3 \end{bmatrix} \qquad B = \begin{bmatrix} 1 & 3 \\ 2 & 4 \end{bmatrix} \qquad C = \begin{bmatrix} 2 & -3 \\ 4 & 1 \end{bmatrix}$$

Verify properties (a) and (b) of Theorem 1.3 for these matrices. To help you get started, note that the formulation $(AB)C$ implies that *first* the matrix product AB is computed and then AB is multiplied times C.
Thus,

$$AB = \begin{bmatrix} 1 \cdot 1 + 2 \cdot 2 & 1 \cdot 3 + 2 \cdot 4 \\ 0 \cdot 1 + 3 \cdot 2 & 0 \cdot 3 + 3 \cdot 4 \end{bmatrix} = \begin{bmatrix} 5 & 11 \\ 6 & 12 \end{bmatrix}$$

Then,

$$(AB)C = \begin{bmatrix} 5 & 11 \\ 6 & 12 \end{bmatrix} \begin{bmatrix} 2 & -3 \\ 4 & 1 \end{bmatrix} = \begin{bmatrix} 5 \cdot 2 + 11 \cdot 4 & 5 \cdot (-3) + 11 \cdot 1 \\ 6 \cdot 2 + 12 \cdot 4 & 6 \cdot (-3) + 12 \cdot 1 \end{bmatrix}$$

Thus,

$$(AB)C = \begin{bmatrix} 54 & -4 \\ 60 & -6 \end{bmatrix}$$

Now complete verifying properties (a) and (b) for the matrices A, B, and C.

— — — — — — — — —

To continue property (a):

$$BC = \begin{bmatrix} 1 \cdot 2 + 3 \cdot 4 & 1 \cdot (-3) + 3 \cdot 1 \\ 2 \cdot 2 + 4 \cdot 4 & 2 \cdot (-3) + 4 \cdot 1 \end{bmatrix} = \begin{bmatrix} 14 & 0 \\ 20 & -2 \end{bmatrix}$$

$$A(BC) = \begin{bmatrix} 1 & 2 \\ 0 & 3 \end{bmatrix} \begin{bmatrix} 14 & 0 \\ 20 & -2 \end{bmatrix} = \begin{bmatrix} 1 \cdot 14 + 2 \cdot 20 & 1 \cdot 0 + 2 \cdot (-2) \\ 0 \cdot 14 + 3 \cdot 20 & 0 \cdot 0 + 3 \cdot (-2) \end{bmatrix}$$

$$= \begin{bmatrix} 54 & -4 \\ 60 & -6 \end{bmatrix}$$

For property (b):

$$A(B + C) = \begin{bmatrix} 1 & 2 \\ 0 & 3 \end{bmatrix} \left(\begin{bmatrix} 1 & 3 \\ 2 & 4 \end{bmatrix} + \begin{bmatrix} 2 & -3 \\ 4 & 1 \end{bmatrix} \right)$$

$$= \begin{bmatrix} 1 & 2 \\ 0 & 3 \end{bmatrix} \begin{bmatrix} 3 & 0 \\ 6 & 5 \end{bmatrix} = \begin{bmatrix} 15 & 10 \\ 18 & 15 \end{bmatrix}$$

Note that *first* B and C were added and then A was multiplied by $(B + C)$. Next, in computing $AB + AC$, *first* the matrix products AB and AC are computed and *then* these products are added.
Thus,

$$AB + AC = \begin{bmatrix} 1 & 2 \\ 0 & 3 \end{bmatrix} \begin{bmatrix} 1 & 3 \\ 2 & 4 \end{bmatrix} + \begin{bmatrix} 1 & 2 \\ 0 & 3 \end{bmatrix} \begin{bmatrix} 2 & -3 \\ 4 & 1 \end{bmatrix}$$

$$= \begin{bmatrix} 5 & 11 \\ 6 & 12 \end{bmatrix} + \begin{bmatrix} 10 & -1 \\ 12 & 3 \end{bmatrix} = \begin{bmatrix} 15 & 10 \\ 18 & 15 \end{bmatrix}$$

Thus, property (b) is verified.

14. Since a vector is a matrix (with one row or one column), the rules for multiplication involving vectors are the same as the general rule for multiplication of a matrix by a matrix. (Refer to the beginning of Section B.) For example, suppose we wish to multiply the $m \times p$ matrix A by a column vector $\mathbf{x}$, to form the product $A\mathbf{x}$. The product will only be defined if $\mathbf{x}$ has p components, in which case $\mathbf{x}$ is the same as a $p \times 1$ matrix.

Example (a): Compute $A\mathbf{x}$ if A and $\mathbf{x}$ are given by

$$A = \begin{bmatrix} 4 & 2 & 3 \\ 1 & 5 & 6 \end{bmatrix}, \qquad \mathbf{x} = \begin{bmatrix} 4 \\ 3 \\ 1 \end{bmatrix}$$

Solution: Notice that $\mathbf{x}$ is a three-component column vector, which is the same as a 3×1 matrix.

$$A\mathbf{x} = \begin{bmatrix} 4 \cdot 4 + 2 \cdot 3 + 3 \cdot 1 \\ 1 \cdot 4 + 5 \cdot 3 + 6 \cdot 1 \end{bmatrix} = \begin{bmatrix} 25 \\ 25 \end{bmatrix}$$

Here, $A\mathbf{x}$ is a 2×1 matrix, or equivalently, a two-component column vector.

Example (b): Suppose that A and $\mathbf{x}$ are given by

$$A = \begin{bmatrix} a_{11} & a_{12} & a_{13} \\ a_{21} & a_{22} & a_{23} \end{bmatrix}, \qquad \mathbf{x} = \begin{bmatrix} x_1 \\ x_2 \\ x_3 \end{bmatrix}$$

This example is like Example (a) except that values have not been assigned to the various terms. Compute $A\mathbf{x}$ for A and $\mathbf{x}$ as given here.

— — — — — — — — — —

$$A\mathbf{x} = \begin{bmatrix} a_{11}x_1 + a_{12}x_2 + a_{13}x_3 \\ a_{21}x_1 + a_{22}x_2 + a_{23}x_3 \end{bmatrix}$$

Thus, $A\mathbf{x}$ is a two-component column vector whose first component is $a_{11}x_1 + a_{12}x_2 + a_{13}x_3$ and whose second component is $a_{21}x_1 + a_{22}x_2 + a_{23}x_3$.

This type of formulation will occur frequently in Chapter Two when we study methods for analyzing systems of m linear equations in n unknowns.

15. A *square matrix* is a matrix that has the same number of rows as columns. It is sometimes referred to as an $n \times n$ matrix, where n is both the number of rows and the number of columns. The general form for an $n \times n$ matrix is

$$A = \begin{bmatrix} a_{11} & a_{12} & \cdots & a_{1n} \\ a_{21} & a_{22} & \cdots & a_{2n} \\ \vdots & & & \vdots \\ a_{n1} & a_{n2} & \cdots & a_{nn} \end{bmatrix}$$

The diagonal containing entries a_{11}, a_{22}, $\ldots$, and a_{nn} is called the *main diagonal* of square matrix A. A square matrix with ones on the main diagonal and zeros elsewhere is called an *identity matrix*. The $n \times n$ identity matrix is denoted by I_n, or just I. For example, the 1×1, 2×2, and 3×3 identity matrices are as follows:

$$I_1 = [1], \qquad I_2 = \begin{bmatrix} 1 & 0 \\ 0 & 1 \end{bmatrix}, \qquad I_3 = \begin{bmatrix} 1 & 0 & 0 \\ 0 & 1 & 0 \\ 0 & 0 & 1 \end{bmatrix}$$

The identity matrix has the property that $I_n A = AI_n = A$ if A is any $n \times n$ matrix. (Here we have matrix multiplication in $I_n A$ and AI_n.) In other words, I_n is analogous to the number 1 of ordinary arithmetic; recall that $1 \cdot a = a \cdot 1 = a$ if a is any number, when we have ordinary multiplication. Because of this property the number 1 is referred to as the *multiplicative identity* of the ordinary number system. In similar fashion we refer to the matrix I_n as the *multiplicative identity matrix* for $n \times n$ matrices.

For example, let us show that $I_2 A = A$ if A is a general 2×2 matrix.

$$I_2 A = \begin{bmatrix} 1 & 0 \\ 0 & 1 \end{bmatrix} \begin{bmatrix} a_{11} & a_{12} \\ a_{21} & a_{22} \end{bmatrix} = \begin{bmatrix} a_{11} + 0 & a_{12} + 0 \\ 0 + a_{21} & 0 + a_{22} \end{bmatrix} = \begin{bmatrix} a_{11} & a_{12} \\ a_{21} & a_{22} \end{bmatrix} = A$$

Show that $AI_2 = A$ if A is a general 2×2 matrix.

— — — — — — — — — —

$$AI_2 = \begin{bmatrix} a_{11} & a_{12} \\ a_{21} & a_{22} \end{bmatrix} \begin{bmatrix} 1 & 0 \\ 0 & 1 \end{bmatrix} = \begin{bmatrix} a_{11} + 0 & 0 + a_{12} \\ a_{21} + 0 & 0 + a_{22} \end{bmatrix} = \begin{bmatrix} a_{11} & a_{12} \\ a_{21} & a_{22} \end{bmatrix} = A$$

16. Now we will consider a practical example that involves multiplying one matrix by another.

Example: A manufacturer makes radios and tape recorders. Each product must go through an assembly process and a finishing process. The times required in these processes (in hours) are presented in matrix A, as follows:

Assembly   Finishing

$$A = \begin{bmatrix} 1 & 2 \\ 4 & 1.5 \end{bmatrix} \begin{array}{l} \text{Radio} \\ \text{Tape recorder} \end{array}$$

For example, it takes 4 hours for the assembly of each tape recorder.

The manufacturer has factories in Massachusetts and New Jersey, where the hourly rates ($/hour) for both processes are given by matrix B:

Mass.   N.J.

$$B = \begin{bmatrix} 10 & 12 \\ 11 & 9 \end{bmatrix} \begin{array}{l} \text{Assembly} \\ \text{Finishing} \end{array}$$

For example, the cost per hour to assemble in New Jersey is $12.

The entries of the product matrix AB have a practical interpretation. For example, if we compute the row 1, column 1 entry of AB, we will be computing the cost of manufacturing a radio (row 1 of A) in Massachusetts (column 1 of B). This is seen, as follows, by canceling of units.

$$(AB)_{11} = 1\frac{\overline{\text{hr. assem.}}}{\text{radio}} \cdot 10\frac{\$}{\overline{\text{hr. assem.}}\ (\text{Mass.})}$$

$$+ 2\frac{\overline{\text{hr. finish}}}{\text{radio}} \cdot 11\frac{\$}{\overline{\text{hr. finish}}\ (\text{Mass.})}$$

$$= 32\frac{\$}{\text{radio (Mass.)}}$$

Compute the remaining entries of AB. Display the matrix AB in a form in which the units of the entries are clearly indicated.

— — — — — — — — — —

The total matrix AB is as follows:

$$AB = \begin{array}{cc} \text{Mass.} & \text{N.J.} \\ \begin{bmatrix} 32 & 30 \\ 56.5 & 61.5 \end{bmatrix} & \begin{array}{l} \text{Radio} \\ \text{Tape recorder} \end{array} \end{array}$$

Note the labeling of the columns and rows. The row labels of AB correspond to the row labels of A, and the column labels of AB correspond to the column labels of B. Each entry of AB gives the cost of manufacturing a radio or tape recorder in either Massachusetts or New Jersey. For example, $(AB)_{22}$ = 61.5 \$/tape recorder is the cost of manufacturing a tape recorder in New Jersey.

17. The concept of powers of a square matrix is useful in several applications (for example, see Markov processes in Chapter 8). Suppose that A is a square matrix. If k is a positive integer, then

$$A^k = A \cdot A \cdot A \cdot \ldots \cdot A \qquad (k \text{ factors on the right}) \tag{1}$$

where the multiplication dots are often omitted in routine calculations.
For A an $n \times n$ matrix, we define

$$A^0 = I_n \tag{2}$$

The following laws of exponents hold, where h and k denote nonnegative integers:

$$A^h \cdot A^k = A^{(h+k)} \tag{3}$$

$$(A^h)^k = A^{hk} \tag{4}$$

Example: Suppose $A = \begin{bmatrix} 1 & 0 \\ 2 & 3 \end{bmatrix}$. Compute A^0, A^1, A^2, and A^3.

Solution: By definition, $A^0 = I_2 = \begin{bmatrix} 1 & 0 \\ 0 & 1 \end{bmatrix}$. $A^1 = A = \begin{bmatrix} 1 & 0 \\ 2 & 3 \end{bmatrix}$.

$$A^2 = AA = \begin{bmatrix} 1 & 0 \\ 2 & 3 \end{bmatrix}\begin{bmatrix} 1 & 0 \\ 2 & 3 \end{bmatrix} = \begin{bmatrix} 1 & 0 \\ 8 & 9 \end{bmatrix}.$$

Now compute A^3. An easy way is to make use of Eq. (3) with $h = 2$ and $k = 1$.

$$A^3 = A^{(2+1)} = A^2A = \begin{bmatrix} 1 & 0 \\ 8 & 9 \end{bmatrix}\begin{bmatrix} 1 & 0 \\ 2 & 3 \end{bmatrix} = \begin{bmatrix} 1 & 0 \\ 26 & 27 \end{bmatrix}$$

C. THE TRANSPOSE OF A MATRIX

18. Often when seeking to understand the characteristics of systems represented in terms of matrices, we perform various manipulations on the matrices. Doing such manipulations often leads to new insights about the properties of such systems. One such manipulation involves the interchanging of the rows and columns of a matrix. Because this manipulation is done quite often in practice, we give a special name to the matrix that arises from such an interchange; it is called the *transpose* of the original matrix.

In forming the transpose of a matrix A, the rows and columns are interchanged in such a way that row i of A becomes column i of the transposed matrix. The transpose of matrix A is denoted by A^t. For example, the transpose of $A = \begin{bmatrix} 1 & 2 \\ 3 & 4 \end{bmatrix}$ is $A^t = \begin{bmatrix} 1 & 3 \\ 2 & 4 \end{bmatrix}$. The entry in row i and column j of A^t is indicated as a_{ij}^t or $(a^t)_{ij}$.

Definition 1.5 (Transpose of a Matrix):

Given an $m \times n$ matrix A. The transpose of matrix A, denoted as A^t, is an $n \times m$ matrix for which

$$a_{ij}^t = a_{ji} \quad \text{for } i = 1, 2, \ldots, n \quad \text{and} \quad j = 1, 2, \ldots, m$$

That is, the row i, column j entry of A^t is equal to the row j, column i entry of A.

Note: To avoid confusion, we agree to use the superscript t in this book to denote only the transpose. Thus, the use of another lowercase letter in the superscript of a matrix will indicate a different quantity, such as an integer power of a matrix (see, for example, frame 17).

Example (a): Give the transpose of a general 2×3 matrix.

Solution: Let

$$A = \begin{bmatrix} a_{11} & a_{12} & a_{13} \\ a_{21} & a_{22} & a_{23} \end{bmatrix}$$

Then

$$A^t = \begin{bmatrix} a_{11} & a_{21} \\ a_{12} & a_{22} \\ a_{13} & a_{23} \end{bmatrix}$$

According to the symbolism discussed previously, we may write

$$A^t = \begin{bmatrix} a^t_{11} & a^t_{12} \\ a^t_{21} & a^t_{22} \\ a^t_{31} & a^t_{32} \end{bmatrix}$$

Thus, $a^t_{12} = a_{21}$, $a^t_{31} = a_{13}$, etc.

Example (b): Determine A^t if A is given by

$$A = \begin{bmatrix} 10 & 8 & 7 & 6 \\ 5 & 9 & 2 & 3 \\ 4 & 5 & 8 & 1 \end{bmatrix}$$

— — — — — — — — —

$$A^t = \begin{bmatrix} 10 & 5 & 4 \\ 8 & 9 & 5 \\ 7 & 2 & 8 \\ 6 & 3 & 1 \end{bmatrix}$$

We observe that row 1 of A is column 1 of A^t, row 2 of A is column 2 of A^t, etc. Also, $a_{24} = 3 = a^t_{42}$, etc.

19. The major properties of the transpose are summarized by the following theorem.

> *Theorem 1.4*
>
> The transpose operation on matrices obeys the following rules:
>
> (a) $(A + B)^t = A^t + B^t$
> (b) $(A^t)^t = A$
> (c) $(kA)^t = kA^t$, where k is a scalar
> (d) $(AB)^t = B^t A^t$

Note: Property (d), which is fairly important, says that the transpose of a product equals the product of the transposes, *but in the opposite order*. Also, the property can be extended. Thus, for example, $(ABC)^t = C^t B^t A^t$.

Example (a): Verify Theorem 1.4, property (d), for the matrices

$$A = \begin{bmatrix} 2 & 3 \\ 1 & 4 \end{bmatrix} \quad \text{and} \quad B = \begin{bmatrix} 1 & 5 \\ 3 & 0 \end{bmatrix}$$

Solution: First, we find the left side, $(AB)^t$.

$$AB = \begin{bmatrix} 2{\cdot}1 + 3{\cdot}3 & 2{\cdot}5 + 3{\cdot}0 \\ 1{\cdot}1 + 4{\cdot}3 & 1{\cdot}5 + 4{\cdot}0 \end{bmatrix} = \begin{bmatrix} 11 & 10 \\ 13 & 5 \end{bmatrix} \tag{1}$$

Thus,

$$(AB)^t = \begin{bmatrix} 11 & 13 \\ 10 & 5 \end{bmatrix} \tag{2}$$

Now find the right side in property (d), and thus complete the verification.

– – – – – – – – – –

Now

$$A^t = \begin{bmatrix} 2 & 1 \\ 3 & 4 \end{bmatrix} \quad \text{and} \quad B^t = \begin{bmatrix} 1 & 3 \\ 5 & 0 \end{bmatrix}$$

Thus,

$$B^t A^t = \begin{bmatrix} 1{\cdot}2 + 3{\cdot}3 & 1{\cdot}1 + 3{\cdot}4 \\ 5{\cdot}2 + 0{\cdot}3 & 5{\cdot}1 + 0{\cdot}4 \end{bmatrix} = \begin{bmatrix} 11 & 13 \\ 10 & 5 \end{bmatrix} \tag{3}$$

Comparing Eqs. (2) and (3), we see that property (d) of Theorem 1.4 is verified.

20. Property (b) of Theorem 1.4 is quite easy to prove. Let us do so for a special case.

Example (a): Prove Theorem 1.4, property (b), for the special case of a 2×3 matrix.

Solution: For a general 2×3 matrix A, we have

$$A = \begin{bmatrix} a & b & c \\ d & e & f \end{bmatrix} \tag{1}$$

where we have labeled the entries as a, b, etc. Thus,

$$A^t = \begin{bmatrix} a & d \\ b & e \\ c & f \end{bmatrix} \tag{2}$$

which is 3×2. Now $(A^t)^t$ is merely the transpose of the matrix A^t, which is given by (2). Thus, we have

$$(A^t)^t = \begin{bmatrix} a & b & c \\ d & e & f \end{bmatrix},$$

which is identical to the starting matrix A.

Example (b): Verify property (a) of Theorem 1.4 for the matrices A and B in Example (a) of frame 19.

Solution:

$$A + B = \begin{bmatrix} 2+1 & 3+5 \\ 1+3 & 4+0 \end{bmatrix} = \begin{bmatrix} 3 & 8 \\ 4 & 4 \end{bmatrix} \tag{1}$$

$$[A + B]^t = \begin{bmatrix} 3 & 4 \\ 8 & 4 \end{bmatrix} \tag{2}$$

$$A^t = \begin{bmatrix} 2 & 1 \\ 3 & 4 \end{bmatrix} \qquad B^t = \begin{bmatrix} 1 & 3 \\ 5 & 0 \end{bmatrix} \tag{3}$$

$$A^t + B^t = \begin{bmatrix} 2+1 & 1+3 \\ 3+5 & 4+0 \end{bmatrix} = \begin{bmatrix} 3 & 4 \\ 8 & 4 \end{bmatrix} \tag{4}$$

Example (c): Verify Theorem 1.4, property (c), for the scalar $k = 3$ and the matrix $A = \begin{bmatrix} -2 & 5 \\ 7 & 4 \end{bmatrix}$.

Solution: First, observe from Definition 1.2 that

$$kA = 3\begin{bmatrix} -2 & 5 \\ 7 & 4 \end{bmatrix} = \begin{bmatrix} 3\cdot(-2) & 3\cdot5 \\ 3\cdot7 & 3\cdot4 \end{bmatrix} = \begin{bmatrix} -6 & 15 \\ 21 & 12 \end{bmatrix} \tag{1}$$

Now finish the verification.

- - - - - - - - - -

Thus,

$$(kA)^t = \begin{bmatrix} -6 & 21 \\ 15 & 12 \end{bmatrix} \tag{2}$$

Now $A^t = \begin{bmatrix} -2 & 7 \\ 5 & 4 \end{bmatrix}$, and hence

$$kA^t = 3\begin{bmatrix} -2 & 7 \\ 5 & 4 \end{bmatrix} = \begin{bmatrix} -6 & 21 \\ 15 & 12 \end{bmatrix} \tag{3}$$

Since the right sides of Eqs. (2) and (3) are the same, the verification is complete.

21. Recall the definitions of a square matrix and the main diagonal of a square matrix (frame 15). A special type of square matrix is a symmetric matrix. Many practical applications involve symmetric matrices [see, for example, Section D of Chapter 7—Diagonalization of Symmetric Matrices].

Definition 1.6 (Symmetric Matrix):

A square matrix A is symmetric if $a_{ij} = a_{ji}$ for all i and j. In other words, $A^t = A$ for a symmetric matrix.

Notes: (a) The reason that $A^t = A$ for a symmetric matrix is that $a_{ij}^t = a_{ji}$, in general. Thus, here $a_{ij}^t = a_{ij}$, since $a_{ij} = a_{ji}$. It follows that $A^t = A$, since $a_{ij}^t = a_{ij}$ for all i and j. (b) If A is symmetric, then the entries of A are symmetric with respect to the main diagonal of A (the diagonal that runs from upper left to lower right).

Example (a): The matrices

$$A = \begin{bmatrix} 3 & 4 \\ 4 & -7 \end{bmatrix} \quad \text{and} \quad B = \begin{bmatrix} 2 & -6 & 8 \\ -6 & 7 & 5 \\ 8 & 5 & 3 \end{bmatrix}$$

are symmetric. For example, in B, $b_{21} = b_{12}$, $b_{31} = b_{13}$, and $b_{32} = b_{23}$. Also, every identity matrix is symmetric. For example, $I_2 = \begin{bmatrix} 1 & 0 \\ 0 & 1 \end{bmatrix}$ is symmetric.

An important type of symmetric matrix is a diagonal matrix.

Definition 1.7 (Diagonal Matrix)

A symmetric matrix for which every term not on the main diagonal is zero is called a diagonal matrix. Thus, $a_{ij} = 0$ for $i \neq j$ for a diagonal matrix.

Example (b): Which of the following matrices are diagonal matrices?

$$A = \begin{bmatrix} -5 & 0 \\ 0 & 3 \end{bmatrix}, \qquad B = \begin{bmatrix} 9 & 0 & 0 \\ 0 & 9 & 0 \\ 0 & 0 & 9 \end{bmatrix},$$

$$C = \begin{bmatrix} 5 & 1 \\ 1 & 5 \end{bmatrix}, \qquad D = \begin{bmatrix} -6 & 0 & 0 \\ 0 & 5 & 0 \\ 0 & 0 & 14 \end{bmatrix}.$$

— — — — — — — — — —

A, B, and D are diagonal matrices.

22. The transpose notation will often be used as a space-saving device throughout this book when we refer to column vectors. Thus, we may indicate the k-component column vector **b** by writing $\mathbf{b}^t = [b_1, b_2, \ldots, b_k]$, or $\mathbf{b} = [b_1, b_2, \ldots, b_k]^t$ instead of writing

$$\mathbf{b} = \begin{bmatrix} b_1 \\ b_2 \\ \vdots \\ b_k \end{bmatrix}.$$

In space-saving fashion, use the transpose notation to indicate the column vector $\mathbf{x} = \begin{bmatrix} 3 \\ -5 \\ 7 \end{bmatrix}$.

— — — — — — — — — —

$$\mathbf{x}^t = [3, \ -5, \ 7] \qquad \text{or} \qquad \mathbf{x} = [3, \ -5, \ 7]^t$$

SELF-TEST

This Self-Test will help you determine whether or not you have mastered the chapter objectives and are ready to go on to the next chapter. Correct answers are given at the end of the test.

1. Identify all the entries of the matrix $A = \begin{bmatrix} 5 & -2 & 3 \\ 7 & 4 & -6 \end{bmatrix}$ using a_{ij} notation.

2. If the matrix $B = \begin{bmatrix} 12 & 7 & 11 \\ 6 & -5 & 13 \end{bmatrix}$ and A is given in question 1, determine the following matrices: (a) $A + B$, (b) $-B$, (c) $4A$, (d) $4A - 2B$.

3. A department store stocks bicycles in three types: Hi-rise, 3-speed, and 10-speed racer, where each type comes in a design for either male (M) or female (F). The inventory at the end of July for the number of bicycles in each category is recorded by the matrix J as follows:

$$J = \begin{matrix} & \text{Hi-r.} & \text{3-sp.} & \text{10-sp.} & \\ & \begin{bmatrix} 12 & 10 & 15 \\ 7 & 16 & 8 \end{bmatrix} & & & \begin{matrix} \text{Male} \\ \text{Fem.} \end{matrix} \end{matrix}$$

Determine the values for entries j_{11}, j_{13}, and j_{22} and interpret.

4. Suppose the department store of question 3 receives shipments of bicycles during the month of August, given by the following matrix B. The number of bikes sold in the different categories during August is given by matrix C.

$$B = \begin{matrix} & \text{Hi-r.} & \text{3-sp.} & \text{10-sp.} & \\ & \begin{bmatrix} 10 & 8 & 12 \\ 8 & 6 & 14 \end{bmatrix} & & & \begin{matrix} \text{Male} \\ \text{Fem.} \end{matrix} \end{matrix}, \quad C = \begin{matrix} & \text{Hi-r} & \text{3-sp} & \text{10-sp} & \\ & \begin{bmatrix} 9 & 8 & 11 \\ 6 & 7 & 5 \end{bmatrix} & & & \begin{matrix} \text{Male} \\ \text{Fem.} \end{matrix} \end{matrix}$$

Let A be the matrix that indicates the inventory at the end of August.

(a) Determine an equation for matrix A in terms of matrices J, B, and C.

(b) Determine the values for the entries of matrix A and display the matrix.

5. Given the following matrices:

$$A = \begin{bmatrix} 3 & -4 & 2 \\ 1 & 0 & 6 \end{bmatrix}, \qquad B = \begin{bmatrix} 5 & 3 & -1 \\ -4 & 2 & 3 \end{bmatrix},$$

$$C = \begin{bmatrix} 4 & 2 & 1 \\ -3 & 2 & 5 \\ 2 & 0 & 3 \end{bmatrix}, \qquad D = \begin{bmatrix} 1 \\ 3 \\ -2 \end{bmatrix}$$

Determine the following matrix products (for those products that are defined): (a) AC, (b) CA, (c) AB, (d) BD, (e) CD

6. Refer to question 5. Show that $(AC)D = A(CD)$, thus verifying property (a) of Theorem 1.3.

7. In an experimental project concerned with studying effects of diet on apes, the matrix A indicates the numbers of different types of apes taking part in the project.

	Young	Old	
$A =$	70	50	Chimpanzees
	30	70	Gibbons

Data for the number of grams per day ingested by each young and old ape are presented by matrix B.

	Protein	Carbohydrate	Fat	
$B =$	25	40	20	Young
	15	30	25	Old

(a) Determine the matrix AB.

(b) Interpret the meaning of entries in the matrix AB. Note that the row labels of AB correspond to the row labels of A and the column labels of AB correspond to the column labels of B.

8. Suppose $B = \begin{bmatrix} 1 & -2 \\ 0 & 3 \end{bmatrix}$. Compute B^0, B^2, and B^3.

9. For the matrices $A = \begin{bmatrix} 6 & -5 \\ 3 & 8 \end{bmatrix}$ and $B = \begin{bmatrix} 9 & 6 \\ 14 & -7 \end{bmatrix}$, verify properties (a) and (d) of Theorem 1.4.

10. Determine the product CD of the two diagonal matrices

$$C = \begin{bmatrix} 5 & 0 & 0 \\ 0 & 3 & 0 \\ 0 & 0 & 2 \end{bmatrix} \qquad \text{and} \qquad D = \begin{bmatrix} -2 & 0 & 0 \\ 0 & 4 & 0 \\ 0 & 0 & 7 \end{bmatrix}$$

Observe that the resulting product matrix CD is also a diagonal matrix. Determine the product matrix DC. Observe that this diagonal matrix is equal to CD. (It can be shown that the product of two general $n \times n$ diagonal matrices C and D is an $n \times n$ diagonal matrix and that $CD = DC$.)

11.   Given a general n component column vector $\mathbf{x}$; that is, $\mathbf{x}^t = [x_1, x_2, \ldots, x_n]$. Determine the expression $\mathbf{x}^t\mathbf{x}$. The expression $\sqrt{\mathbf{x}^t\mathbf{x}}$ is sometimes called the *length* of the vector $\mathbf{x}$. We shall encounter this term in Chapter Four.

ANSWERS TO SELF-TEST

If your answers to the test questions do not agree with the ones given here, review the frames indicated in parentheses after each answer before you go to the next chapter.

1. $a_{11} = 5, a_{12} = -2, a_{13} = 3; a_{21} = 7, a_{22} = 4, a_{23} = -6.$ (frame 2)

2.   (a) $A + B = \begin{bmatrix} 17 & 5 & 14 \\ 13 & -1 & 7 \end{bmatrix}$,   (b) $-B = \begin{bmatrix} -12 & -7 & -11 \\ -6 & 5 & -13 \end{bmatrix}$,

(c) $4A = \begin{bmatrix} 20 & -8 & 12 \\ 28 & 16 & -24 \end{bmatrix}$,

(d) $4A - 2B = \begin{bmatrix} -4 & -22 & -10 \\ 16 & 26 & -50 \end{bmatrix}$ (frames 6, 8–10)

3. $j_{11} = 12$ high-rise, male; $j_{13} = 15$ 10-speed, male; $j_{22} = 16$ 3-speed, female.   (frame 3)

4.   (a) $A = J + B - C,$
(b)

$$A = \begin{bmatrix} \text{Hi-r.} & \text{3-sp.} & \text{10-sp.} \\ 13 & 10 & 16 \\ 9 & 15 & 17 \end{bmatrix} \begin{matrix} \\ \text{Male} \\ \text{Fem.} \end{matrix}$$ (frames 7–11)

5.   (a) $AC = \begin{bmatrix} 28 & -2 & -11 \\ 16 & 2 & 19 \end{bmatrix}$,   (b)   not defined,   (c)   not defined,

(d) $BD = \begin{bmatrix} 16 \\ -4 \end{bmatrix}$,   (e) $CD = \begin{bmatrix} 8 \\ -7 \\ -4 \end{bmatrix}$.   (frame 12)

6. $(AC)D = A(CD) = \begin{bmatrix} 44 \\ -16 \end{bmatrix}$.   (frame 13)

7.   (a)

$$AB = \begin{bmatrix} \text{Prot.} & \text{Carb.} & \text{Fat} \\ 2500 & 4300 & 2650 \\ 1800 & 3300 & 2350 \end{bmatrix} \begin{matrix} \\ \text{Chimpanzees} \\ \text{Gibbons} \end{matrix}$$

(b)   For example, $(AB)_{11} = 2500$ gm of protein per day consumed by all chimpanzees, and $(AB)_{23} = 2350$ gm of fat consumed by all gibbons. (frame 14)

8. $B^0 = I_2 = \begin{bmatrix} 1 & 0 \\ 0 & 1 \end{bmatrix}$, $B^2 = \begin{bmatrix} 1 & -8 \\ 0 & 9 \end{bmatrix}$, $B^3 = \begin{bmatrix} 1 & -26 \\ 0 & 27 \end{bmatrix}$.   (frame 17)

9.   (a) $(A + B)^t = A^t + B^t = \begin{bmatrix} 15 & 17 \\ 1 & 1 \end{bmatrix}$

    (b) $(AB)^t = B^t A^t = \begin{bmatrix} -16 & 139 \\ 71 & -38 \end{bmatrix}$ (frames 18–20)

10. $CD = DC = \begin{bmatrix} -10 & 0 & 0 \\ 0 & 12 & 0 \\ 0 & 0 & 14 \end{bmatrix}$ (frame 21)

11. $\mathbf{x}^t \mathbf{x} = x_1^2 + x_2^2 + \ldots + x_n^2.$ (frames 12, 14, 22)

CHAPTER TWO

Linear Systems

In this chapter we shall study methods for analyzing a system of m linear equations in n unknowns. Our approach will lead us to a matrix–vector equation for representing such a system and to the Gauss–Jordan elimination method. From this will follow the concepts of rank and inverse of a matrix and major theorems for analyzing homogeneous and nonhomogeneous systems of linear equations.

The methods and results of this chapter will form the foundation for our future analysis and study of linear algebra.

OBJECTIVES

When you complete this chapter you should be able to

- Represent a system of m linear equations in n unknowns by the simple matrix–vector equation $A\mathbf{x} = \mathbf{b}$.
- Use the Gauss–Jordan elimination method to find solutions (if they exist) of a linear system.
- Determine the special characteristics of homogeneous linear systems.
- Apply the matrix rank concept to analyzing nonhomogeneous and homogeneous systems.
- Compute the inverse of a square matrix (if it exists).
- Do calculations using the Leontief input–output economic model.

A. INTRODUCTION TO LINEAR SYSTEMS

1. A linear equation in the two variables x and y has the general form

$$a_1 x + a_2 y = b, \tag{1}$$

where a_1, a_2, and b are real number constants. This equation can be interpreted as the equation of a straight line in the xy plane (provided that at least one of a_1 and a_2 is unequal to zero). Remember that the slope–intercept form for a straight line is $y = mx + B$, where m is the slope and B is the y-intercept.

Thus, from (1), we obtain the following "slope–intercept" form when we divide through by a_2 (assuming $a_2 \neq 0$):

$$y = -\frac{a_1}{a_2}x + \frac{b}{a_2} \tag{2}$$

Here the slope is $-a_1/a_2$ and the y-intercept is b/a_2. If $a_2 = 0$ (and $a_1 \neq 0$) we obtain the equation $x = b/a_1$ after dividing through by a_1 in Eq. (1). This is the equation of a line parallel to the Y axis (or coincident with the Y axis if $b = 0$).

Example (a): Determine the slope and y-intercept of the straight line in the xy plane given by the following equation: $-x + 2y = 4$.

Solution: Since we can write this as $y = \frac{1}{2}x + 2$, this represents a straight line with slope 1/2 and y-intercept 2.

It will be more efficient for our later purposes if we label our variables as x_1, x_2, etc. Thus, our general linear equation in two variables is rewritten as

$$a_1x_1 + a_2x_2 = b \tag{1'}$$

Thus, we can rewrite the Example (a) equation as $-x_1 + 2x_2 = 4$.

A *solution* of a single linear equation in two variables x_1 and x_2 is a pair of values $x_1 = k_1$, $x_2 = k_2$ that causes Eq. (1') to be satisfied. Thus, we see that each solution corresponds to a *point* on the line whose equation is Eq. (1'). Thus, for a single linear equation in two variables, there is an *infinite* number of solutions. These solutions correspond to the *infinite* number of points on the line.

Example (b): Determine several solutions (points) for the single equation $-x_1 + 2x_2 = 4$.

Solution: Suppose we solve for x_2 in terms of x_1 (we could equally well solve for x_1 in terms of x_2) as in Example (a):

$$x_2 = \tfrac{1}{2}x_1 + 2.$$

Thus, for $x_1 = 0$, $x_2 = 2$, and this is one solution. Now find solutions for $x_1 = 1, 2,$ and 3.

— — — — — — — —

For $x_1 = 1$, $x_2 = 2.5$; for $x_1 = 2$, $x_2 = 3$; for $x_1 = 3$, $x_2 = 3.5$, etc. Here $(0, 2)$, $(1, 2.5)$, $(2, 3)$, $(3, 3.5)$ represent four solutions, or, equivalently, four points on the line with equation $-x_1 + 2x_2 = 4$.

2.   A *general linear equation* in the n variables $x_1, x_2, \ldots, x_n$ has the form

$$a_1x_1 + a_2x_2 + \ldots + a_nx_n = b,$$

where $a_1, a_2, \ldots, a_n$, and b are real number constants. If $n = 3$, we have the equation

$$a_1x_1 + a_2x_2 + a_3x_3 = b$$

This can be interpreted as the equation of a plane in three-dimensional space (provided that at least one of a_1, a_2, and a_3 is unequal to zero).

To sketch a plane, first we lay out the mutually perpendicular X_1, X_2, and X_3 axes as in Fig. 2.1; clearly, some distortion of the perpendicularity requirement will occur, since our sketch is on a two-dimensional page. For sketching purposes it is often useful to determine intercepts of the plane on the three axes. For example, to find the x_1-intercept, denoted by $(x_1)_i$, we set x_2 and x_3 equal to zero and solve for x_1. Thus,

$$(x_1)_i = \frac{b}{a_1} \qquad \text{if} \quad a_1 \neq 0.$$

If $a_1 = 0$, then $(x_1)_i$ does not exist. We denote the other intercepts (if they exist) as $(x_2)_i$ and $(x_3)_i$. They are found in similar fashion.

Example (a):   The equation of a plane in three-dimensional space is $2x_1 + 4x_2 + 3x_3 = 12$. Determine the intercepts of the plane with the X_1, X_2, and X_3 axes.

Solution: If $x_2 = x_3 = 0$, then $(x_1)_i = 6$.

Now find the other intercepts. For example, to find $(x_2)_i$, set $x_1 = x_3 = 0$.

If $x_1 = x_3 = 0$, then $(x_2)_i = 3$. If $x_1 = x_2 = 0$, then $(x_3)_i = 4$.

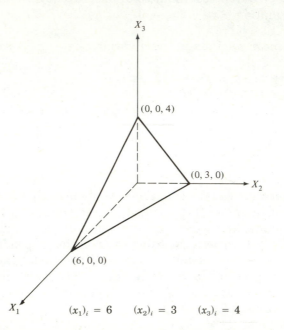

X_3

$(0, 0, 4)$

$(0, 3, 0)$

X_2

$(6, 0, 0)$

X_1 $(x_1)_i = 6$ $(x_2)_i = 3$ $(x_3)_i = 4$

FIGURE 2.1   The plane $2x_1 + 4x_2 + 3x_3 = 12$ in three-dimensional space.

3.   The sketch of the portion of the plane in the first octant (the octant where x_1, x_2, and x_3 are all positive) is given in Figure 2.1.

Consider again the general linear equation in three variables:

$$a_1x_1 + a_2x_2 + a_3x_3 = b \tag{1}$$

A solution of such an equation consists of a triple of values $x_1 = k_1, x_2 = k_2, x_3 = k_3$ that causes Eq. (1) to be satisfied. Thus, we see that each solution corresponds to a *point* on the plane whose equation is given by Eq. (1).

Example (a):   Determine several solutions (points) for the equation $2x_1 + 4x_2 + 3x_3 = 12$.

Solution: Three solutions are given by the three intercept points for this equation, listed in Example (a) of frame 2. For other solutions (points), let us solve our equation for x_1 in terms of x_2 and x_3:

$$x_1 = 6 - 2x_2 - \tfrac{3}{2}x_3$$

To obtain other solutions, we vary x_2 and x_3 freely and solve for x_1. For example, if we let $x_2 = 1$ and $x_3 = 0$, we obtain $x_1 = 4$. Obtain several other solutions in this fashion.

Thus, $(4, 1, 0)$, $(4.5, 0, 1)$, $(2.5, 1, 1)$, $(3, 0, 2)$, and $(3, 1.5, 0)$ are five solutions obtained in this way. Note that in listing solutions, we first indicate the x_1 value, then the x_2 value, and finally the x_3 value. This will be our approach throughout this book. Thus, the solution $(4.5, 0, 1)$ means $x_1 = 4.5$, $x_2 = 0$, and $x_3 = 1$.

Clearly, the set of all solutions to a single linear equation in three variables consists of the infinite number of solutions that correspond to the infinite number of points on the plane represented by the equation.

4. One of the most important problems in elementary linear algebra is to find a solution (or solutions) of a system of linear equations. A system of linear equations consists of one or more (usually more) linear equations.

Example (a): Determine a solution (or solutions) of the system of two linear equations given by

$$x_1 + 2x_2 = 4 \tag{1}$$

$$2x_1 - x_2 = 3 \tag{2}$$

Solution: First, observe that by a solution to the system of equations (1) and (2) we mean a pair of values for x_1 and x_2, i.e., $x_1 = k_1$, $x_2 = k_2$, such that *both* equations are satisfied. Now each of the preceding equations is the equation of a straight line in two-dimensional space. Thus, a solution $x_1 = k_1$, $x_2 = k_2$ corresponds to a point (k_1, k_2) that lies on *both* lines. In the current situation, the slopes of the two lines differ ($m_1 = -\frac{1}{2}$; $m_2 = 2$), and thus the two lines are not parallel. Therefore the lines intersect at a single point. Hence, there is a *unique* (i.e., only one) solution of the preceding system of two equations, and this occurs at the x_1 and x_2 values for the single point of intersection.

There are many methods of solution for a system of linear equations. The method that we will employ here is closely related to the main computational method of this chapter, the Gauss–Jordan elimination method.* This method, which we shall call the *elimination method*, involves eliminating x_2 from the first equation, and x_1 from the second equation so that we end up with one equation containing x_1 alone, and the other equation containing x_2 alone. Eliminating an unknown is done mainly by adding an appropriate multiple of one equation to the other equation. For the preceding system we begin by

---

* Carl Friedrich Gauss (1777–1855), one of the greatest mathematicians, made significant contributions to probability–statistics, the theory of equations, number theory, astronomy, and magnetism. Camille Jordan (1838–1922) started his career as an engineer. He did memorable work in many areas of mathematics, including matrix theory, the theory of functions, measure theory, and what later became known as topology. He is famous for the Jordan curve theorem.

eliminating x_1 from the second equation. To do this we subtract 2 times the first equation from the second equation, as follows:

$$2x_1 - x_2 = 3$$

$$-2(x_1 + 2x_2 = 4)$$

Thus, we end up with $-5x_2 = -5$. The system of equations becomes the following, where Eq. (1a) is the same as Eq. (1), and Eq. (2) is replaced by the equation just obtained.

$$x_1 + 2x_2 = \quad 4 \tag{1a}$$

$$-5x_2 = -5 \tag{2a}$$

Next we divide Eq. (2a) by -5 to get

$$x_1 + 2x_2 = 4 \tag{1b}$$

$$x_2 = 1 \tag{2b}$$

Now it is an easy matter to eliminate x_2 from Eq. (1b). We subtract 2 times the second equation from Eq. (1b), thus converting the first equation to

$$x_1 + 2x_2 - 2x_2 = 4 - 2{\cdot}1, \quad \text{or} \quad x_1 = 2$$

Thus, our system has been reduced to

$$x_1 \quad = 2 \tag{1c}$$

$$x_2 = 1 \tag{2c}$$

This indicates the unique solution of our system of two equations in two unknowns. Geometrically, we have the two lines corresponding to the two equations intersecting at the single point (2, 1). Refer to Figure 2.2.

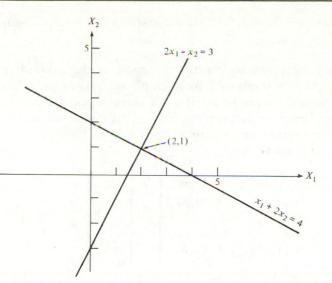

FIGURE 2.2   Lines with equations $x_1 + 2x_2 = 4$ and $2x_1 - x_2 = 3$ intersect at $(2, 1)$.

In general, we shall be interested in finding solutions for a general system of m linear equations in n unknowns. We indicate this as follows, where the a_{ij}'s and b_j's denote real number constants:

$$a_{11}x_1 + a_{12}x_2 + \ldots + a_{1n}x_n = b_1 \tag{1}$$

$$a_{21}x_1 + a_{22}x_2 + \ldots + a_{2n}x_n = b_2 \tag{2}$$

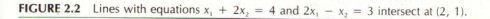

$$a_{m1}x_1 + a_{m2}x_2 + \ldots + a_{mn}x_n = b_m \tag{m}$$

FIGURE 2.3   General system of m linear equations in n unknowns.

By a solution to this system we mean a collection of values for the x_j's, namely, $x_1 = k_1, x_2 = k_2, \ldots, x_n = k_n$, which cause *each* of the preceding m equations to be satisfied. In our work all of the three possibilities $m < n$ (read m less than n), $m = n$, and $m > n$ (read m greater than n) will occur. The $m = n$ case is probably very familiar to the reader (same number of equations as unknowns), and the $m < n$ case (fewer equations than unknowns) will occur frequently in linear algebra.

Example (b):   Identify m and n and the a_{ij}'s and b_j's for the system of Example (a).

In this system, $m = n = 2$ (two equations in two unknowns), and $a_{11} = 1$, $a_{12} = 2$, $b_1 = 4$; $a_{21} = 2$, $a_{22} = -1$, and $b_2 = 3$.

5. It is a relatively easy matter to represent the general system of linear equations given in Figure 2.3 by a simple matrix–vector equation.

Now, two vectors are equal if their corresponding entries are equal, and conversely. Thus, we can replace Eqs. (1) through (m) of Figure 2.3 by the following *vector equation*, where on each side we have an $m \times 1$ matrix, or m component column vector:

$$\begin{bmatrix} a_{11}x_1 + a_{12}x_2 + \ldots + a_{1n}x_n \\ a_{21}x_1 + a_{22}x_2 + \ldots + a_{2n}x_n \\ \ldots \quad . \quad \ldots \quad . \quad \ldots \quad . \quad \ldots \\ \ldots \quad . \quad \ldots \quad . \quad \ldots \quad . \quad \ldots \\ a_{m1}x_1 + a_{m2}x_2 + \ldots + a_{mn}x_n \end{bmatrix} = \begin{bmatrix} b_1 \\ b_2 \\ . \\ . \\ b_m \end{bmatrix} \tag{I}$$

Observe again that the left side of this equation is an m-component column vector; for example, the second component is the sum $a_{21}x_1 + \ldots + a_{2n}x_n$.

We now define the *coefficient matrix* A, the *unknowns vector* $\mathbf{x}$, and the *right-side vector* $\mathbf{b}$, as follows:

$$A = \begin{bmatrix} a_{11} & a_{12} & \ldots & a_{1n} \\ a_{21} & a_{22} & \ldots & a_{2n} \\ . . & . . & \ldots & . . \\ . . & . . & \ldots & . . \\ a_{m1} & a_{m2} & \ldots & a_{mn} \end{bmatrix}, \quad \mathbf{x} = \begin{bmatrix} x_1 \\ x_2 \\ . \\ . \\ x_n \end{bmatrix}, \quad \mathbf{b} = \begin{bmatrix} b_1 \\ b_2 \\ . \\ . \\ b_m \end{bmatrix}$$

We thus see that the matrix equation (I) may be written in the concise form

$$A\mathbf{x} = \mathbf{b} \tag{I'}$$

The m terms of the product $A\mathbf{x}$ are given on the left side of Eq. (I). Note that $A\mathbf{x}$ is an $m \times 1$ matrix ($m \times n$ matrix times $n \times 1$ matrix). For example, the second component of $A\mathbf{x}$ is $a_{21}x_1 + a_{22}x_2 + \ldots + a_{2n}x_n$, obtained by forming products of the entries of row 2 of A with the components of $\mathbf{x}$ and then summing. The remaining $(m - 1)$ components of the $m \times 1$ matrix $A\mathbf{x}$ are computed similarly. Refer to frame 14 of Chapter 1 for the multiplication of a 2×3 matrix times a three-component column vector $\mathbf{x}$.

Theorem 2.1

The matrix–vector form of the general system of m linear equations in n unknowns is $A\mathbf{x} = \mathbf{b}$, where A, $\mathbf{x}$, and $\mathbf{b}$ are defined above.

Example (a): Indicate the matrix A and the vectors $\mathbf{x}$ and $\mathbf{b}$ for the system of Example (a) of frame 4.

$$A = \begin{bmatrix} 1 & 2 \\ 2 & -1 \end{bmatrix}, \qquad \mathbf{x} = \begin{bmatrix} x_1 \\ x_2 \end{bmatrix}, \qquad \mathbf{b} = \begin{bmatrix} 4 \\ 3 \end{bmatrix}$$

It is useful to compare A and $\mathbf{b}$ here, with the data given in Example (b) of frame 4. Observe that the solution values for x_1 and x_2 that we obtained in Example (a) of frame 4 can be indicated by the *solution column vector* $\mathbf{x}_s$, where

$$\mathbf{x}_s = \begin{bmatrix} 2 \\ 1 \end{bmatrix}, \qquad \text{or equivalently} \quad (\mathbf{x}_s)^t = [2, \ 1], \quad \text{or} \quad \mathbf{x}_s = [2, \ 1]^t$$

6. For the system of three linear equations in three unknowns:

$$x_1 - x_2 + x_3 = \quad 1 \tag{1}$$

$$2x_1 + x_2 + 3x_3 = \quad 4 \tag{2}$$

$$3x_1 - x_2 - x_3 = -5 \tag{3}$$

identify A, $\mathbf{x}$, and $\mathbf{b}$.

$$A = \begin{bmatrix} 1 & -1 & 1 \\ 2 & 1 & 3 \\ 3 & -1 & -1 \end{bmatrix}, \qquad \mathbf{x} = \begin{bmatrix} x_1 \\ x_2 \\ x_3 \end{bmatrix}, \qquad \mathbf{b} = \begin{bmatrix} 1 \\ 4 \\ -5 \end{bmatrix}$$

Note: Later on in this chapter (frame 18) we shall find that the unique solution for this system is $x_1 = -1$, $x_2 = 0$, and $x_3 = 2$. Interpreting the three preceding equations as representing three planes in three-dimensional space, we can interpret the solution as indicating that the three planes intersect at the single point $(-1, 0, 2)$. Remember, the solution satisfies *each* equation; equivalently, the point cited lies on *each* plane.

7. In Example (a) of frame 4 in effect we used the elimination method to determine the solution for a simple system of two equations in two unknowns. By applying the steps we used, we replaced our original system of linear equations by a new (and simpler) system that had the same solution but that was easier to analyze. In general, three types of operations, known as *elementary operations*, will be used on our starting system of equations, and on systems derived from it, in order to eliminate unknowns systematically.

Definition 2.1 (Elementary Operations on a System of Equations)

The three possible elementary operations are

(a)   Interchange two equations.
(b)   Multiply an equation by a nonzero constant.
(c)   Add to any equation a (nonzero) multiple of another equation.

In Example (a) of frame 4, we used a type (c) operation in going from Eqs. (1) and (2) to (1a) and (2a); the multiple was -2. We used a type (b) operation in going from Eqs. (1a) and (2a) to (1b) and (2b); the multiple was $-\frac{1}{5}$ [recall we divided by -5]. We again used a type (c) operation in going from Eqs. (1b) and (2b) to (1c) and (2c); the multiple was again -2. (Note that subtracting h times an equation is the same as adding on $-h$ multiplied by an equation.) On occasion, in future calculations, we shall use type (a) operations.

The elimination method works, since each of the elementary operations results in a system that is *equivalent* to the previous system in the sense that any collection of values for x_1, x_2, and so on, which is a solution of the previous system, will also be a solution of the new system, and conversely.

It should be clear that interchanging any two equations in a system will leave the solutions of the system unchanged. Also, multiplying an equation by a nonzero constant will not change the solutions.

Thus, application of elementary operations (a) and (b) will yield an equivalent system. It is not hard to show that elementary operation (c) has this property also.

In the next stage of our work we shall pave the way for the development of a matrix method for finding solutions to a system of m linear equations in n unknowns. This matrix method, known as the Gauss–Jordan elimination method, is the counterpart of the elimination method illustrated in frame 4.

It is useful to define the augmented matrix $A|\mathbf{b}$. Refer to Figure 2.3 (frame 4), Theorem 2.1 (frame 5), and the equations for A and $\mathbf{b}$ that precede Theorem 2.1.

Definition 2.2

The augmented coefficient matrix $A|\mathbf{b}$ for the general system of m linear equations in n unknowns is given by

$$A|\mathbf{b} = \left[\begin{array}{cccc|c} a_{11} & a_{12} & \ldots & a_{1n} & b_1 \\ a_{21} & a_{22} & \ldots & a_{2n} & b_2 \\ \ldots & \ldots & \ldots & \ldots & . \\ \ldots & \ldots & \ldots & \ldots & . \\ a_{m1} & a_{m2} & \ldots & a_{mn} & b_m \end{array}\right]$$

The matrix $A|\mathbf{b}$ is of size $m \times (n + 1)$.

Note: It is important to observe that the purpose of the vertical line is to set off the right-side coefficients b_1 through b_m. Thus, we would still have the same augmented matrix if the vertical line were omitted. Some authors use a broken line $\left(\vdots\right)$ and others use no line at all to set off the right-side coefficients.

Example: Determine the 2×3 matrix $A|\mathbf{b}$ for the system of linear equations in Example (a) of frame 4.

- - - - - - - - - - -

$$A|\mathbf{b} = \begin{bmatrix} 1 & 2 & 4 \\ 2 & -1 & 3 \end{bmatrix}$$

8. Observe that the rows in the augmented matrix $A|\mathbf{b}$ correspond to the equations in the related system of linear equations (except that now the variables $x_1, x_2, \ldots, x_n$ are omitted). Thus, it should be possible to find the solution(s) of a system by working solely with the augmented matrix, and on matrices derived from it, through execution of elementary operations.

We restate the elementary operations, now called elementary *row* operations, as they would apply to the augmented matrix $A|\mathbf{b}$ and to matrices derived from it.

Definition 2.3 (Elementary Row Operations on a Matrix)

The three possible elementary row operations are

(a) Interchange two rows.
(b) Multiply a row by a nonzero constant.
(c) Add to any row a (nonzero) multiple of another row.

Note: It is useful to compare the statements in Definitions 2.3 and 2.1. The only difference is in the replacement of the word *equation* by *row*.

Example: Refer to the augmented matrix for the system of Example (a) of frame 4 (see the answer to frame 7). Apply matrix elementary row operations that correspond to those used in Example (a) of frame 4.

Solution: We start with

$$\begin{bmatrix} 1 & 2 & 4 \\ 2 & -1 & 3 \end{bmatrix} \tag{I}$$

Our first goal is to transform the coefficient 2 in row 2 to a zero (this is equivalent to "eliminating" the x_1 term from the second equation).

We use a type (c) row operation on row 2 in which we add (-2) times the first row to the second row. (This is equivalent to *subtracting* 2 times the first row.) This will transform the coefficient 2 in row 2 to a zero. Note that the row being transformed is the second row, not the first row! The new entries for the second row are given by

$$2 + (-2)\cdot 1 = 0; \qquad -1 + (-2)\cdot 2 = -5; \qquad 3 + (-2)\cdot 4 = -5.$$

Thus, augmented matrix (I) is transformed to

$$\left[\begin{array}{cc|c} 1 & 2 & 4 \\ 0 & -5 & -5 \end{array}\right], \tag{Ia}$$

which corresponds directly to Eqs. (1a) and (2a) of Example (a) of frame 4.

A convenient symbolic way of representing the change from matrix (I) to matrix (Ia) is by the equation

$$r_2' = r_2 + (-2)r_1 \qquad \text{or} \qquad r_2' = r_2 - 2r_1 .$$

This says that the *new* row 2, r_2', is equal to the *old* row 2, r_2, plus (-2) times the *old* row 1, r_1.

Next, we transform the leading -5 in row 2 to a 1 by employing a type (b) row operation in which the multiple is $(-\frac{1}{5})$. Here we obtain

$$\left[\begin{array}{cc|c} 1 & 2 & 4 \\ 0 & 1 & 1 \end{array}\right], \tag{Ib}$$

which corresponds directly to Eqs. (1b) and (2b) of Example (a) of frame 4. We can indicate the change from matrix (Ia) to matrix (Ib) by the symbolic equation $r_2' = -\frac{1}{5}r_2$. (Again, the prime symbol indicates *new* and nonprime indicates *old*, or former.)

Now we can transform the 2 in row 1, column 2 to a zero (equivalently, we eliminate the x_2 term from the first equation) by using a type (c) row operation on row 1, in which we add (-2) times the second row to the first row. Determine the new entries for row 1 and the new augmented matrix.

— — — — — — — — — — —

The new entries for row 1 are

$$1 + (-2) \cdot 0 = 1; \qquad 2 + (-2) \cdot 1 = 0; \qquad 4 + (-2) \cdot 1 = 2$$

Row 2 is not altered. Thus, we obtain

$$\left[\begin{array}{cc|c} 1 & 0 & 2 \\ 0 & 1 & 1 \end{array}\right], \tag{Ic}$$

which corresponds directly to Eqs. (1c) and (2c) of Example (a) of frame 4. The symbolic equation indicating the change from matrix (Ib) to matrix (Ic) is

$$r_1' = r_1 + (-2)r_2 = r_1 - 2r_2$$

9. It is important to be able to interpret transformed augmented matrices in terms of equations involving the original variables $x_1, x_2, \ldots$, etc. A key point is that when we perform an elementary row operation on a matrix $A|\mathbf{b}$ or on an augmented matrix derived from it, the resulting new augmented matrix corresponds to a system of equations that is *equivalent* to the original system of equations. In particular, any collection of values for x_1, x_2, etc., that is a solution of the original system will also be a solution to any new system of equations, and conversely.

The concept of equivalent systems of linear equations was previously discussed in frame 7, in the text between Definitions 2.1 and 2.2. For example, suppose we apply an elementary row operation to the augmented matrix $A|\mathbf{b}$ (which corresponds to the original system $A\mathbf{x} = \mathbf{b}$) and obtain $A'|\mathbf{b}'$. The new system of equations given by $A'\mathbf{x} = \mathbf{b}'$ is *equivalent* to the original system. Referring to (Ia) of frame 8, we see that the following is equivalent to the original system of equations, where $A' = \begin{bmatrix} 1 & 2 \\ 0 & -5 \end{bmatrix}$ and $\mathbf{b}' = \begin{bmatrix} 4 \\ 5 \end{bmatrix}$, here. That is, we have

$$\begin{bmatrix} 1 & 2 \\ 0 & -5 \end{bmatrix} \mathbf{x} = \begin{bmatrix} 4 \\ -5 \end{bmatrix} \quad \text{or} \quad \begin{bmatrix} 1 & 2 \\ 0 & -5 \end{bmatrix}\begin{bmatrix} x_1 \\ x_2 \end{bmatrix} = \begin{bmatrix} 4 \\ -5 \end{bmatrix} \tag{1}$$

Carrying out the matrix multiplication on the left of the equation (2×2 matrix times 2×1 matrix) and then equating corresponding components on left and right leads to

$$\begin{bmatrix} x_1 + 2x_2 \\ 0 - 5x_2 \end{bmatrix} = \begin{bmatrix} 4 \\ -5 \end{bmatrix} \quad \text{and} \quad \begin{aligned} x_1 + 2x_2 &= 4 \\ -5x_2 &= -5 \end{aligned} \qquad \begin{aligned} &\text{(1a)} \\ &\text{(1b)} \end{aligned}$$

Thus, we arrive at Eqs. (1a) and (1b) of Example (a) of frame 4 again; the latter are, of course, equivalent to the original system of equations.

Note: There is a simple schematic way of determining a set of linear equations that corresponds to an augmented matrix. The method consists first of listing the variables x_1, x_2, and so on, consecutively, in a horizontal margin (called a *top margin*) above the coefficient matrix part of the augmented matrix. Thus, for $A'|\mathbf{b}' = \begin{bmatrix} 1 & 2 & 4 \\ 0 & -5 & -5 \end{bmatrix}$ from the preceding, we would write

$$x_1 \qquad x_2$$

$$\begin{bmatrix} 1 & 2 & 4 \\ 0 & -5 & -5 \end{bmatrix}$$

To determine the linear equation corresponding to the first row of the augmented matrix, we first multiply the first row entries to the left of the dividing line by the margin variables x_1 and x_2, respectively, and then sum the products. Next, after treating the vertical dividing line as an "equals sign," we obtain $1 \cdot x_1 + 2 \cdot x_2 = 4$, or

$$x_1 + 2x_2 = 4,$$

as the equation corresponding to row 1 of $A'|\mathbf{b}'$. This is Eq. (1a) given above.

The equation corresponding to the second row of $A'|\mathbf{b}'$ is obtained in similar fashion. Thus, we have

$$0 \cdot x_1 + (-5) \cdot x_2 = -5, \qquad \text{or} \qquad -5x_2 = -5,$$

and this is Eq. (1b). Here we first multiplied 0 and -5 in row 2 by x_1 and x_2, respectively; then we equated the sum of the two products thus obtained to the entry in row 2 of $A'|\mathbf{b}'$ to the right of the dividing line.

Example: From the augmented matrix (Ic) of frame 8, determine the solution of the original system of equations.

Solution: First, observe that we can write (Ic) as $\left[I_2 \;\middle|\; \begin{matrix} 2 \\ 1 \end{matrix} \right]$ since I_2, the 2×2 identity matrix, is given by $I_2 = \begin{bmatrix} 1 & 0 \\ 0 & 1 \end{bmatrix}$. The identity matrices were introduced in frame 15 of Chapter One. Thus, a corresponding system of equations is $I_2\mathbf{x} = \begin{bmatrix} 2 \\ 1 \end{bmatrix}$, which can be written as follows:

$$\begin{bmatrix} 1 & 0 \\ 0 & 1 \end{bmatrix}\begin{bmatrix} x_1 \\ x_2 \end{bmatrix} = \begin{bmatrix} 2 \\ 1 \end{bmatrix} \qquad \text{or} \qquad \begin{bmatrix} x_1 \\ x_2 \end{bmatrix} = \begin{bmatrix} 2 \\ 1 \end{bmatrix}$$

Now finish the problem.

– – – – – – – – – –

Equating corresponding components on the left and right of the final equation leads to $x_1 = 2$ and $x_2 = 1$. These values for x_1 and x_2 constitute the solution of the original system of equations. [Refer to Example (a) of frame 4.]

10. Let us focus for a while on the very important case of a system of n linear equations in n unknowns. Thus, suppose A is $n \times n$ and $A|\mathbf{b}$ is $n \times (n + 1)$. A result we strive for (but don't always achieve) is the obtaining of a *unique* solution for $x_1, x_2, \ldots, x_n$. When we say unique solution, we mean only one solution. The following theorem generalizes the result of the Example of frame 9. Also, I_n denotes the $n \times n$ identity matrix and $\mathbf{k}$ denotes an n-component column vector with components $k_1, k_2, \ldots, k_n$.

> *Theorem 2.2 (n Equations in n Unknowns Case)*
>
>   If the augmented matrix $A|\mathbf{b}$ can be transformed by means of elementary row operations to the form $I_n|\mathbf{k}$, then the unique solution to the original system is given by $x_1 = k_1, x_2 = k_2, \ldots, x_n = k_n$.

Notes: (a) We can also say that the unique solution vector $\mathbf{x}_s$ is given by $\mathbf{x}_s = \mathbf{k}$. (b) If it is not possible to transform the matrix portion on the left of the dividing line to I_n, then this indicates that a unique solution for $x_1, x_2, \ldots, x_n$ does not exist. We shall deal with situations of this type later in this chapter (Section C) and in other chapters.

Discussion of Theorem: The key aspect to justifying the theorem is to realize that the solution(s) for $A\mathbf{x} = \mathbf{b}$ are identical to the solution(s) for $A^*\mathbf{x} = \mathbf{b}^*$ if $A^*|\mathbf{b}^*$ is an augmented matrix derived from $A|\mathbf{b}$ through a sequence of elementary row operations. The idea of equivalent systems (see frame 7) applies here.
  Now if we obtain $I_n|\mathbf{k}$ from $A|\mathbf{b}$ through a sequence of elementary row operations, this means that the solution(s) for $A\mathbf{x} = \mathbf{b}$ are identical to the solution(s) for $I_n\mathbf{x} = \mathbf{k}$. But $I_n\mathbf{x} = \mathbf{x}$ and, thus, we obtain the unique result $\mathbf{x} = \mathbf{k}$, which means $x_1 = k_1, x_2 = k_2, \ldots, x_n = k_n$, if we equate corresponding components on both sides of the vector equation $\mathbf{x} = \mathbf{k}$.

Illustrate this theorem for the $n = 2$ case. A specific situation for $n = 2$ is provided by the example of frame 9.

— — — — — — — — — —

For example, for the $n = 2$ case, we would have

$$\begin{bmatrix} 1 & 0 \\ 0 & 1 \end{bmatrix} \begin{bmatrix} x_1 \\ x_2 \end{bmatrix} = \begin{bmatrix} k_1 \\ k_2 \end{bmatrix}$$

as in the example of frame 9. The left side, which equals $I_2\mathbf{x}$, reduces to $\begin{bmatrix} x_1 \\ x_2 \end{bmatrix}$, which is $\mathbf{x}$. Thus, here $\mathbf{x} = \mathbf{k}$, or, equivalently, $x_1 = k_1$ and $x_2 = k_2$.

B. THE GAUSS–JORDAN ELIMINATION METHOD

11.   Shortly, we will present the Gauss–Jordan elimination method. The method involves a systematic procedure, using elementary row operations, for transforming the original augmented matrix into a form from which the solution(s) of the original system can be easily obtained. For example, in frame 8, we generated the augmented matrix

$$\begin{bmatrix} 1 & 0 & | & 2 \\ 0 & 1 & | & 1 \end{bmatrix} \tag{Ic}$$

from which we easily obtained the unique solution $x_1 = 2$, $x_2 = 1$ (frame 9).

Matrix (Ic) is an example of a matrix that is said to be in *reduced row echelon form*. The Gauss–Jordan elimination method consists of a sequence of steps (each step using an elementary row operation) that converts a matrix into reduced row echelon form. The general definition of reduced row echelon form is now presented.

Definition 2.4

A matrix is in *reduced row echelon form* if it satisfies the following properties:

(1)   All rows, consisting entirely of zeros, if any, are at the bottom of the matrix.

(2)   For a row that does not consist entirely of zeros, the first nonzero entry of the row is a 1. We call this entry the *leading entry* of the row. Often we shall underline the leading entry of the row (as in 1). Also, for the leading entry of row i, we identify the column location (or label) of the leading entry as column J_i.

(3)   If row i and $(i + 1)$ are two successive rows that do not consist entirely of zeros, then the leading entry of row $(i + 1)$ is to the right of the leading entry of row i. That is, $J_{(i+1)} > J_i$. [Recall that row $(i + 1)$ is positionally lower than row i.]

(4)   For a column that has a leading entry, *all other* entries in that column are *zeros*. Such a column will often be referred to as a *cleared-out column*. Thus, a cleared-out column has a 1 for one entry, and 0's for all other entries.

In our discussions on whether matrices are in reduced row echelon form we will not distinguish between whether a matrix is augmented or not. For example, both $\begin{bmatrix} \underline{1} & 0 & | & 2 \\ 0 & \underline{1} & | & 1 \end{bmatrix}$ and $\begin{bmatrix} \underline{1} & 0 & 2 \\ 0 & \underline{1} & 1 \end{bmatrix}$ are in reduced row echelon form.

Example: The following matrices are in reduced row echelon form:

$$
\text{(a)} \begin{bmatrix} 1 & 0 & 0 & | & -1 \\ 0 & 1 & 0 & | & 0 \\ 0 & 0 & 1 & | & 2 \end{bmatrix}, \quad
\text{(b)} \begin{bmatrix} 1 & 0 & 0 \\ 0 & 1 & 0 \\ 0 & 0 & 1 \end{bmatrix}, \quad
\text{(c)} \begin{bmatrix} 1 & 0 & 5 & 0 \\ 0 & 1 & 3 & 0 \\ 0 & 0 & 0 & 1 \end{bmatrix},
$$

$$
\text{(d)} \begin{bmatrix} 1 & 3 & 0 & | & 0 \\ 0 & 0 & 1 & | & 6 \\ 0 & 0 & 0 & | & 0 \end{bmatrix}, \quad
\text{(e)} \begin{bmatrix} 1 & 4 & 0 & | & 0 \\ 0 & 0 & 1 & | & 0 \\ 0 & 0 & 0 & | & 1 \end{bmatrix}, \quad
\text{(f)} \begin{bmatrix} 1 & 0 & 0 & -2 \\ 0 & 1 & 8 & 6 \\ 0 & 0 & 0 & 0 \\ 0 & 0 & 0 & 0 \end{bmatrix}
$$

To illustrate the J_i notation of part (2) of Definition 2.4 (recall that J_i is the column location of the leading entry in row i), determine the list of J_i's for each of the preceding matrices.

Solution: (a) and (b): $J_1 = 1$, $J_2 = 2$, $J_3 = 3$. Now do parts (c) through (f).

— — — — — — — —

(c) $J_1 = 1$, $J_2 = 2$, $J_3 = 4$; (d) $J_1 = 1$, $J_2 = 3$; (e) $J_1 = 1$, $J_2 = 3$, $J_3 = 4$; (f) $J_1 = 1$, $J_2 = 2$.

Note that each cleared-out column has an underlined 1 in its leading entry position and zeros for all other entries.

12.

Example (a): None of the following matrices is in reduced row echelon form, since each fails to satisfy all the properties (1), (2), (3), and (4) of Definition 2.4.

$$
\text{(a)} \begin{bmatrix} 1 & 0 & 5 & | & -2 \\ 0 & 1 & -4 & | & 6 \\ 0 & 0 & 1 & | & 2 \end{bmatrix}, \quad
\text{(b)} \begin{bmatrix} 1 & 3 & 0 & 8 \\ 0 & 0 & 0 & 0 \\ 0 & 0 & 1 & 4 \end{bmatrix},
$$

$$
\text{(c)} \begin{bmatrix} 1 & 0 & 0 & | & 2 \\ 0 & 1 & 0 & | & 2 \\ 0 & 0 & 3 & | & 7 \\ 0 & 0 & 0 & | & 0 \end{bmatrix}, \quad
\text{(d)} \begin{bmatrix} 1 & 0 & 0 & 7 & 2 \\ 0 & 0 & 1 & 4 & 3 \\ 0 & 1 & 0 & -2 & 7 \\ 0 & 0 & 0 & 0 & 0 \end{bmatrix}
$$

Example (b): Suppose we wish to find solutions for the system

$$x_1 - x_2 + x_3 = 1 \tag{1}$$

$$2x_1 + x_2 + 3x_3 = 4 \tag{2}$$

$$3x_1 - x_2 - x_3 = -5 \tag{3}$$

This system represents three planes in three-dimensional space. The augmented matrix for this system is

$$A|\mathbf{b} = \begin{bmatrix} 1 & -1 & 1 & 1 \\ 2 & 1 & 3 & 4 \\ 3 & -1 & -1 & -5 \end{bmatrix} \tag{4}$$

The system itself in matrix–vector form is $A\mathbf{x} = \mathbf{b}$, with A and $\mathbf{b}$ as given by (4) (A to the left of the dividing line and $\mathbf{b}$ to the right). By employing the Gauss–Jordan elimination method (as we shall do in frame 18), we can obtain the following matrix in reduced row echelon form:

$$A^*|\mathbf{b}^* = \begin{bmatrix} \underline{1} & 0 & 0 & -1 \\ 0 & \underline{1} & 0 & 0 \\ 0 & 0 & \underline{1} & 2 \end{bmatrix} \tag{5}$$

Now the system $A^*\mathbf{x} = \mathbf{b}^*$ is equivalent to the original system. Here $A^*\mathbf{x} = I_3\mathbf{x} = \mathbf{x}$. From (5) determine the unique solution of the original system.

_____

Since (5) is equivalent to $\mathbf{x} = \mathbf{b}$, with $\mathbf{b}$ given by $[-1, 0, 2]^t$, we have $[x_1, x_2, x_3]^t = [-1, 0, 2]^t$. Thus, the unique solution is given by $x_1 = -1, x_2 = 0, x_3 = 2$. The geometrical interpretation is that the three planes (1), (2), and (3) intersect at the single point $(-1, 0, 2)$.

13.   Let us continue now with our preparatory discussion of the Gauss–Jordan elimination method. The method may be summarized in rough form, as follows. (Stage 1 applies to row 1, Stage 2 to row 2, etc.)

Stage 1:   Determine a leading entry for row 1, which is as far to the left as possible by using, if necessary, elementary row operations of types (a) and (b). (Henceforth, such a leading entry will be called a left-most leading entry.) Let the column label for the leading entry of row 1 be J_1.

For example, if

$$A|\mathbf{b} = \begin{bmatrix} 0 & 0 & 2 & 6 \\ 0 & 0 & 1 & 3 \\ 0 & 1 & 0 & 2 \end{bmatrix}$$

we would interchange rows 1 and 3 to obtain

$$A^*|\mathbf{b}^* = \begin{bmatrix} 0 & \underline{1} & 0 & 2 \\ 0 & 0 & 1 & 3 \\ 0 & 0 & 2 & 6 \end{bmatrix}$$

Then $J_1 = 2$, since the leading entry for row 1 is in column 2. Thus, only a type (a) row operation is needed here. The operation can be indicated here symbolically by $r_1 \leftrightarrow r_3$.

The next step in Stage 1 would be to clear out column J_1. For this purpose we would use, if necessary, elementary row operations of type (c). [In $A^*|\mathbf{b}^*$, immediately preceding, column 2 is already cleared out, so no type (c) operation is needed.]

Stage 2:   In the submatrix determined by rows 2, 3, ..., m; determine a left-most leading entry for row 2 [by using, if necessary, elementary row operations of types (a) and (b)]. Let the column label for the leading entry of row 2 be J_2.

Clear out the *entire* column J_2. (This includes generating a zero in the row 1 position.) Here again we use elementary row operations of type (c).

Then we do Stages 3, 4, etc., on rows 3, 4, etc., until our matrix is in reduced row echelon form. Initially, in Stage 3, we focus on the submatrix determined by rows 3, 4, ..., m.

The most unusual aspect of applying the Gauss–Jordan elimination method involves the clearing out of a column.

For example, suppose that $A|\mathbf{b}$ is 3×4 (as are all subsequent matrices), and we have completed Stage 1 and the leading entry part of Stage 2. For a particular case where $J_1 = 1$ and $J_2 = 2$, say, the augmented matrix would appear as follows:

$$\left[\begin{array}{ccc|c} \underline{1} & a_{12} & a_{13} & b_1 \\ 0 & \underline{1} & a_{23} & b_2 \\ 0 & a_{32} & a_{33} & b_3 \end{array}\right]$$

(Note that the symbols a_{12}, a_{13}, b_1 denote current entries in row 1 and that they are undoubtedly different in value from the initial values for these entries. The same is true for the other subscripted entries.)

Now we wish to clear out column 2. That is, we wish to transform the a_{12} in row 1 and the a_{32} in row 3 to zeros. The correct way to transform the a_{12} to a zero is by using the type (c) row operation indicated by $r_1' = r_1 + (-a_{12})r_2 = r_1 - a_{12}r_2$. That is, the *new* row 1 is equal to the old row 1 minus a_{12} times the old row 2. Thus, the new entries in row 1 will be

$$1 - a_{12} \cdot 0 = 1, \qquad a_{12} - a_{12} \cdot 1 = 0,$$

$$a_{13} - a_{12} \cdot a_{23}, \quad \text{and} \quad b_1 - a_{12} \cdot b_2 ,$$

respectively. Similarly, the correct transformation for row 3 to cause the new a_{32} value to become zero is $r_3' = r_3 - a_{32}r_2$.

The ideas of the preceding discussion are best illustrated with an example.

Example: Suppose that after completing Stage 1 and the leading entry part of Stage 2, we have the following matrix:

$$\left[\begin{array}{ccc|c} \underline{1} & -1 & 1 & 1 \\ 0 & \underline{1} & \frac{1}{3} & \frac{2}{3} \\ 0 & 2 & -4 & -8 \end{array}\right]$$

Use type (c) elementary row operations to clear out column 2.

Solution: Symbolically, the proper row operations are $r_1' = r_1 - a_{12}r_2 = r_1 - (-1)r_2 = r_1 + r_2$, and $r_3' = r_3 - a_{32}r_2 = r_3 - 2r_2$. Note that for both rows, we shall be adding on a multiple of row 2 that will cause the new values for a_{12} and a_{32} to be zero. The multiples are then $-a_{12}$ and $-a_{32}$, respectively.

Now complete the calculations.

_ _ _ _ _ _ _ _ _ _

The computations for the new row 1 entries are as follows: $1 + 0 = 1$, $-1 + 1 = 0$, $1 + \frac{1}{3} = \frac{4}{3}$, $1 + \frac{2}{3} = \frac{5}{3}$.

The computations for the new row 3 entries are as follows: $0 - 2 \cdot 0 = 0$, $2 - 2 \cdot 1 = 0$, $-4 - 2(\frac{1}{3}) = -\frac{14}{3}$, $-8 - 2(\frac{2}{3}) = -\frac{28}{3}$.

The resulting matrix is as follows:

$$\left[\begin{array}{ccc|c} \underline{1} & 0 & \frac{4}{3} & \frac{5}{3} \\ 0 & \underline{1} & \frac{1}{3} & \frac{2}{3} \\ 0 & 0 & -\frac{14}{3} & -\frac{28}{3} \end{array}\right]$$

14. We are now ready to illustrate all the aspects of the Gauss–Jordan elimination method. It will be used to transform any matrix (augmented or not) to reduced row echelon form.

To make the method more easy to understand we shall apply it to the augmented matrix for the following system of three equations in four unknowns:

$$7x_3 + 14x_4 = -7 \tag{1}$$

$$2x_1 - 8x_2 + 4x_3 + 18x_4 = 0 \tag{2}$$

$$3x_1 - 12x_2 - x_3 + 13x_4 = 7 \tag{3}$$

Here

$$A|\mathbf{b} = \left[\begin{array}{cccc|c} 0 & 0 & 7 & 14 & -7 \\ 2 & -8 & 4 & 18 & 0 \\ 3 & -12 & -1 & 13 & 7 \end{array}\right]$$

In the illustration of the Gauss–Jordan elimination method (Method 2.1), the *general* steps are bracketed (as in [   ]) and are located to the right of the central dividing line.

Method 2.1 (Gauss–Jordan Elimination Method)

Stage 1 (for Row 1)

$$\begin{bmatrix} 0 & 0 & 7 & 14 & -7 \\ 2 & -8 & 4 & 18 & 0 \\ 3 & -12 & -1 & 13 & 7 \end{bmatrix}$$
↑
Left-most nonzero column.

[*Step 1:* Locate the left-most column that does not contain all zeros. This locates column J_1.]

Here the left-most nonzero column is column 1. Thus, $J_1 = 1$.

$$\begin{bmatrix} 2 & -8 & 4 & 18 & 0 \\ 0 & 0 & 7 & 14 & -7 \\ 3 & -12 & -1 & 13 & 7 \end{bmatrix}$$

[*Step 2:* If necessary, interchange the top row with another row such that the entry at top of column J_1 is unequal to zero.]

Here we interchange rows 1 and 2 [type (a) row operation]. In symbols, $r_1 \leftrightarrow r_2$.

$$\begin{bmatrix} 1 & -4 & 2 & 9 & 0 \\ 0 & 0 & 7 & 14 & -7 \\ 3 & -12 & -1 & 13 & 7 \end{bmatrix}$$

[*Step 3:* Let the entry now at the top of the column found at step 1 be a. If $a \neq 1$, multiply the first row by $1/a$ in order to generate a leading entry for row 1. [Type (b) row operation: $r_1' = \left(\dfrac{1}{a}\right) r_1$.] If $a = 1$, no multiplication is needed. Underline leading entry in row 1.]

Here $r_1' = \frac{1}{2}r_1 = r_1/2$, since we multiply row 1 by $\frac{1}{2}$.

$$\begin{bmatrix} 1 & -4 & 2 & 9 & 0 \\ 0 & 0 & 7 & 14 & -7 \\ 0 & 0 & -7 & -14 & 7 \end{bmatrix}$$

[*Step 4:* Use type (c) row operations on rows $2, 3, \ldots, m$, when necessary, to clear out column J_1. This completes Stage 1.]

Here leave row 2 alone ($r_2' = r_2$), and use $r_3' = r_3 - 3r_1$ on row 3.

Stage 2 (for Row 2)

$$\begin{bmatrix} 1 & -4 & 2 & 9 & 0 \\ 0 & 0 & 7 & 14 & -7 \\ 0 & 0 & -7 & -14 & 7 \end{bmatrix}$$
↑
Left-most nonzero column in submatrix.

[*Step 1:* Focus on the submatrix consisting of rows 2, 3, etc. For this submatrix locate the left-most column that does not consist of all zeros. This locates column J_2.]

Here the left-most nonzero column is column 3, and thus, $J_2 = 3$.

$$\begin{bmatrix} \underline{1} & -4 & 2 & 9 & | & 0 \\ 0 & 0 & 7 & 14 & | & -7 \\ 0 & 0 & -7 & -14 & | & 7 \end{bmatrix}$$

Stage 2 (for Row 2)

$\begin{bmatrix} \text{\textit{Step 2:} Repeat Step 2 of Stage 1 for} \\ \text{submatrix referred to in Step 1 of} \\ \text{Stage 2.} \end{bmatrix}$

Here *no change* is necessary, since $a_{23} \neq 0$ ($a_{23} = 7$ here).

$$\begin{bmatrix} \underline{1} & -4 & 2 & 9 & | & 0 \\ 0 & 0 & \underline{1} & 2 & | & -1 \\ 0 & 0 & -7 & -14 & | & 7 \end{bmatrix}$$

$\begin{bmatrix} \text{\textit{Step 3:} Repeat Step 3 of Stage 1 for} \\ \text{submatrix referred to in Step 1 of} \\ \text{Stage 2.} \end{bmatrix}$

Here $r_2' = r_2/7$, since we multiply row 2 by $1/7$.

$$\begin{bmatrix} \underline{1} & -4 & 0 & 5 & | & 2 \\ 0 & 0 & \underline{1} & 2 & | & -1 \\ 0 & 0 & 0 & 0 & | & 0 \end{bmatrix}$$

$\begin{bmatrix} \text{\textit{Step 4:} Now focus again on the \textit{entire}} \\ \text{matrix. Use type (c) row operations} \\ \text{on rows } 1, 3, \ldots, m, \text{ when necessary,} \\ \text{to clear out column } J_2. \end{bmatrix}$

Here we use $r_1' = r_1 - 2r_2$ on row 1 and $r_3' = r_3 + 7r_2$ on row 3. Our matrix is now in reduced row echelon form, since rows 1 and 2 have leading entries, and row 3 consists entirely of zeros. Method 2.1 is thus terminated.

If the matrix is not in reduced row echelon form after two stages, then Stage 3 is used. In steps 1, 2, and 3 of Stage 3 we would focus on the submatrix consisting of rows $3, 4, \ldots, m$. Thus, we see that we have a similar pattern for all stages.

We terminate after a particular stage if the matrix has been transformed into reduced row echelon form.

Note: The procedure outlined in Method 2.1 is well suited for computer programming. (In fact, many computer programs are currently in existence that are similar in structure. We postpone our discussion on computer usage until Chapter 9.) In particular, the definition of the variable J_i, which locates the column label of the leading entry of row i, is very useful for computational purposes. (Observe that $J_1 < J_2 < J_3 < \ldots$, in general.) So also is the r_i, r_i' notation.

Example: Interpret the reduced row echelon form just determined (in the sample calculation pertaining to the presentation of Method 2.1).

Solution: Letting

$$A^* | \mathbf{b}^* = \begin{bmatrix} \underline{1} & -4 & 0 & 5 & | & 2 \\ 0 & 0 & \underline{1} & 2 & | & -1 \\ 0 & 0 & 0 & 0 & | & 0 \end{bmatrix},$$

our original system of equations is equivalent to $A^*\mathbf{x} = \mathbf{b}^*$, that is, to

$$\begin{bmatrix} \underline{1} & -4 & 0 & 5 \\ 0 & 0 & \underline{1} & 2 \\ 0 & 0 & 0 & 0 \end{bmatrix} \begin{bmatrix} x_1 \\ x_2 \\ x_3 \\ x_4 \end{bmatrix} = \begin{bmatrix} 2 \\ -1 \\ 0 \end{bmatrix}$$

Now multiply on the left and then equate corresponding components to obtain scalar equations from the preceding matrix–vector equation.

--- --- --- --- --- --- ---

We are led to the equations

$$x_1 - 4x_2 \qquad\qquad + 5x_4 = \quad 2 \tag{1}$$

$$x_3 + 2x_4 = -1 \tag{2}$$

We discard the last equation $0x_1 + 0x_2 + 0x_3 + 0x_4 = 0$, since the left side is automatically equal to zero, regardless of the values of the x_j's.

15. Variables that correspond to leading entries are called *leading variables*. In the example of frame 14, the leading variables are x_1 and x_3 [underlined in Eqs. (1) and (2)]. Variables that are not leading variables are called *nonleading variables*. Thus, in the example of frame 14, the nonleading variables are x_2 and x_4. In Eqs. (1) and (2) in the answer for frame 14, let us now solve for the leading variables x_1 and x_3 in terms of the nonleading variables.

$$x_1 \quad = \quad 2 + 4x_2 - 5x_4 \tag{1a}$$

$$x_3 = -1 \qquad\quad - 2x_4 \tag{2a}$$

If we set x_2 and x_4 equal to the symbols r and s, respectively, we can represent the solutions to our linear system by

$$x_1 \qquad = \quad 2 + 4r - 5s \tag{1b}$$

$$x_3 \quad = -1 \qquad - 2s \tag{2b}$$

$$x_2 \quad = \qquad\qquad r \tag{3b}$$

$$x_4 = \qquad\qquad\quad s \tag{4b}$$

Now if r and s are assigned arbitrary real values (i.e., any real number values whatsoever) and if we then solve for x_1, x_3, x_2 ($= r$), and x_4 ($= s$) from Eqs. (1b), (2b), (3b), and (4b), we will have a solution to our original linear system. Thus, it is clear that our original linear system has *infinitely many* solutions. For example, one solution is as follows: With $r = 0$ and $s = 0$, we have $x_1 = 2$, $x_3 = -1$, $x_2 = 0$, $x_4 = 0$.

Now determine solutions for (a) $r = 1$ and $s = 0$, and (b) $r = 0$ and $s = 1$.

_ _ _ _ _ _ _ _ _ _

(a)   With $r = 1$ and $s = 0$, we have $x_1 = 6$, $x_3 = -1$, $x_2 = 1$, $x_4 = 0$.

(b)   With $r = 0$ and $s = 1$, we have $x_1 = -3$, $x_3 = -3$, $x_2 = 0$, $x_4 = 1$.

We can continue in this fashion, and list other solutions for arbitrary choices of real number values for r and s.

Note: Often in our work we shall replace nonleading variables by arbitrary value symbols, such as r, s, and so on.

C. GEOMETRICAL INTERPRETATIONS; HOMOGENEOUS SYSTEMS

16.   One of the most powerful features of the Gauss–Jordan elimination method is that it enables us to interpret solutions of systems of linear equations in a geometrically plausible way. This is readily apparent for systems with two and three variables. For example, we can interpret a system of linear equations in two variables as representing lines in two-dimensional space (i.e., in a plane). For three-variable systems we would have planes lying in three-dimensional space. Since our "real life" experience is geared to visualizing phenomena in two and three dimensions, we should have a sound intuition about the different cases that can occur in two and three dimensions. In fact, we shall see that there are three possible cases; these are (a) unique solution, (b) infinitely many solutions, and (c) no solution.

Of great importance is the fact that these three cases, which are easy to visualize in two and three dimensions, apply also to the general n-dimensional (or n variable) case.

Let us now proceed to use Method 2.1 to analyze familiar type systems. We shall focus, for a while, on simple systems involving n equations in n unknowns. (Note that we loosely interchange the words *variable* and *unknown* in our discussions.)

Let us consider some examples dealing with two equations in two unknowns. (We already considered a situation with two equations in two unknowns that led to a unique solution, in a series of calculations starting with frame 4 and ending in frame 9.)

Example (a):   For the problem first cited in Example (a) of frame 4, namely,

$$x_1 + 2x_2 = 4 \tag{1}$$

$$2x_1 - x_2 = 3 \tag{2}$$

we obtained the reduced row echelon form

$$\left[\begin{array}{cc|c} 1 & 0 & 2 \\ 0 & 1 & 1 \end{array}\right]$$

in frame 8. This indicated that the unique solution was $x_1 = 2$, $x_2 = 1$. Geometrically, the two lines with equations given by (1) and (2) intersected at a single point (2, 1). See Figure 2.2 in frame 4.

Example (b): Consider the system

$$2x_1 + x_2 = 5 \tag{1}$$

$$2x_1 + x_2 = 2 \tag{2}$$

The starting augmented matrix is thus

$$\left[\begin{array}{cc|c} 2 & 1 & 5 \\ 2 & 1 & 2 \end{array}\right]$$

Show that application of Method 2.1 to this augmented matrix leads to

$$\left[\begin{array}{cc|c} 1 & \frac{1}{2} & \frac{5}{2} \\ 0 & 0 & -3 \end{array}\right]$$

- - - - - - - - - -

$$\left[\begin{array}{cc|c} 2 & 1 & 5 \\ 2 & 1 & 2 \end{array}\right] \xrightarrow{r_1' \;=\; r_1/2} \left[\begin{array}{cc|c} 1 & \frac{1}{2} & \frac{5}{2} \\ 2 & 1 & 2 \end{array}\right]$$

$$\xrightarrow{r_2' \;=\; r_2 \,-\, 2r_1} \left[\begin{array}{cc|c} 1 & \frac{1}{2} & \frac{5}{2} \\ 0 & 0 & -3 \end{array}\right]$$

Note that the last form is not in reduced row echelon form. The equation corresponding to the second row is $0x_1 + 0x_2 = -3$, or $0 = -3$.

The absurdity $0 = -3$ indicates that there is *no* solution to our system of equations. Geometrically (see Figure 2.4), the original two equations correspond to two distinct, parallel lines. (Such lines do not intersect.)

17.   Observe that the reduced row echelon form for the preceding system is

$\left[\begin{array}{cc|c} 1 & \frac{1}{2} & 0 \\ 0 & 0 & 1 \end{array}\right]$. Working with this form leads to the same conclusion as obtained

earlier. Here the absurd equation $0 = 1$ corresponds to the second row.

Note: In general, if at any point in the application of Method 2.1, we obtain a row with all zeros to the left of the dividing line and a nonzero number

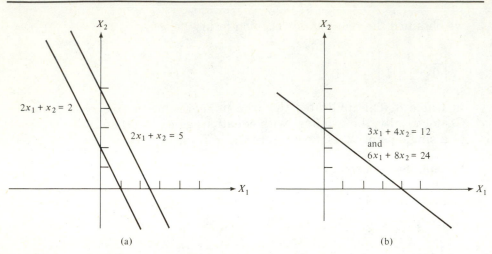

FIGURE 2.4   Special cases for two linear equations in two unknowns. (a) No solution. (b) Infinitely many solutions.

to the right of the dividing line, this indicates that there is no solution to the system.

Example: Consider the system

$$3x_1 + 4x_2 = 12 \tag{1}$$

$$6x_1 + 8x_2 = 24 \tag{2}$$

Apply Method 2.1 to determine the reduced row echelon form for this system, and interpret.

- - - - - - - - - -

$$\begin{bmatrix} 3 & 4 & | & 12 \\ 6 & 8 & | & 24 \end{bmatrix} \xrightarrow{\ r_1' = r_1/3\ } \begin{bmatrix} 1 & \frac{4}{3} & | & 4 \\ 6 & 8 & | & 24 \end{bmatrix}$$

$$\xrightarrow{\ r_2' = r_2 - 6r_1\ } \begin{bmatrix} 1 & \frac{4}{3} & | & 4 \\ 0 & 0 & | & 0 \end{bmatrix}$$

The last form is the reduced row echelon form.

The row of zeros indicates that the two lines are in fact the same line (i.e., they *coincide*). In other words, there are infinitely many solutions. See Figure 2.4. This is indicated by writing the equation corresponding to row 1 with the leading variable x_1 solved for in terms of the nonleading variable x_2. Thus, $x_1 = -\frac{4}{3}x_2 + 4$, or equivalently,

$$x_1 = -\tfrac{4}{3}r + 4 \tag{3}$$

$$x_2 = r \tag{4}$$

where r is arbitrary (i.e., r can take on any real number value).

18.   Actually, the three cases illustrated in the last three examples apply to all linear systems of m equations in n unknowns.

Theorem 2.3

Given a linear system of m linear equations in n unknowns represented by the matrix–vector equation $Ax = b$ (see Figure 2.3 in frame 4, and the discussion that follows Figure 2.3), there are three possible cases with respect to solutions of $Ax = b$:

   (a)   A unique solution.
   (b)   Infinitely many solutions.
   (c)   No solution.

It is not difficult to see that the only possibilities, in general, would be (a), (b), and (c). All one has to do is consider the possible reduced row equivalent forms that result from application of the Gauss–Jordan elimination method (Method 2.1) to any starting augmented matrix $A|b$.

We have fairly easy geometrical interpretations if $n = 3$ (three-dimensional space). If $m = 3$ also, the diagrams of Figure 2.5 indicate some of the geometrical configurations that apply for cases (a), (b), and (c).

Note: Some authors group cases (a) and (b) of the Theorem 2.3 together and refer to a system that has either a unique solution or infinitely many solutions as a *consistent system*. A system for which there is no solution [case (c) above] is then referred to as an *inconsistent system*.

Now we consider the problem discussed previously in frame 6 and in Example (b) of frame 12.

Example: Solve by using the Gauss–Jordan elimination method.

$$x_1 - x_2 + x_3 = 1 \tag{1}$$

$$2x_1 + x_2 + 3x_3 = 4 \tag{2}$$

$$3x_1 - x_2 - x_3 = -5 \tag{3}$$

Solution: The starting augmented matrix for this $m = n = 3$ system is

$$A|b = \begin{bmatrix} 1 & -1 & 1 & 1 \\ 2 & 1 & 3 & 4 \\ 3 & -1 & -1 & -5 \end{bmatrix}$$

Now use the Gauss–Jordan elimination method (Method 2.1 of frame 14).

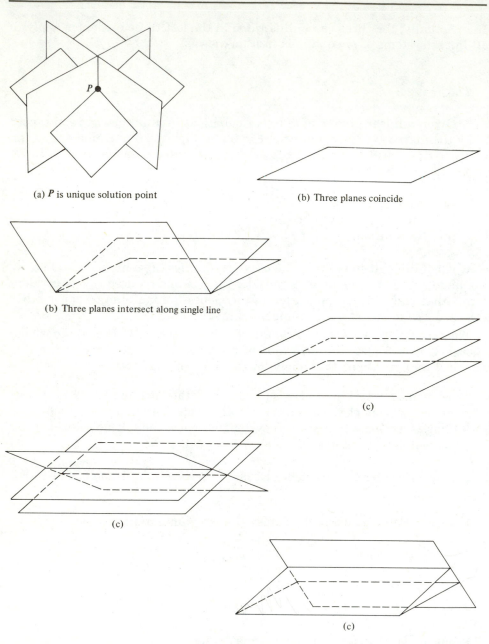

(a) *P* is unique solution point

(b) Three planes coincide

(b) Three planes intersect along single line

(c)

(c)

(c)

FIGURE 2.5   Three equations in three unknowns. (a) Unique solution. (b) Infinitely many solutions. (c) No solution.

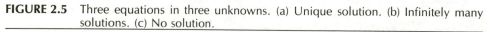

$$\begin{bmatrix} \underline{1} & -1 & 1 & | & 1 \\ 0 & 3 & 1 & | & 2 \\ 0 & 2 & -4 & | & -8 \end{bmatrix}$$

$\uparrow$

Column for Stage 2

Application of Stage 1 involves the row operations $r_2' = r_2 - 2r_1$ and $r_3' = r_3 - 3r_1$ to clear out column 1. Stage 1 is then complete.

$$\begin{bmatrix} \underline{1} & -1 & 1 & | & 1 \\ 0 & \underline{1} & \frac{1}{3} & | & \frac{2}{3} \\ 0 & 2 & -4 & | & -8 \end{bmatrix}$$

Application of Stage 2 first involves row operation $r_2' = r_2/3$ to transform the 3 in the row 2, column 2 position to 1.

$$\begin{bmatrix} \underline{1} & 0 & \frac{4}{3} & | & \frac{5}{3} \\ 0 & \underline{1} & \frac{1}{3} & | & \frac{2}{3} \\ 0 & 0 & -\frac{14}{3} & | & -\frac{28}{3} \end{bmatrix}$$

$\uparrow$

Column for Stage 3

Stage 2 continues with row operations $r_1' = r_1 + r_2$ and $r_3' = r_3 - 2r_2$ to clear out column 2. Stage 2 is then complete.

$$\begin{bmatrix} \underline{1} & 0 & \frac{4}{3} & | & \frac{5}{3} \\ 0 & \underline{1} & \frac{1}{3} & | & \frac{2}{3} \\ 0 & 0 & \underline{1} & | & 2 \end{bmatrix}$$

Application of Stage 3 first involves row operation $r_3' = (-\frac{3}{14})r_3$ to transform the $-\frac{14}{3}$ in the row 3, column 3 position to 1.

$$\begin{bmatrix} \underline{1} & 0 & 0 & | & -1 \\ 0 & \underline{1} & 0 & | & 0 \\ 0 & 0 & \underline{1} & | & 2 \end{bmatrix}$$

Stage 3 continues with row operations $r_1' = r_1 - \frac{4}{3}r_3$ and $r_2' = r_2 - \frac{1}{3}r_3$ to clear out column 3. Stage 3 is then complete.

Our final augmented matrix is in reduced row echelon form. We read off the unique solution from the final augmented matrix as follows:

$$x_1 = -1, \qquad x_2 = 0, \qquad x_3 = 2$$

See Figure 2.5, case (a) for a corresponding three-dimensional picture. Observe that $J_1 = 1, J_2 = 2$, and $J_3 = 3$ and that all three variables are leading variables in this problem.

19.   Example (a) of frame 16 and the example of frame 18 reveal what happens for an n equations–n unknowns system if the unique solution case occurs. The next example illustrates the case in which infinitely many solutions occur [case (b)].

Example: Solve by the Gauss–Jordan elimination method.

$$2x_2 - 2x_3 = -8 \qquad (1)$$

$$x_1 + x_2 + x_3 = 2 \qquad (2)$$

$$x_1 + 2x_2 = -2 \qquad (3)$$

_ _ _ _ _ _ _ _ _

The augmented matrix for this $m = n = 3$ system is

$$A|\mathbf{b} = \begin{bmatrix} 0 & 2 & -2 & -8 \\ 1 & 1 & 1 & 2 \\ 1 & 2 & 0 & -2 \end{bmatrix}$$

↑
Column for Stage 1

$$\begin{bmatrix} \underline{1} & 1 & 1 & 2 \\ 0 & 2 & -2 & -8 \\ 1 & 2 & 0 & -2 \end{bmatrix}$$

Application of Stage 1 first involves the type (a) row operation $r_1 \leftrightarrow r_2$. That is, rows 1 and 2 are interchanged.

$$\begin{bmatrix} \underline{1} & 1 & 1 & 2 \\ 0 & 2 & -2 & -8 \\ 0 & 1 & -1 & -4 \end{bmatrix}$$

↑
Column for Stage 2

Stage 1 continues with the type (c) row operation $r_3' = r_3 - r_1$ to clear out column 1. Stage 1 is complete.

$$\begin{bmatrix} \underline{1} & 1 & 1 & 2 \\ 0 & \underline{1} & -1 & -4 \\ 0 & 1 & -1 & -4 \end{bmatrix}$$

Application of Stage 2 first involves the row operation $r_2' = r_2/2$.

$$\begin{bmatrix} \underline{1} & 0 & 2 & 6 \\ 0 & \underline{1} & -1 & -4 \\ 0 & 0 & 0 & 0 \end{bmatrix}$$

Stage 2 continues with row operations $r_1' = r_1 - r_2$ and $r_3' = r_3 - r_2$ to clear out column 2. Stage 2 is then complete.

The matrix is now in reduced row echelon form, since row 3 consists entirely of zeros. Here $J_1 = 1$, $J_2 = 2$, and J_3 does not exist. From the final augmented matrix, we obtain the following system of equations (the leading variables x_1 and x_2, which correspond to the leading entries, are underlined):

$$\underline{x_1} + 2x_3 = 6$$

$$\underline{x_2} - x_3 = -4$$

After replacing the nonleading variable x_3 by r and transposing the r terms to the right sides, we obtain

$$x_1 \quad\;\; = -2r + 6$$

$$x_2 \;\; = \quad r - 4$$

$$x_3 = \quad r$$

Here r can be any real number. This situation, in which we have infinitely many solutions, corresponds to three planes intersecting along a common straight line. See case (b), left diagram in Figure 2.5.

20.   Theorem 2.3 of frame 18 indicates that a system of linear equations either has a unique solution [case (a)], infinitely many solutions [case (b)], or no solution at all [case (c)]. In this and the next frame, we shall study homogeneous systems; in such systems the no solution case cannot occur.

Definition 2.5

A linear system of m equations in n unknowns is said to be *homogeneous* if all the right-side constants (i.e., the b_i's) are zero. Thus, a homogeneous system has the form

$$a_{11}x_1 + a_{12}x_2 + \ldots + a_{1n}x_n = 0 \tag{1}$$
$$a_{21}x_1 + a_{22}x_2 + \ldots + a_{2n}x_n = 0 \tag{2}$$
$$\cdots\cdots\cdots\cdots\cdots\cdots\cdots\cdots\cdots\cdots\cdots\cdots\cdots\cdots$$
$$\cdots\cdots\cdots\cdots\cdots\cdots\cdots\cdots\cdots\cdots\cdots\cdots\cdots\cdots$$
$$a_{m1}x_1 + a_{m2}x_2 + \ldots + a_{mn}x_n = 0 \tag{m}$$

The matrix–vector form is $A\mathbf{x} = \mathbf{0}$, and the augmented coefficient matrix is $A|\mathbf{0}$.

Example (a):   The system given by

$$4x_1 + 12x_2 = 0 \tag{1}$$

$$2x_1 - 3x_2 = 0 \tag{2}$$

is a homogeneous system since both right-side constants equal zero. Here $m = 2$ (two equations) and $n = 2$ (two variables).

Observe that any homogeneous system always has a solution, namely, the solution for which $x_1 = 0$, $x_2 = 0, \ldots, x_n = 0$. The reason for this is that this collection of values causes Eqs. (1) through (m) in Definition 2.5 to be satisfied. This solution is called the *trivial solution* of a homogeneous linear system. [For example, in Example (a), $x_1 = 0$ and $x_2 = 0$ cause both Eqs. (1) and (2) to be satisfied in the form $0 = 0$. Thus, the trivial solution in Example (a) is given by $x_1 = 0$, $x_2 = 0$.]

In vector form, the trivial solution is expressed as $\mathbf{x} = \mathbf{0}$, where both column vectors have n components.

If solutions besides the trivial solution exist for a particular homogeneous system, such solutions are called *nontrivial solutions*.

Referring again to Theorem 2.3 in frame 18, we see that the no solution case [case (c)] cannot occur for a homogeneous system. Another fact is that if a homogeneous system has a unique solution, then that solution has to be the trivial solution.

At this point our method of solution for homogeneous systems will be the Gauss–Jordan elimination method (Method 2.1 in frame 14).

Example (b): Determine the solution(s) for the homogeneous system of Example (a).

_ _ _ _ _ _ _ _ _ _

Solution: First, we observe that the starting augmented coefficient matrix is $A|\mathbf{0} = \begin{bmatrix} 4 & 12 & 0 \\ 2 & -3 & 0 \end{bmatrix}$. Applying Method 2.1 to this matrix yields the following sequence of augmented matrices:

$$\begin{bmatrix} 4 & 12 & 0 \\ 2 & -3 & 0 \end{bmatrix} \xrightarrow{r_1' = r_1/4} \begin{bmatrix} 1 & 3 & 0 \\ 2 & -3 & 0 \end{bmatrix} \xrightarrow{r_2' = r_2 - 2r_1} \begin{bmatrix} 1 & 3 & 0 \\ 0 & -9 & 0 \end{bmatrix}$$

$$\xrightarrow{r_2' = r_2/(-9)} \begin{bmatrix} 1 & 3 & 0 \\ 0 & 1 & 0 \end{bmatrix} \xrightarrow{r_1' = r_1 - 3r_2} \begin{bmatrix} 1 & 0 & 0 \\ 0 & 1 & 0 \end{bmatrix}$$

The final augmented matrix, which is in the reduced row echelon form $I_2|\mathbf{0}$, reveals the unique (trivial) solution $x_1 = 0$, $x_2 = 0$.

21. The following notes pertain to homogeneous systems.

Notes: (a) If the unique (trivial) solution occurs for an n equations–n unknowns homogeneous system, then the reduced row echelon form is $I_n|\mathbf{0}$. The converse is valid also. (b) For a homogeneous system the numbers to the right of the dividing line will stay fixed at zero for any possible row operation. This is clearly shown in Example (b) of frame 20. (c) Geometrically, the two lines with Eqs. (1) and (2) in Example (a) of frame 20 intersect solely at the origin.

Example: Determine the solution(s) of the homogeneous system:

$$2x_2 - 2x_3 = 0 \qquad (1)$$

$$x_1 + x_2 + x_3 = 0 \qquad (2)$$

$$x_1 + 2x_2 \qquad = 0 \qquad (3)$$

Solution: Observe that the coefficient matrix A here is identical to that for the Example of frame 19. Thus, it follows that both Examples will have identical reduced row echelon forms to the *left* of the dividing line.

Now finish the problem.

- - - - - - - - - -

For the current example, the numbers to the right of the dividing line will, of course, be zeros in the reduced row echelon form [see Note (b)]. Thus, the reduced row echelon form for our current example is

$$\left[\begin{array}{ccc|c} 1 & 0 & 2 & 0 \\ 0 & 1 & -1 & 0 \\ 0 & 0 & 0 & 0 \end{array}\right] \qquad (4)$$

The equations corresponding to (4) are

$$\underline{x_1} \qquad = -2x_3$$

$$\underline{x_2} = \qquad x_3$$

after transposing the nonleading variable x_3 to the right and thereby solving for the leading variables x_1 and x_2. Replacing x_3 by r we have

$$x_1 \qquad = -2r$$

$$x_2 \qquad = \qquad r$$

$$x_3 = \qquad r$$

where r is arbitrary (i.e., any real number). This indicates the system has infinitely many solutions. For example, letting $r = 1$, we have $x_1 = -2$, $x_2 = 1$, and $x_3 = 1$, and for $r = 0$, we have $x_1 = x_2 = x_3 = 0$, namely, the trivial solution. Observe that we can check an alleged solution by substituting back into the original equations [here, Eqs. (1), (2), and (3)] and seeing whether they are satisfied.

D. THE RANK OF A MATRIX

22. The concept of *rank of a matrix* will prove useful in determining the solution characteristics of systems of linear equations.

Definition 2.6

The rank of a matrix is the number of leading entries in its reduced row echelon form.

Thus, we have an easy way of determining the rank of a matrix. All we have to do is apply Method 2.1 (frame 14) and then count the number of leading entries in the resulting reduced row echelon form.

Example (a): Determine the rank of the augmented coefficient matrix $A|\mathbf{b}$

$= \begin{bmatrix} 1 & 2 & | & 4 \\ 2 & -1 & | & 3 \end{bmatrix}$, which occurs in frame 8, at the beginning of frame 11, and

in Example (a) of frame 16.

Solution: The corresponding reduced row echelon form is $\begin{bmatrix} 1 & 0 & | & 2 \\ 0 & 1 & | & 1 \end{bmatrix}$, as is

noted in Example (a) of frame 16. Since there are two leading entries here, it follows that the rank of $A|\mathbf{b}$ is 2.

Henceforth, we shall often denote the rank of any matrix C by writing

$$\text{Rank } (C)$$

Thus, in Example (a), we can write Rank $(A|\mathbf{b}) = 2$.

Example (b): Determine the rank of $A|\mathbf{b}$ for Example (b) of frame 16.

Solution: Here $A|\mathbf{b} = \begin{bmatrix} 2 & 1 & | & 5 \\ 2 & 1 & | & 2 \end{bmatrix}$, and the reduced row echelon form, dis-

played at the beginning of frame 17, is $\begin{bmatrix} 1 & \frac{1}{2} & | & 0 \\ 0 & 0 & | & 1 \end{bmatrix}$.

Now finish the problem.

— — — — — — — — — — —

Since the second matrix has two leading entries, Rank $(A|\mathbf{b}) = 2$. Recall that for this problem there is no solution for the system. This is related to the fact that the leading entry in row 2 is to the right of the dividing line (indicating an absurd equation of the form $0 = 1$ corresponding to row 2).

23. Suppose that we know the reduced row echelon form of an augmented matrix. For example, we know that the reduced row echelon form of

$\begin{bmatrix} 1 & 2 & | & 4 \\ 2 & -1 & | & 3 \end{bmatrix}$ is $\begin{bmatrix} 1 & 0 & | & 2 \\ 0 & 1 & | & 1 \end{bmatrix}$, as noted in Example (a) of frame 22. Observe that it automatically follows that the reduced row echelon form of the matrix $\begin{bmatrix} 1 & 2 & 4 \\ 2 & -1 & 3 \end{bmatrix}$ is $\begin{bmatrix} 1 & 0 & 2 \\ 0 & 1 & 1 \end{bmatrix}$. Note that all we have done is remove the dividing line in the first two matrices. This type of result applies in general. Thus, to find the reduced row echelon form of any given matrix, augmented or not, we merely apply Method 2.1 (frame 14) to the given matrix. Further, we say that the rank of either $\begin{bmatrix} 1 & 2 & | & 4 \\ 2 & -1 & | & 3 \end{bmatrix}$ or $\begin{bmatrix} 1 & 2 & 4 \\ 2 & -1 & 3 \end{bmatrix}$ is 2 because the number of leading entries in the resulting reduced row echelon form is 2.

Still another useful fact is the following. Suppose we know the reduced row echelon form for some augmented matrix $A|\mathbf{b}$. We can then automatically obtain the reduced row echelon form for the matrix A by merely *deleting the last column* of the reduced row echelon form of $A|\mathbf{b}$. This is so because of the nature of the elementary row operations (Definition 2.3 in frame 8). Then, knowing the reduced row echelon form of matrix A, we can determine the rank of A by counting the number of leading entries in the corresponding reduced row echelon form.

Example (a): Determine the ranks of matrices $A|\mathbf{b}$ and A, respectively, in Example (a) of frame 22.

Solution: Since the reduced row echelon form of $A|\mathbf{b} = \begin{bmatrix} 1 & 2 & | & 4 \\ 2 & -1 & | & 3 \end{bmatrix}$ is $\begin{bmatrix} 1 & 0 & | & 2 \\ 0 & 1 & | & 1 \end{bmatrix}$, it follows that the reduced row echelon form of $A = \begin{bmatrix} 1 & 2 \\ 2 & -1 \end{bmatrix}$ is $\begin{bmatrix} 1 & 0 \\ 0 & 1 \end{bmatrix}$, the identity matrix I_2.

Thus, Rank $(A|\mathbf{b}) = 2$ and Rank $(A) = 2$.

Example (b): Determine the ranks of matrices $A|\mathbf{b}$ and A, respectively, in Example (b) of frame 22.

Solution: Since the reduced row echelon form of $A|\mathbf{b} = \begin{bmatrix} 2 & 1 & | & 5 \\ 2 & 1 & | & 2 \end{bmatrix}$ is $\begin{bmatrix} 1 & \frac{1}{2} & | & 0 \\ 0 & 0 & | & 1 \end{bmatrix}$, it follows that the reduced row echelon form of $A = \begin{bmatrix} 2 & 1 \\ 2 & 1 \end{bmatrix}$ is $\begin{bmatrix} 1 & \frac{1}{2} \\ 0 & 0 \end{bmatrix}$.

Now finish the problem.

- - - - - - - - -

Rank $(A|\mathbf{b}) = 2$, and Rank $(A) = 1$.

24. In Example (b) of the preceding two frames, we encountered a linear system for which there is no solution. Such a system is also called an inconsistent system (see Note in frame 18). For every no solution system, the reduced row echelon form of $A|\mathbf{b}$ has a single row of the form $0\ 0\ \dots\ 0\,|\,1$.

The presence of such a row can be related to rank. In fact, it is always true that a linear system will have no solution precisely when the ranks of $A|\mathbf{b}$ and A differ, because it is in this case that a row of the form $0\ 0\ \dots\ 0\,|\,1$ occurs in the reduced row echelon form of $A|\mathbf{b}$. In fact, Rank $(A|\mathbf{b})$ = Rank (A) + 1, in this case.

To illustrate, in Example (b) of the preceding two frames, Rank $(A|\mathbf{b})$ = 2 and Rank (A) = 1.

The other two cases described in Theorem 2.3 of frame 18—namely, (a) unique solution and (b) infinitely many solutions—occur precisely when Rank $(A|\mathbf{b})$ = Rank (A). Recall that a linear system that has either a unique solution or infinitely many solutions is referred to as a consistent system. Thus, for a consistent system, Rank $(A|\mathbf{b})$ = Rank (A), and conversely.

Now let us consider how to describe further the solution properties of the two types of consistent systems in terms of rank.

Example (a): In Example (a) of the previous two frames, we have the unique solution case, where the unique solution is given by x_1 = 2 and x_2 = 1 [see also Example (a) in frame 16]. Observe that Rank $(A|\mathbf{b})$ = n = 2 here and that both variables are leading variables. Recall that n is our symbol for the total number of variables (also called unknowns).

Example (b): Another simple consistent system occurs in the example of frame 17. There the system consists of the following two equations in two variables:

$$3x_1 + 4x_2 = 12 \tag{1}$$

$$6x_1 + 8x_2 = 24 \tag{2}$$

The reduced row echelon form of $A|\mathbf{b}$ is $\begin{bmatrix} 1 & \frac{4}{3} & 4 \\ 0 & 0 & 0 \end{bmatrix}$, indicating that

Rank $(A|\mathbf{b})$ = 1. In addition, there is one leading variable here, namely x_1; the variable x_2 is a nonleading variable. The infinitely many solutions for the preceding system are determined by the equations

$$x_1 = -\tfrac{4}{3}r + 4 \tag{3}$$

$$x_2 = r \tag{4}$$

where r is arbitrary.

In both the preceding examples, the rank of $A|\mathbf{b}$ is equal to the number of leading variables. (By the way, it is *always true* for a consistent system that the rank of $A|\mathbf{b}$ is equal to the number of leading variables.) In the first of the preceding examples the number of leading variables equals the total number of variables, so that each variable has its value completely determined; thus, there is a unique solution.

In the second example there is a total of two variables, and there is one leading variable. This means that there is a solution for each possible choice of value for the nonleading variable (x_2). Since there are infinitely many possible values for the nonleading variable, the system has infinitely many solutions.

The ideas illustrated in the preceding example and discussion can be generalized. We do so now in the following master theorem:

Theorem 2.4

Given a linear system of m equations with a total of n variables, with augmented coefficient matrix $A|\mathbf{b}$.

(I)   If Rank $(A|\mathbf{b})$ = Rank (A) + 1, there is no solution (also called inconsistent case).

(II)   If Rank $(A|\mathbf{b})$ = Rank (A), we have a consistent system, where
    (a)   there is a unique solution if Rank $(A|\mathbf{b})$ = n,
and
    (b)   there are infinitely many solutions if Rank $(A|\mathbf{b})$ < n.

Example (c): Verify Theorem 2.4 with respect to (i) the system referred to in frame 14, and (ii) the system referred to in the Example of frame 18.

Solution: (i) In this system, Rank $(A|\mathbf{b})$ = Rank (A) = 2, indicating we have a consistent system (with two leading variables). Since $n = 4$, it follows that there are infinitely many solutions.

Now you do part (ii).

— — — — — — — — —

(ii) Rank $(A|\mathbf{b})$ = Rank (A) = 3, indicating a consistent system. Since $n = 3$, there is a unique solution.

25. We should not forget some commonsense facts about the rank concept. One is that for a consistent system, we cannot have a leading entry in column $(n + 1)$ of the reduced row echelon form. [Recall that column $(n + 1)$ is the $\mathbf{b}$ column.] Moreover, there is at most one leading entry per column in each of the remaining n columns. Thus, for a consistent system,

Rank $(A|\mathbf{b}) \leqslant n$ (1)

We could also write Rank $(A) \leqslant n$ here, since Rank $(A|\mathbf{b})$ = Rank (A) for a consistent system.

Another fact is that there is at most one leading entry per row in each of the m rows of the reduced row echelon form. That is, in general,

$$\text{Rank } (A|\mathbf{b}) \leqslant m \tag{2}$$

Now suppose we have a consistent system with fewer equations than unknowns (i.e., $m < n$). From (2) above, and $m < n$, we see that we must have

$$\text{Rank } (A|\mathbf{b}) < n \tag{3}$$

in this case. But this means that the system falls into category (IIb) of Theorem 2.4 and that there are infinitely many solutions. Thus, we have a familiar result for a linear system, namely, that if we have fewer equations than unknowns (or variables), then there will be infinitely many solutions (provided we have a consistent system).

Let us return to the special case of a homogeneous system (see frames 20 and 21). We know that for homogeneous systems the no solution (or inconsistent) case cannot occur. Thus, Rank $(A|\mathbf{b})$ = Rank (A) for any homogeneous linear system, which, of course, indicates a consistent system. Recall that the trivial solution $x_1 = 0$, $x_2 = 0$, $\dots$, $x_n = 0$ is a solution of any homogeneous linear system. We summarize the important characteristics of homogeneous linear systems in the following. Henceforth, we replace $A|\mathbf{b}$ by $A|\mathbf{0}$ for homogeneous systems.

Corollary to Theorem 2.4 (for Homogeneous Linear Systems)

Given a homogeneous linear system of m equations with n variables, with augmented coefficient matrix $A|\mathbf{0}$. Then Rank $(A|\mathbf{0})$ = Rank (A), which indicates the system of equations is consistent. Also,
   (a)   there is a unique solution (namely, the trivial solution) if Rank $(A|\mathbf{0})$ = n, and
   (b)   there are infinitely many solutions if Rank $(A|\mathbf{0}) < n$.

Note: It follows from the prior discussion of this frame that a homogeneous linear system with fewer equations than variables will automatically have infinitely many solutions.

Example: Verify the Corollary to Theorem 2.4 for the homogeneous systems analyzed in frame 20 and frame 21.

Solution: *Frame 20:* For the system given by

$$4x_1 + 12x_2 = 0 \tag{1}$$

$$2x_1 - 3x_2 = 0 \tag{2}$$

we find that Rank $(A|0)$ = Rank (A) = 2, n = 2, and we have the unique (trivial) solution $x_1 = 0$, $x_2 = 0$.

Now verify for the system of frame 21.

— — — — — — — — — —

Frame 21: For this frame we see that Rank $(A|0)$ = Rank (A) = 2, n = 3, and there are infinitely many solutions.

26. It is useful to return now to a discussion of linear systems that have n equations in n unknowns. Let us reconsider Theorem 2.2 of frame 10. In the following, observe that A is an $n \times n$ matrix, and $A|b$ is $n \times (n + 1)$.

Theorem 2.2 (n Equations in n Unknowns Case)

If the augmented matrix $A|b$ can be transformed, by means of elementary row operations, to the form $I_n|k$, then the unique solution to the original system is given by $x_1 = k_1$, $x_2 = k_2$, ..., $x_n = k_n$.

First of all, observe that if $A|b$ can be transformed, by elementary row operations, to $I_n|k$, that also means that the reduced row echelon form of $A|b$ is $I_n|k$. Then it follows that Rank $(A|b)$ = Rank (A) = n, since I_n has n leading entries, one in each row and one in each column. From this we have the following result:

(i)   If the reduced row echelon form of $A|b$ is $I_n|k$, then Rank $(A|b)$ = Rank (A) = n.

Also, the converse of (i) is valid. It can be stated as follows:

(ii)   If Rank $(A|b)$ = Rank (A) = n, then the reduced row echelon form of $A|b$ is $I_n|k$.

From the preceding discussion it follows that expressions (iii) and (iv) following are valid for an $n \times n$ matrix A.

(iii)   If the reduced row echelon form of A is I_n, then Rank (A) = n.
(iv)   If Rank (A) = n, then the reduced row echelon form of A is I_n.

Example (a): Refer to the Example of frame 18. There,

$$A|b = \begin{bmatrix} 1 & -1 & 1 & 1 \\ 2 & 1 & 3 & 4 \\ 3 & -1 & -1 & -5 \end{bmatrix},$$

and application of Method 2.1 led to the following reduced row echelon form:

$$
\left[
\begin{array}{ccc|c}
\underline{1} & 0 & 0 & -1 \\
0 & \underline{1} & 0 & 0 \\
0 & 0 & \underline{1} & 2
\end{array}
\right]
$$

Here, the identity matrix I_3 appears to the left, and $\mathbf{k} = [-1, 0, 2]^t$ appears to the right of the dividing line. Also, Rank $(A|\mathbf{b})$ = Rank (A) = 3. This discussion verifies expressions (i) and (ii).

Now if we had applied Method 2.1 to matrix A alone, where

$$
A = \left[
\begin{array}{ccc}
1 & -1 & 1 \\
2 & 1 & 3 \\
3 & -1 & -1
\end{array}
\right],
$$

we would have found that the reduced row echelon form of A was

$$
I_3 = \left[
\begin{array}{ccc}
\underline{1} & 0 & 0 \\
0 & \underline{1} & 0 \\
0 & 0 & \underline{1}
\end{array}
\right],
$$

thereby indicating that Rank (A) = 3. This verifies expressions (iii) and (iv).

From expressions (iii) and (iv), it follows that statements (a) and (b) are equivalent for an $n \times n$ matrix A:

(a)   Rank $(A) = n$.
(b)   The reduced row echelon form of A is I_n.

We can interpret a list of equivalent statements as follows: if any of the statements in such a list is true, then all the statements in the list are true; if any one is false, then all are false. Thus, in the preceding, where we have a list of two equivalent statements, if, say, (a) is true for some $n \times n$ matrix A, then (b) is also true. We shall return to statements (a) and (b), and to an extended equivalence list at the end of Section F (Theorem 2.9 in frame 32).

If two statements are equivalent, then we can represent this by means of an "if and only if" expression. Thus, for example, we can indicate the equivalence of (a) and (b) (for an $n \times n$ matrix A) by writing

Rank $(A) = n$ if and only if the reduced row echelon form of A is I_n.

We could just as well reverse the order here. The preceding comments on terminology apply, in general, and are used extensively in the wording of theorems.

Thus, using the symbols P and Q to represent statements, if statements P and Q are equivalent, we can indicate this by the expression

P if and only if Q, (I)

(or, Q if and only if P).

Still another way of indicating the equivalence of statements P and Q is by writing the two expressions:

If P, then Q, (IIa)

and,

If Q, then P. (IIb)

Now let us return to Theorem 2.2 (see the beginning of the current frame). It is useful to restate Theorem 2.2 in the following form, which incorporates the concepts of reduced row echelon form and rank.

Theorem 2.2' (n Equations in n Unknowns Case)

If the reduced row echelon form of $A|b$ is $I_n|k$, then Rank $(A|b)$ = Rank (A) = n, and the unique solution of the original system is given by $x_1 = k_1, x_2 = k_2, \ldots, x_n = k_n$, where the k_i's are the components of the column vector $\mathbf{k}$.

Example (b):   Illustrate Theorem 2.2' by finding the reduced row echelon form corresponding to the system

$$5x_1 + 2x_2 = 14 \tag{1}$$

$$3x_1 - 4x_2 = 24 \tag{2}$$

Solution: Let us apply Method 2.1 to $A|b = \begin{bmatrix} 5 & 2 & | & 14 \\ 3 & -4 & | & 24 \end{bmatrix}$. Now finish the

problem.

-- -- -- -- -- -- -- --

The reduced row echelon form is $\begin{bmatrix} 1 & 0 & | & 4 \\ 0 & 1 & | & -3 \end{bmatrix}$, indicating that Rank

$(A|b)$ = Rank (A) = 2. Also, the unique solution is $x_1 = 4$ and $x_2 = -3$, which can be easily checked by substitution into Eqs. (1) and (2).

E. THE INVERSE OF A SQUARE MATRIX

27. The inverse of a square matrix is a very important quantity. Suppose $A\mathbf{x}$ = $\mathbf{b}$ represents a linear system with n equations in n unknowns (A is $n \times n$). Then if the so-called inverse of A (denoted as A^{-1}) exists, we can solve for the solution vector $\mathbf{x}$ by means of the simple equation $\mathbf{x} = A^{-1}\mathbf{b}$, as will be shown later. This is particularly useful if we are given one matrix for A and several right-side vectors, say, $\mathbf{b}_1$, $\mathbf{b}_2$, etc. This type of calculation will be illustrated in Sections F and G.

The inverse of a matrix is also encountered in many practical applications, one of which will be illustrated in Section G.

Suppose that A is an $n \times n$ matrix and I_n is the $n \times n$ identity matrix.

Definition 2.7

If there exists an $n \times n$ matrix B such that $AB = I_n$ and $BA = I_n$, then we say that matrix A has an *inverse* and an inverse is B.

Notes: (a) If matrix A has an inverse, then A is also called a *nonsingular* or *invertible* matrix. If A has no inverse, then A is called a *singular* or *noninvertible* matrix. (b) Suppose A and B are both $n \times n$ matrices. It can be proved that if $BA = I_n$, then $AB = I_n$, and thus, A is invertible. (It is also true that $AB = I_n$ implies $BA = I_n$.) The proof is given in Lipschutz (1968, p. 58), and in Shields (1980, p. 104).* This result is useful to us in the following way: If we wish to show that B is an inverse of A, all we have to do is show that $BA = I_n$ (or that $AB = I_n$). (c) It is clear that if B is an inverse of A, then A is an inverse of B.

Theorem 2.5

An invertible $n \times n$ matrix A has exactly one inverse. Thus, if an inverse of a matrix exists, it is unique.

Proof: Suppose the $n \times n$ matrix A has two inverses, say B and C. Then

$$AB = BA = I_n \quad \text{and} \quad AC = CA = I_n$$

Thus, from the multiplicative property of I_n and the associative property of matrix multiplication [Theorem 1.3, Property (a)], we have

$$B = BI_n = B(AC) = (BA)C = I_nC = C$$

Thus, $B = C$, and the proof is complete.

---

* References are listed at the back of the book. Our method of referring to references is by listing the name(s) of the author(s), followed by the date of publication in parentheses.

Because of this result, we now can speak of *the* inverse of an invertible matrix; we denote the inverse by A^{-1}, which is read "A inverse." Also, we can write $AA^{-1} = I_n$ and $A^{-1}A = I_n$.

Example: Given that $A = \begin{bmatrix} 1 & 3 \\ 2 & 4 \end{bmatrix}$ and $B = \begin{bmatrix} -2 & \frac{3}{2} \\ 1 & -\frac{1}{2} \end{bmatrix}$. Verify that B is the inverse of A.

Solution: It is sufficient to show that $AB = I_2$ because of Note (b) following Definition 2.7 and Theorem 2.5. Now finish the problem.

- - - - - - - - - - -

$$AB = \begin{bmatrix} 1 & 3 \\ 2 & 4 \end{bmatrix}\begin{bmatrix} -2 & \frac{3}{2} \\ 1 & -\frac{1}{2} \end{bmatrix} = \begin{bmatrix} -2+3 & \frac{3}{2}-\frac{3}{2} \\ -4+4 & 3-2 \end{bmatrix}$$

$$= \begin{bmatrix} 1 & 0 \\ 0 & 1 \end{bmatrix} = I_2$$

28. Certain matrices do not have inverses, as the following example illustrates.

Example (a): Show that $A = \begin{bmatrix} 2 & 1 \\ 6 & 3 \end{bmatrix}$ has no inverse.

Solution: If A had an inverse, then there would exist a matrix

$$B = \begin{bmatrix} b_{11} & b_{12} \\ b_{21} & b_{22} \end{bmatrix}$$

such that $AB = I_2$, i.e., such that

$$\begin{bmatrix} 2 & 1 \\ 6 & 3 \end{bmatrix}\begin{bmatrix} b_{11} & b_{12} \\ b_{21} & b_{22} \end{bmatrix} = \begin{bmatrix} 1 & 0 \\ 0 & 1 \end{bmatrix}$$

Carrying out the matrix multiplication, this becomes

$$\begin{bmatrix} 2b_{11} + b_{21} & 2b_{12} + b_{22} \\ 6b_{11} + 3b_{21} & 6b_{12} + 3b_{22} \end{bmatrix} = \begin{bmatrix} 1 & 0 \\ 0 & 1 \end{bmatrix}$$

Equating the corresponding column one entries yields the following equations:

$$2b_{11} + b_{21} = 1 \tag{1}$$

$$6b_{11} + 3b_{21} = 0 \tag{2}$$

If we attempt to solve these two equations for b_{11} and b_{21}, we would be led to no solution. For example, applying Method 2.1 would lead to an augmented matrix with second row given by $0 \ \ 0 | -3$. Thus, since there is no solution for b_{11} and b_{21}, it follows that matrix B does not exist. Thus, this particular matrix A has no inverse.

The matrix inverse has many useful properties, several of which are summarized in the following theorem.

Theorem 2.6 (Properties of the Inverse)

(a)   If A is an invertible matrix, then A^{-1} is also invertible and its inverse is A. That is, $(A^{-1})^{-1} = A$.

(b)   If A is invertible, then its transpose A^t is invertible. The inverse of A^t is given by $(A^t)^{-1} = (A^{-1})^t$.

(c)   If A is invertible with inverse A^{-1}, then kA (where k is a nonzero scalar) is invertible, and $(kA)^{-1} = \dfrac{1}{k}A^{-1}$.

(d)   If A and B are invertible $n \times n$ matrices, then AB is an invertible $n \times n$ matrix whose inverse $(AB)^{-1}$ is given by $(AB)^{-1} = B^{-1}A^{-1}$.

Notes: (i) Part (b) is read "the inverse of A transpose equals the transpose of A inverse." (ii) A result comparable to part (d) holds for the transpose [see Theorem 1.4, part (d)]. (iii) Also, part (d) can be extended to any number of matrices. Thus, for example, for three invertible $n \times n$ matrices A, B, and C, the product ABC is also an invertible $n \times n$ matrix, and $(ABC)^{-1} = C^{-1}B^{-1}A^{-1}$.

Example (b):   Prove part (d) of Theorem 2.6.

Proof: If we can show that $(AB)(B^{-1}A^{-1}) = I_n$, then we will have shown, first, that AB is invertible, and, second, that $(AB)^{-1} = B^{-1}A^{-1}$. We shall employ the associative property of matrix multiplication and the definition of a matrix inverse. Thus,

$$(AB)(B^{-1}A^{-1}) = ((AB)B^{-1})A^{-1} = (A(BB^{-1}))A^{-1}$$
$$= (AI_n)A^{-1} = AA^{-1} = I_n$$

Initially, we think of (AB) as being a separate matrix C, and we employ $C(B^{-1}A^{-1}) = (CB^{-1})A^{-1}$. Then, later, we use the fact that $BB^{-1} = I_n$.

Example (c):   The inverse of $A = \begin{bmatrix} 1 & 3 \\ 2 & 4 \end{bmatrix}$ is given by $A^{-1} = \begin{bmatrix} -2 & \frac{3}{2} \\ 1 & -\frac{1}{2} \end{bmatrix}$, as was verified in the example of frame 27. Illustrate parts (a) and (b) of Theorem 2.6 with respect to matrix A. Employ part (c) of Theorem 2.6 to find the inverse of matrix $C = \begin{bmatrix} 4 & 12 \\ 8 & 16 \end{bmatrix}$.

Solution: Part (a): Since $A^{-1} = \begin{bmatrix} -2 & \frac{3}{2} \\ 1 & -\frac{1}{2} \end{bmatrix}$, the inverse of $\begin{bmatrix} -2 & \frac{3}{2} \\ 1 & -\frac{1}{2} \end{bmatrix}$ is

$\begin{bmatrix} 1 & 3 \\ 2 & 4 \end{bmatrix}$. That is, $\begin{bmatrix} -2 & \frac{3}{2} \\ 1 & -\frac{1}{2} \end{bmatrix}^{-1} = \begin{bmatrix} 1 & 3 \\ 2 & 4 \end{bmatrix}$.

Part (b): Since $A^t = \begin{bmatrix} 1 & 2 \\ 3 & 4 \end{bmatrix}$, it follows that

$$\begin{bmatrix} 1 & 2 \\ 3 & 4 \end{bmatrix}^{-1} = (A^{-1})^t = \begin{bmatrix} -2 & \frac{3}{2} \\ 1 & -\frac{1}{2} \end{bmatrix}^t = \begin{bmatrix} -2 & 1 \\ \frac{3}{2} & -\frac{1}{2} \end{bmatrix}.$$

Now complete the problem and find the inverse of matrix C.

— — — — — — — — —

First, we note that $C = 4A$, and the scalar k in part (c) equals 4. Thus, $C^{-1} = \frac{1}{4}A^{-1}$, and hence,

$$C^{-1} = \frac{1}{4}\begin{bmatrix} -2 & \frac{3}{2} \\ 1 & -\frac{1}{2} \end{bmatrix} = \begin{bmatrix} -\frac{1}{2} & \frac{3}{8} \\ \frac{1}{4} & -\frac{1}{8} \end{bmatrix}.$$

F. A METHOD FOR CALCULATING A^{-1}

29. We shall now focus on developing a practical method for calculating the inverse of a matrix. Our method will turn out to be a variation of the Gauss–Jordan elimination method (Method 2.1).

Let us digress for a while and consider a type of example that will later prove to be helpful toward the ultimate development of our method for calculating the inverse of a matrix.

Example (a): Solve each of the following two linear systems of three equations in three unknowns.

$$\begin{aligned} x_1 - x_2 + x_3 &= 0 \\ 2x_1 + x_2 + 3x_3 &= 2 \\ 3x_1 - x_2 - x_3 &= 6 \end{aligned} \tag{1}$$

$$\begin{aligned} x_1 - x_2 + x_3 &= 6 \\ 2x_1 + x_2 + 3x_3 &= 7 \\ 3x_1 - x_2 - x_3 &= 10 \end{aligned} \tag{2}$$

Solution: Observe that these two systems differ only in their right-hand sides. That is, the coefficient matrix A is the same for both. (In fact, we have the same A here as in the example of frame 18.) Thus, if we use Method 2.1 twice (once for each system) to find the solutions, the calculations to the left of the dividing line will be the same for both systems. This means that we can solve both systems at the same time by using Method 2.1 on the following augmented matrix.

$$\begin{bmatrix} 1 & -1 & 1 & | & 0 & 6 \\ 2 & 1 & 3 & | & 2 & 7 \\ 3 & -1 & -1 & | & 6 & 10 \end{bmatrix} \tag{3}$$

This matrix has both right-side vectors $\mathbf{b}_1 = [0, 2, 6]^t$ and $\mathbf{b}_2 = [6, 7, 10]^t$, respectively, listed in order, to the right of the dividing line. Symbolically, the matrix in (3) can be indicated as $A|\mathbf{b}_1\,\mathbf{b}_2$. Now let us use Method 2.1 to convert (3) to reduced row echelon form. The sequence of elementary row operations is the same as that used in the example of frame 18. The operations are as follows, in the order indicated:

(a) First, $r_2' = r_2 - 2r_1$, and then $r_3' = r_3 - 3r_1$ to clear out column 1.

(b) Next, $r_2' = r_2/3$, and then $r_1' = r_1 + r_2$ and $r_3' = r_3 - 2r_2$ to clear out column 2.

(c) Next, $r_3' = \left(-\frac{3}{14}\right)r_3$, and then $r_1' = r_1 - \left(\frac{4}{3}\right)r_3$ and $r_2' = r_2 - \left(\frac{1}{3}\right)r_3$ to clear out column 3.

Thus, the starting augmented matrix $A|\mathbf{b}_1\,\mathbf{b}_2$ is transformed into the following reduced row echelon form:

$$\begin{bmatrix} \underline{1} & 0 & 0 & | & 2 & 3 \\ 0 & \underline{1} & 0 & | & 1 & -2 \\ 0 & 0 & \underline{1} & | & -1 & 1 \end{bmatrix} \tag{4}$$

Symbolically, this is in the form $I_3|\mathbf{k}_1\,\mathbf{k}_2$, where $\mathbf{k}_1$ and $\mathbf{k}_2$ are, respectively, the unique solution column vectors of systems (1) and (2). Thus, for systems (1) and (2), the unique solutions are as follows:

$$x_1 = 2, x_2 = 1, x_3 = -1 \quad \text{[system (1)]}$$

$$x_1 = 3, x_2 = -2, x_3 = 1 \quad \text{[system (2)]}$$

We shall now apply the approach used in Example (a) to find the inverse of an $n \times n$ matrix. To be specific, let us work with a definite matrix.

Example (b): Find the inverse of the matrix $A = \begin{bmatrix} 1 & 3 \\ 2 & 4 \end{bmatrix}$ of the example of frame 27.

Solution: Let the inverse (if it exists) be denoted by

$$B = \begin{bmatrix} b_{11} & b_{12} \\ b_{21} & b_{22} \end{bmatrix}$$

Thus, we are interested in calculating b_{11}, b_{12}, b_{21}, and b_{22}. Now B satisfies the matrix equation $AB = I_2$, that is,

$$\begin{bmatrix} 1 & 3 \\ 2 & 4 \end{bmatrix} \begin{bmatrix} b_{11} & b_{12} \\ b_{21} & b_{22} \end{bmatrix} = \begin{bmatrix} 1 & 0 \\ 0 & 1 \end{bmatrix}$$

Carrying out the matrix multiplication on the left leads to the matrix equation

$$\begin{bmatrix} b_{11} + 3b_{21} & b_{12} + 3b_{22} \\ 2b_{11} + 4b_{21} & 2b_{12} + 4b_{22} \end{bmatrix} = \begin{bmatrix} 1 & 0 \\ 0 & 1 \end{bmatrix}$$

Equating the corresponding column 1 entries and then the corresponding column 2 entries leads to the following two systems of equations, one for column 1 entries and the other for column 2 entries:

$$\begin{aligned} b_{11} + 3b_{21} &= 1 \\ 2b_{11} + 4b_{21} &= 0 \end{aligned} \qquad \text{(column 1)} \tag{1}$$

$$\begin{aligned} b_{12} + 3b_{22} &= 0 \\ 2b_{12} + 4b_{22} &= 1 \end{aligned} \qquad \text{(column 2)} \tag{2}$$

Each of systems (1) and (2) is a system of two linear equations in two unknowns. Also, the systems have the same coefficient matrix, namely,

$\begin{bmatrix} 1 & 3 \\ 2 & 4 \end{bmatrix}$, which is the matrix A, whose inverse we seek. The systems differ

only in their right-side terms. Therefore, we may solve the two systems at the same time by using the approach of Example (a). We thus apply Method 2.1 to transform the following augmented matrix to reduced row echelon form:

$$\left[\begin{array}{cc|cc} 1 & 3 & 1 & 0 \\ 2 & 4 & 0 & 1 \end{array} \right] \tag{3}$$

Note that a symbolic form for (3) is $A|I_2$. We will have unique solutions for the b_{ij}'s provided that the reduced row echelon form is $I_2|\mathbf{k}_1\ \mathbf{k}_2$. Here $\mathbf{k}_1$ and $\mathbf{k}_2$ are the unique solution column vectors of systems (1) and (2). Thus, here they are the columns of the inverse matrix B (or A^{-1}).

We now employ Method 2.1 to find the reduced row echelon form of $A|I_2$ for this example.

$$\left[\begin{array}{cc|cc} 1 & 3 & 1 & 0 \\ 2 & 4 & 0 & 1 \end{array}\right] \xrightarrow{r_2' = r_2 - 2r_1}$$

$$\left[\begin{array}{cc|cc} 1 & 3 & 1 & 0 \\ 0 & -2 & -2 & 1 \end{array}\right] \xrightarrow{r_2' = r_2/(-2)}$$

Now finish the calculations.

— — — — — — — — — —

Continuing the application of Method 2.1, we obtain

$$\left[\begin{array}{cc|cc} 1 & 3 & 1 & 0 \\ 0 & 1 & 1 & -\frac{1}{2} \end{array}\right] \xrightarrow{r_1' = r_1 - 3r_2} \left[\begin{array}{cc|cc} 1 & 0 & -2 & \frac{3}{2} \\ 0 & 1 & 1 & -\frac{1}{2} \end{array}\right] \qquad (4)$$

From (4), which is in the desired form (I_2 is to the left of the dividing line), we read off $b_{11} = -2$, $b_{21} = 1$, $b_{12} = \frac{3}{2}$, and $b_{22} = -\frac{1}{2}$. Thus,

$$A^{-1} = \left[\begin{array}{cc} -2 & \frac{3}{2} \\ 1 & -\frac{1}{2} \end{array}\right],$$

which checks out with the inverse cited in the example of frame 27.

30. We summarize the general procedure for finding A^{-1}, if it exists, in the following theorem:

Theorem 2.7

Let A be an $n \times n$ matrix and suppose the $n \times 2n$ augmented matrix $A|I_n$ is formed. If the reduced row echelon form of $A|I_n$ is $I_n|K$ (obtained by using the Gauss–Jordan elimination method, or some equivalent approach), then the matrix K is the inverse of A (i.e., $K = A^{-1}$). If the reduced row echelon form of $A|I_n$ is not in the form $I_n|K$ (i.e., I_n does not occur to the left of the dividing line in the reduced row echelon form), then matrix A has no inverse.

Notes: (a) Similar comments apply to an $n \times n$ matrix A by itself. If the reduced row echelon form of A is I_n, then the inverse of A exists (i.e., A is invertible). If the reduced row echelon form of A is not I_n, then matrix A has no inverse.
  (b) One indication that A has no inverse is if at any step of the Gauss–Jordan elimination method a row of zeros occurs to the left of the dividing line ($|$). Such an occurrence clearly indicates that $A|I_n$ cannot be transformed to $I_n|K$.

(c) Another way of indicating that A has no inverse involves the J_i function. Recall (Definition 2.4 of frame 11) that J_i is the column label of the leading entry of row i. We note that the inverse of A exists if $J_1 = 1$, $J_2 = 2, \ldots, J_n = n$, and conversely. The reason is that this condition is equivalent to $A|I_n$ having $I_n|K$ as its reduced row echelon form. In Example (b) of frame 29 we see that $J_1 = 1$ and $J_2 = 2$.

If in applying Method 2.1 to $A|I_n$ it turns out for a row numbered $\hat{i}$, that J_i exceeds $\hat{i}$ (in general, $J_i \geqslant i$, if J_i exists), then this indicates that A has no inverse. Thus, the J_i function is very well suited for the writing of a computer program to determine first whether an inverse exists, and, second, to calculate it if it does exist.

Example (a): Find the inverse of matrix A below, if it exists.

$$A = \begin{bmatrix} 3 & -9 & 1 \\ -1 & 3 & 2 \\ 2 & -6 & -1 \end{bmatrix}$$

Solution: We apply Method 2.1 to $A|I_3$ until we reach a definite conclusion about the inverse.

$$\left[\begin{array}{ccc|ccc} 3 & -9 & 1 & 1 & 0 & 0 \\ -1 & 3 & 2 & 0 & 1 & 0 \\ 2 & -6 & -1 & 0 & 0 & 1 \end{array}\right] \xrightarrow{r_1' = r_1/(3)} \left[\begin{array}{ccc|ccc} 1 & -3 & \frac{1}{3} & \frac{1}{3} & 0 & 0 \\ -1 & 3 & 2 & 0 & 1 & 0 \\ 2 & -6 & -1 & 0 & 0 & 1 \end{array}\right]$$

$$\xrightarrow[\text{Then } r_3' = r_3 - 2r_1]{r_2' = r_2 + r_1.} \left[\begin{array}{ccc|ccc} 1 & -3 & \frac{1}{3} & \frac{1}{3} & 0 & 0 \\ 0 & 0 & \frac{7}{3} & \frac{1}{3} & 1 & 0 \\ 0 & 0 & -\frac{5}{3} & -\frac{2}{3} & 0 & 1 \end{array}\right] \xrightarrow{r_2' = r_2/(7/3)}$$

$$\left[\begin{array}{ccc|ccc} 1 & -3 & \frac{1}{3} & \frac{1}{3} & 0 & 0 \\ 0 & 0 & 1 & \frac{1}{7} & \frac{3}{7} & 0 \\ 0 & 0 & -\frac{5}{3} & -\frac{2}{3} & 0 & 1 \end{array}\right] \xrightarrow[\text{Then } r_3' = r_3 + (5/3)r_2]{r_1' = r_1 - (1/3)r_2.} \left[\begin{array}{ccc|ccc} 1 & -3 & 0 & \frac{2}{7} & -\frac{1}{7} & 0 \\ 0 & 0 & 1 & \frac{1}{7} & \frac{3}{7} & 0 \\ 0 & 0 & 0 & -\frac{3}{7} & \frac{5}{7} & 1 \end{array}\right]$$

Since row 3 of the last augmented matrix has only zeros to the left of the dividing line, the matrix A has no inverse.

Actually, we can tell from the third augmented preceding matrix that A has no inverse, since the entries to the left of the dividing line in row 2 (i.e., $0\ 0\ \frac{7}{3}$) indicate that $J_2 = 3$. (Thus, $J_i > \hat{i}$ here, since $3 > 2$.)

Example (b): Find the inverse of the matrix A, if it exists.

$$A = \begin{bmatrix} 2 & 3 & 4 \\ 1 & 2 & 1 \\ 1 & 2 & 3 \end{bmatrix}$$

Solution: We apply Method 2.1 to $A|I_3$ until we reach a definite conclusion about the inverse.

$$
\begin{bmatrix}
2 & 3 & 4 & | & 1 & 0 & 0 \\
1 & 2 & 1 & | & 0 & 1 & 0 \\
1 & 2 & 3 & | & 0 & 0 & 1
\end{bmatrix}
\xrightarrow{r_1' = r_1/2}
\begin{bmatrix}
1 & \frac{3}{2} & 2 & | & \frac{1}{2} & 0 & 0 \\
1 & 2 & 1 & | & 0 & 1 & 0 \\
1 & 2 & 3 & | & 0 & 0 & 1
\end{bmatrix}
\begin{matrix}
r_2' = r_2 - r_1; \\
\xrightarrow{\hspace{1cm}} \\
r_3' = r_3 - r_1
\end{matrix}
$$

$$
\begin{bmatrix}
1 & \frac{3}{2} & 2 & | & \frac{1}{2} & 0 & 0 \\
0 & \frac{1}{2} & -1 & | & -\frac{1}{2} & 1 & 0 \\
0 & \frac{1}{2} & 1 & | & -\frac{1}{2} & 0 & 1
\end{bmatrix}
\xrightarrow{r_2' = r_2/(1/2) = 2r_2}
\begin{bmatrix}
1 & \frac{3}{2} & 2 & | & \frac{1}{2} & 0 & 0 \\
0 & 1 & -2 & | & -1 & 2 & 0 \\
0 & \frac{1}{2} & 1 & | & -\frac{1}{2} & 0 & 1
\end{bmatrix}
$$

Now finish the calculations. The next step involves clearing out column 2.

$$
\begin{matrix}
r_1' = r_1 - (3/2)r_2; \\
\xrightarrow{\hspace{1cm}} \\
r_3' = r_3 - (1/2)r_2
\end{matrix}
\begin{bmatrix}
1 & 0 & 5 & | & 2 & -3 & 0 \\
0 & 1 & -2 & | & -1 & 2 & 0 \\
0 & 0 & 2 & | & 0 & -1 & 1
\end{bmatrix}
$$

$$
\xrightarrow{r_3' = r_3/2}
\begin{bmatrix}
1 & 0 & 5 & | & 2 & -3 & 0 \\
0 & 1 & -2 & | & -1 & 2 & 0 \\
0 & 0 & 1 & | & 0 & -\frac{1}{2} & \frac{1}{2}
\end{bmatrix}
$$

$$
\begin{matrix}
r_1' = r_1 - 5r_3; \\
\xrightarrow{\hspace{1cm}} \\
r_2' = r_2 + 2r_3
\end{matrix}
\begin{bmatrix}
1 & 0 & 0 & | & 2 & -\frac{1}{2} & -\frac{5}{2} \\
0 & 1 & 0 & | & -1 & 1 & 1 \\
0 & 0 & 1 & | & 0 & -\frac{1}{2} & \frac{1}{2}
\end{bmatrix}
$$

Since I_3 appears to the left of the final dividing line, the matrix A^{-1} appears to the right. (Observe that $J_1 = 1$, $J_2 = 2$, and $J_3 = 3$ here.) Thus,

$$
A^{-1} =
\begin{bmatrix}
2 & -\frac{1}{2} & -\frac{5}{2} \\
-1 & 1 & 1 \\
0 & -\frac{1}{2} & \frac{1}{2}
\end{bmatrix}
$$

31.   In this frame we will show that the unique solution for **x** in the matrix–vector equation $A\mathbf{x} = \mathbf{b}$ is $\mathbf{x} = A^{-1}\mathbf{b}$ if A^{-1} exists. We know, from frames 4–6, that we can represent a system of n linear equations in n unknowns by the matrix equation

$$A\mathbf{x} = \mathbf{b} , \tag{1}$$

where A is the $n \times n$ coefficient matrix. Suppose A^{-1} exists. If we premultiply (i.e., multiply on the left) on both sides of Eq. (1) by A^{-1}, we obtain

$$A^{-1}(A\mathbf{x}) = A^{-1}\mathbf{b} . \tag{2}$$

Because of the associative property of matrix multiplication [Theorem 1.3, Property (a)], the definition of A^{-1}, and the multiplicative property of I_n, the left side of Eq. (2) can be rewritten as

$$A^{-1}(Ax) = (A^{-1}A)x = I_n x = x .\tag{3}$$

Substitution from Eq. (3) into Eq. (2) yields the unique result $x = A^{-1}b$ for x. We summarize the preceding discussion in the following theorem.

Theorem 2.8

If $Ax = b$ represents a linear system of n equations in n unknowns, and A^{-1} exists, then the unique solution for the column vector x is given by $x = A^{-1}b$ for any right-side vector b.

If we are given a system of n linear equations in n unknowns represented by $Ax = b$, and A^{-1} exists, then we can solve for the unknowns vector x either by using the Gauss–Jordan elimination method, or by first finding A^{-1} and then using $x = A^{-1}b$. The former method usually involves less work. Suppose we are asked to solve several systems $Ax = b_1$, $Ax = b_2$, etc., where A is the same for each system, but the right-side vectors b_1, b_2, etc., vary. In this situation it is sometimes best to first find A^{-1} and then use Theorem 2.8 repeatedly for each of the b_i vectors.

Example: Solve the following linear systems for the unknowns vectors.

(a) $x_1 + 3x_2 = -3$ (b) $x_1 + 3x_2 = 11$ (c) $x_1 + 3x_2 = 5$

      $2x_1 + 4x_2 = -2$ $2x_1 + 4x_2 = 14$ $2x_1 + 4x_2 = 7$

Solution: For each of these systems, $A = \begin{bmatrix} 1 & 3 \\ 2 & 4 \end{bmatrix}$, and the right-side vectors

are

$$b_1 = \begin{bmatrix} -3 \\ -2 \end{bmatrix}, \qquad b_2 = \begin{bmatrix} 11 \\ 14 \end{bmatrix}, \qquad b_3 = \begin{bmatrix} 5 \\ 7 \end{bmatrix},$$

respectively. The matrix inverse A^{-1} has been determined in Example (b) of frame 29, and

$$A^{-1} = \begin{bmatrix} -2 & \frac{3}{2} \\ 1 & -\frac{1}{2} \end{bmatrix}.$$

Now we apply Theorem 2.8 three times in succession to find the respective unknowns vectors:

(a) $\mathbf{x} = A^{-1}\mathbf{b}_1 = \begin{bmatrix} -2 & \frac{3}{2} \\ 1 & -\frac{1}{2} \end{bmatrix} \begin{bmatrix} -3 \\ -2 \end{bmatrix} = \begin{bmatrix} 3 \\ -2 \end{bmatrix}$. Thus, $x_1 = 3$, $x_2 = -2$.

Now do parts (b) and (c).

- - - - - - - - - -

(b) $\mathbf{x} = A^{-1}\mathbf{b}_2 = \begin{bmatrix} -2 & \frac{3}{2} \\ 1 & -\frac{1}{2} \end{bmatrix} \begin{bmatrix} 11 \\ 14 \end{bmatrix} = \begin{bmatrix} -1 \\ 4 \end{bmatrix}$. Thus, $x_1 = -1$, $x_2 = 4$.

(c) $\mathbf{x} = A^{-1}\mathbf{b}_3 = \begin{bmatrix} -2 & \frac{3}{2} \\ 1 & -\frac{1}{2} \end{bmatrix} \begin{bmatrix} 5 \\ 7 \end{bmatrix} = \begin{bmatrix} \frac{1}{2} \\ \frac{3}{2} \end{bmatrix}$. Thus, $x_1 = \frac{1}{2}$, $x_2 = \frac{3}{2}$.

32. It is useful at this point to try to tie together some of the main concepts covered in this chapter thus far. Let us focus on properties associated with an $n \times n$ matrix. For example, we can describe a linear system of n equations in n unknowns (variables) by the matrix–vector equation $A\mathbf{x} = \mathbf{b}$, where A is an $n \times n$ matrix. In frame 26, we indicated the equivalence of the following two statements:

(a)   Rank $(A) = n$.
(b)   The reduced row echelon form of A is I_n.

We now know that A^{-1}, the inverse of matrix A, will exist if the reduced row echelon form of A is I_n. Also, A^{-1} will not exist if the reduced row echelon form of A is not I_n. [See Note (a) after Theorem 2.7 in frame 30.] We also know that if A^{-1} exists, then the system will have a unique solution for $\mathbf{x}$ (Theorem 2.8 of frame 31). If A^{-1} does not exist, then $A\mathbf{x} = \mathbf{b}$ will not have a unique solution, in which case there will be either no solution for $\mathbf{x}$ or infinitely many solutions. We summarize the preceding discussion as follows, in Theorem 2.9.

Theorem 2.9

If A is an $n \times n$ matrix, then the following statements are equivalent:

(a)   Rank $(A) = n$.
(b)   The reduced row echelon form of A is I_n.
(c) A is invertible (i.e., A^{-1} exists).
(d)   The system $A\mathbf{x} = \mathbf{b}$ has a unique solution (given by $\mathbf{x} = A^{-1}\mathbf{b}$) for any right-side vector $\mathbf{b}$.

Notes: (i) As indicated in frame 26, the meaning of "equivalent" is that if any statement of the list is true, then all the statements are true, and, similarly, if any statement is false, then all are false. (ii) Also, as discussed in frame 26, we can relate any two statements in a list of equivalent statements through an "if and only if" expression, and vice versa. Thus, we can relate parts (c) and (d) above by writing "A is invertible if and only if the system $A\mathbf{x} = \mathbf{b}$ has a unique solution given by $\mathbf{x} = A^{-1}\mathbf{b}$ for any $\mathbf{b}$." (iii) In part (d), if we have a homogeneous system, then $\mathbf{b} = \mathbf{0}$. The unique solution in this case is the trivial solution, i.e., $\mathbf{x} = \mathbf{0}$, since A^{-1}, (or for that matter, any $n \times n$ matrix) times $\mathbf{0}$ (the zero vector) equals $\mathbf{0}$. (iv) We shall add more equivalent statements to the list of Theorem 2.9 when we encounter an expression known as the determinant (Chapter 3) and the concept of independence (Chapter 5).

Example (a): Interpret all the statements of Theorem 2.9 with respect to the 2×2 matrix $A = \begin{bmatrix} 1 & 3 \\ 2 & 4 \end{bmatrix}$.

Solution: In Example (b) of frame 29 we determined that the inverse of A exists; in fact, we found that $A^{-1} = \begin{bmatrix} -2 & \frac{3}{2} \\ 1 & -\frac{1}{2} \end{bmatrix}$. This means that statement (c) of the list is true, and thus the remaining three statements are also true. In other words, Rank $(A) = 2$ from (a), and the reduced row echelon form of A is I_2 from (b). Also, from (d), the system $A\mathbf{x} = \mathbf{b}$ has a unique solution vector $\mathbf{x}$, regardless of the right-side vector $\mathbf{b}$.

Example (b): Interpret all statements of Theorem 2.9 with respect to the 3×3 matrix

$$A = \begin{bmatrix} 3 & -9 & 1 \\ -1 & 3 & 2 \\ 2 & -6 & -1 \end{bmatrix}$$

Hint: Refer to Example (a) of frame 30.

— — — — — — — — — —

Solution: In Example (a) of frame 30, we determined that A^{-1} does not exist for this matrix. This means that statement (c) of the list in Theorem 2.9 is false, and thus the remaining three statements are also false.

From statement (b), we know that the reduced row echelon form of A is *not* I_3; from Example (a) of frame 30, we observe that the reduced row echelon form of A is

$$\begin{bmatrix} 1 & -3 & 0 \\ 0 & 0 & 1 \\ 0 & 0 & 0 \end{bmatrix}.$$

To see this, observe the matrix to the left of the dividing line in the last augmented matrix of Example (a) of frame 30.

From statement (a), we know that Rank $(A) \neq 3$; in fact, Rank $(A) = 2$, as can be seen from the reduced row echelon form given above since there are two leading entries. [In general, for an $n \times n$ matrix A, Rank $(A) \leq n$. Thus, here we know that Rank (A) has to be less than n.]

From statement (d), we know that the linear system of three equations in three unknowns given by $A\mathbf{x} = \mathbf{b}$ will not have a unique solution. If $\mathbf{b} \neq \mathbf{0}$, then there will be either no solution or infinitely many solutions (depending on $\mathbf{b}$). If $\mathbf{b} = \mathbf{0}$, there will be infinitely many solutions.

G. AN APPLICATION OF THE MATRIX INVERSE

33.   Input–output analysis was developed by W. W. Leontief in 1936 as a method for studying the interactions of various parts of an economy.* Here we present a simplified form of the input–output model.

Suppose we have an economy consisting of the three production sectors, agriculture, manufacturing, and services, and also a consumer sector. Each of the three production sectors (or industries) requires inputs from each of the production sectors so as to produce its output.

Consider the following hypothetical table, which shows how the total output of each production sector is used by the different production sectors and the consumer sector. The entries are in millions of dollars. We label agriculture, manufacturing, and services as production sectors 1, 2, and 3, respectively.

TABLE 2.1   Input–Output Table for a Hypothetical Economy

	Agriculture (1)	Manufacturing (2)	Services (3)	Consumer Demand	Total Output
Agriculture (1)	12	40	15	53	120
Manufacturing (2)	36	60	25	79	200
Services (3)	15	50	20	65	150

For example, the total output of the manufacturing sector is worth $200 million. Of this, $36 million is used by the agriculture sector, $60 million is used by the manufacturing sector itself, $25 million is used by the services sector, and $79 million is used by the consumer sector.

Suppose we wish to know the input, in dollars, of each production sector for each dollar of output of a particular production sector. From Table 2.1 we see that for $200 million of output in the manufacturing sector, there are required $40 million, $60 million, and $50 million (see *column* labeled "Manufacturing")

---

* Leontief won the Nobel Prize in 1973 for Economics (*Newsweek*, Oct. 29, 1973, p. 94).

from agriculture, manufacturing, and services, respectively. Thus, for $1 of output for the manufacturing sector, there are required $0.20 ($=40/200$), $0.30 ($=60/200$), and $0.25 ($=50/200$) from the three sectors, respectively.

These latter numbers are called *input–output coefficients*. If we use the labels 1, 2, and 3, as introduced in Table 2.1, and let the input–output coefficient a_{ij} denote the amount of input from production sector i (in dollars) used to produce one dollar's worth of output in production sector j, we have the following input–output coefficients with respect to output in the manufacturing sector:

$$a_{12} = 0.20 \, \frac{\$ \text{ input } 1}{\$ \text{ output } 2},$$

$$a_{22} = 0.30 \, \frac{\$ \text{ input } 2}{\$ \text{ output } 2},$$

$$a_{32} = 0.25 \, \frac{\$ \text{ input } 3}{\$ \text{ output } 2}.$$

Interpreting a_{12} in words, there is an input of $0.20 from the agriculture sector for each dollar of output in the manufacturing sector. In similar fashion we obtain the entries in Table 2.2 for all the input–output coefficients.

TABLE 2.2   Entries of the Input–Output Matrix A.

	Inputs per $1.00 of Output of		
	Agriculture (1)	Manufacturing (2)	Services (3)
Agriculture (1)	$a_{11} = 0.10$	$a_{12} = 0.20$	$a_{13} = 0.10$
Manufacturing (2)	0.30	0.30	0.1667
Services (3)	0.125	0.25	0.1333

The collection of the a_{ij} entries constitutes the *input–output matrix A*, where, for the case of three production sectors,

$$A = \begin{bmatrix} a_{11} & a_{12} & a_{13} \\ a_{21} & a_{22} & a_{23} \\ a_{31} & a_{32} & a_{33} \end{bmatrix} \tag{1}$$

Again, the entries in a particular column of Table 2.2 are obtained by dividing the corresponding column entries of Table 2.1 by the total output of the sector which is indicated in the column label.

Example: Illustrate the calculation of the a_{ij}'s in Table 2.2.

Solution: We already did the calculation of the a_{ij}'s for column 2. Now for column 1 we have

$$a_{11} = \frac{12}{120} = 0.10 \frac{\$ \text{ input 1}}{\$ \text{ output 1}} ,$$

$$a_{21} = \frac{36}{120} = 0.30 \frac{\$ \text{ input 2}}{\$ \text{ output 1}} ,$$

$$a_{31} = \frac{15}{120} = 0.125 \frac{\$ \text{ input 3}}{\$ \text{ output 1}} .$$

Now do the calculations of the input–output coefficients for column 3.

– – – – – – – – – –

$$a_{13} = \frac{15}{150} = 0.10 \frac{\$ \text{ input 1}}{\$ \text{ output 3}} ,$$

$$a_{23} = \frac{25}{150} = 0.1667 \frac{\$ \text{ input 2}}{\$ \text{ output 3}} ,$$

$$a_{33} = \frac{20}{150} = 0.1333 \frac{\$ \text{ input 3}}{\$ \text{ output 3}} .$$

34. In input–output analysis, one assumes that the a_{ij}'s remain *constant* over some period of time. With this assumption one can then compute total outputs in the various production sectors needed to satisfy different consumer demands on the production sectors. It is convenient to define x_j as the total output, in millions of dollars, of sector j, and c_j as the consumer demand, in millions of dollars, on sector j. The column vectors **x** (total output vector) and **c** (consumer demand vector) are defined as follows:

$$\mathbf{x} = \begin{bmatrix} x_1 \\ x_2 \\ x_3 \end{bmatrix}, \qquad \mathbf{c} = \begin{bmatrix} c_1 \\ c_2 \\ c_3 \end{bmatrix}. \tag{2a), (2b}$$

(For example, for Table 2.1, $\mathbf{x} = [120, \ 200, \ 150]^t$, and $\mathbf{c} = [53, \ 79, \ 65]^t$.)

Now let us derive equations relating the x_j's, the c_j's, and the a_{ij}'s.

Let us focus on sector 1, the agriculture sector, for example. We have the following equation for the total output of sector 1 (symbolized by x_1):

$$\begin{pmatrix} \text{Total output} \\ \text{of sector 1} \end{pmatrix} = \begin{pmatrix} \text{sum of inputs from} \\ \text{sector 1 going to} \\ \text{sectors 1, 2, and 3} \end{pmatrix} + \begin{pmatrix} \text{consumer de-} \\ \text{mand on sector 1} \end{pmatrix} \tag{3}$$

In symbols, we have

$$x_1 = a_{11}x_1 + a_{12}x_2 + a_{13}x_3 + c_1 \tag{4a}$$

For example, the second term on the right gives the input in millions of dollars (denoted by $\$$) from production sector 1 to production sector 2, as is indicated by the canceling of terms in the following:

$$a_{12} \frac{\$ \text{ input } 1}{\$ \ \cancel{\text{output 2}}} \cdot x_2 \ \$ \ \cancel{\text{output 2}} = a_{12}x_2 \ \$ \text{ input } 1$$

Likewise, $a_{11}x_1$ and $a_{13}x_3$ represent the inputs from production sector 1, in millions of dollars, to sectors 1 and 3, respectively.

We can likewise obtain equations for x_2 and x_3, respectively, which are similar to Eq. (4a):

$$x_2 = a_{21}x_1 + a_{22}x_2 + a_{23}x_3 + c_2 \tag{4b}$$

$$x_3 = a_{31}x_1 + a_{32}x_2 + a_{33}x_3 + c_3 \tag{4c}$$

Next we combine Eqs. (4a), (4b), and (4c) into the single column vector equation:

$$\begin{bmatrix} x_1 \\ x_2 \\ x_3 \end{bmatrix} = \begin{bmatrix} a_{11}x_1 + a_{12}x_2 + a_{13}x_3 \\ a_{21}x_1 + a_{22}x_2 + a_{23}x_3 \\ a_{31}x_1 + a_{32}x_2 + a_{33}x_3 \end{bmatrix} + \begin{bmatrix} c_1 \\ c_2 \\ c_3 \end{bmatrix} \tag{5}$$

The first and last terms in Eq. (5) are the vectors $\mathbf{x}$ and $\mathbf{c}$, respectively, and the second term is none other than the product of input–output matrix A [see Eq. (1) of frame 33] times column vector $\mathbf{x}$. Thus, we have the following matrix–vector equation:

$$\mathbf{x} = A\mathbf{x} + \mathbf{c} \tag{6}$$

Now let us attempt to develop an equation where $\mathbf{x}$ is solved for in terms of other quantities. From Eq. (6), we have

$$\mathbf{x} - A\mathbf{x} = \mathbf{c} \tag{7}$$

Since $\mathbf{x} = I\mathbf{x}$, where I is the appropriate identity matrix (I_3 for three production sectors, I_n for n production sectors), we have

$$I\mathbf{x} - A\mathbf{x} = \mathbf{c} \tag{8}$$

Now complete the job. *Hints:* First use Theorem 1.3, part (c) distributive law to "factor out" **x**. Then multiply the resulting equation by the appropriate matrix inverse, and thus solve for **x**.

– – – – – – – – – –

From Eq. (8), we have

$$(I - A)\mathbf{x} = \mathbf{c} \tag{9}$$

Then premultiplying both sides by the inverse $(I - A)^{-1}$ (assuming it exists) leads to

$$\mathbf{x} = (I - A)^{-1}\mathbf{c} \tag{10}$$

35. Equation (10) of frame 34 is useful if we wish to calculate total outputs (x_j's) from the various production sectors corresponding to a set of consumer demands (c_j's) on the sectors. Of course, we have to compute the matrix inverse first.

Example: For the input–output matrix A given in Table 2.2 of frame 33, compute $(I_3 - A)^{-1}$.

Solution: First, subtracting A from the 3×3 identity matrix yields

$$(I_3 - A) = \begin{bmatrix} 1 & 0 & 0 \\ 0 & 1 & 0 \\ 0 & 0 & 1 \end{bmatrix} - \begin{bmatrix} 0.10 & 0.20 & 0.10 \\ 0.30 & 0.30 & 0.1667 \\ 0.125 & 0.25 & 0.1333 \end{bmatrix}$$

$$= \begin{bmatrix} 0.90 & -0.20 & -0.10 \\ -0.30 & 0.70 & -0.1667 \\ -0.125 & -0.25 & 0.8667 \end{bmatrix} \tag{11}$$

Now we use the technique of Section F to obtain $(I_3 - A)^{-1}$. For example, refer to frames 29 and 30. Thus, we apply Method 2.1 to the following augmented matrix

$$\begin{bmatrix} 0.90 & -0.20 & -0.10 & \vert & 1 & 0 & 0 \\ -0.30 & 0.70 & -0.1667 & \vert & 0 & 1 & 0 \\ -0.125 & -0.25 & 0.8667 & \vert & 0 & 0 & 1 \end{bmatrix} \tag{12}$$

Observe that matrix $(I_3 - A)$ is to the left of the dividing line. Now finish the problem.

– – – – – – – – – –

The reduced row echelon form for (12) is

$$
\begin{bmatrix}
\underline{1} & 0 & 0 & 1.2956 & 0.45481 & 0.23696 \\
0 & \underline{1} & 0 & 0.64399 & 1.7600 & 0.41277 \\
0 & 0 & \underline{1} & 0.37263 & 0.57328 & 1.3071
\end{bmatrix},
\tag{13}
$$

which indicates that

$$
(I_3 - A)^{-1} =
\begin{bmatrix}
1.2956 & 0.45481 & 0.23696 \\
0.64399 & 1.7600 & 0.41277 \\
0.37263 & 0.57328 & 1.3071
\end{bmatrix}
\tag{14}*
$$

36. Now that we have calculated $(I_3 - A)^{-1}$, we can do calculations for total outputs needed to satisfy a set of consumer demands.

Example (a): Determine the values for total outputs x_1, x_2, and x_3 if the consumer demands, in millions of dollars, are $c_1 = 10$, $c_2 = 146.67$, and $c_3 = 85.833$.

Solution: Substituting $c = [10, 146.67, 85.833]^t$ and $(I_3 - A)^{-1}$ from Eq. (14) into Eq. (10) of frame 34 yields

$$
x = (I_3 - A)^{-1}c
$$

$$
=
\begin{bmatrix}
1.2956 & 0.45481 & 0.23696 \\
0.64399 & 1.7600 & 0.41277 \\
0.37263 & 0.57328 & 1.3071
\end{bmatrix}
\begin{bmatrix}
10.00 \\
146.67 \\
85.833
\end{bmatrix}
$$

$$
=
\begin{bmatrix}
100. \\
300. \\
200.
\end{bmatrix},
$$

which means that the individual total outputs, in millions of dollars, are $x_1 = 100$ (for agriculture), $x_2 = 300$ (for manufacturing), and $x_3 = 200$ (for services).*

Example (b): Determine the values for total outputs x_1, x_2, and x_3 if the consumer demands, in millions of dollars, are $c_1 = 70$, $c_2 = 105$, and $c_3 = 48.75$.

_ _ _ _ _ _ _ _ _ _

----

* These calculations were checked by using the Master Library Module of the Texas Instrument 58C (or 59) Programmable Calculator. Checks were also made of other calculations involving matrix multiplication, matrix inverses, and solutions of linear systems of n equations in n unknowns. Programmable calculator usage is discussed further in Section C of Chapter 9.

Solution: Proceeding as in Example (a), we solve for the total output vector **x** from

$$\mathbf{x} = (I_3 - A)^{-1}\mathbf{c}$$

$$= \begin{bmatrix} 1.2956 & 0.45481 & 0.23696 \\ 0.64399 & 1.7600 & 0.41277 \\ 0.37263 & 0.57328 & 1.3071 \end{bmatrix} \begin{bmatrix} 70. \\ 105. \\ 48.75 \end{bmatrix} = \begin{bmatrix} 150 \\ 250 \\ 150 \end{bmatrix}$$

Thus, the individual total outputs, in millions of dollars, are $x_1 = 150$, $x_2 = 250$, and $x_3 = 150$.*

SELF-TEST

This Self-Test will help you determine whether or not you have mastered the chapter objectives and are ready to go on to the next chapter. Correct answers are given at the end of the test.

1. Given the following system of three equations in three variables (unknowns):

$$2x_1 - 4x_2 + 3x_3 = -8 \qquad (1)$$

$$3x_1 + 5x_2 - 4x_3 = 4 \qquad (2)$$

$$4x_1 - 6x_2 + 5x_3 = -12 \qquad (3)$$

Determine (a) the coefficient matrix A, (b) the right-side vector **b**, (c) the expression for the unknowns vector **x**, and (d) the augmented coefficient matrix $A|\mathbf{b}$ for this system.

2. Which of the following matrices are in reduced row echelon form? If a matrix is in reduced row echelon form, determine the J_i's (J_i is column location of leading entry in row i).

(a) $\begin{bmatrix} 1 & 6 & 0 & | & 3 \\ 0 & 0 & 1 & | & -5 \\ 0 & 0 & 0 & | & 0 \end{bmatrix}$ (b) $\begin{bmatrix} 1 & 0 & -3 & 0 \\ 0 & 1 & 2 & 0 \\ 0 & 0 & 0 & 4 \end{bmatrix}$ (c) $\begin{bmatrix} 1 & 0 & 2 & 0 \\ 0 & 1 & 5 & 0 \\ 0 & 0 & 0 & 1 \end{bmatrix}$

(d) $\begin{bmatrix} 1 & 0 & 7 & 2 \\ 0 & 1 & -5 & 0 \\ 0 & 0 & 0 & 1 \end{bmatrix}$ (e) $\begin{bmatrix} 1 & 0 & 0 & | & 0 \\ 0 & 1 & 0 & | & 8 \\ 0 & 0 & 1 & | & 0 \\ 0 & 0 & 0 & | & 0 \end{bmatrix}$ (f) $\begin{bmatrix} 1 & 0 & 0 & -4 \\ 0 & 0 & 1 & 6 \\ 0 & 1 & 0 & 5 \end{bmatrix}$

---

* These calculations were checked by using the Master Library Module of the Texas Instrument 58C (or 59) Programmable Calculator. Checks were also made of other calculations involving matrix multiplication, matrix inverses, and solutions of linear systems of n equations in n unknowns. Programmable calculator usage is discussed further in Section C of Chapter 9.

3.   Employ the Gauss–Jordan elimination method (Method 2.1 in frame 14) to obtain the reduced row echelon form for the augmented matrix $A|\mathbf{b}$ of question (1). List the augmented matrices that result at the ends of all stages. Determine solutions to the linear system of question (1), if they exist. (There may be only one solution.) Interpret geometrically.

4.   Find solution(s), if they exist, for the following system, by using Method 2.1.

$$
\begin{aligned}
2x_2 - 2x_3 &= -8 \\
x_1 + x_2 + x_3 &= 2 \\
x_1 + 2x_2 &= 4
\end{aligned}
$$

5.   Find solution(s), if they exist, for the following system, by using Method 2.1.

$$
\begin{aligned}
4x_1 - 8x_2 + x_3 &= 18 \\
x_1 - 2x_2 - x_3 &= 2 \\
3x_1 - 6x_2 - 2x_3 &= 8
\end{aligned}
$$

6.   Find solution(s) for the following homogeneous systems:

(a) $2x_1 - 3x_2 = 0$ (b) $4x_1 - 8x_2 + x_3 = 0$

$\qquad -4x_1 + 5x_2 = 0$ $\qquad x_1 - 2x_2 - x_3 = 0$

$\qquad\qquad\qquad\qquad\qquad\qquad 3x_1 - 6x_2 - 2x_3 = 0$

7.   (a)   For the linear system of questions (1) and (3), determine Rank $(A|\mathbf{b})$, and Rank (A), and verify Theorem 2.4 of frame 24.

     (b)   Repeat part (a) for the linear system of question (4).

     (c)   Repeat part (a) for the linear system of question (5).

8.   For the following system of three equations in four unknowns, you are given that Rank $(A|\mathbf{b})$ = Rank (A).

$$
\begin{aligned}
x_1 - 3x_2 + 8x_3 - 6x_4 &= 5 \\
2x_2 + 10x_3 + 3x_4 &= 2 \\
x_2 - 6x_3 &= 4
\end{aligned}
$$

Use Theorem 2.4 and common sense (that is, don't do any further calculations) to determine the solution characteristics of this system.

9.   For the homogeneous linear systems of questions (6a) and (6b), determine Rank $(A|\mathbf{0})$, and verify the Corollary to Theorem 2.4 (frame 25).

10. (a) Use the method suggested in Theorem 2.7 (frame 30) to determine the inverse of matrix B, if $B = \begin{bmatrix} 2 & 6 \\ 3 & 6 \end{bmatrix}$. Check your result by computing BB^{-1} and $B^{-1}B$.

(b) Then use Theorem 2.6, part (b), to determine the inverse of $B^t = \begin{bmatrix} 2 & 3 \\ 6 & 6 \end{bmatrix}$.

(c)

From Example (b) of frame 29, $A = \begin{bmatrix} 1 & 3 \\ 2 & 4 \end{bmatrix}$. Thus, using B from part (a), we have $AB = \begin{bmatrix} 11 & 24 \\ 16 & 36 \end{bmatrix}$ (verify this). Determine $(AB)^{-1}$ by using the known results for A^{-1} and B^{-1}, and Theorem 2.6, part (d).

11. Find the matrix inverse, if it exists, for each of the following:

(a) $C = \begin{bmatrix} 2 & 1 & 2 \\ 0 & 0 & 1 \\ 1 & 0 & 2 \end{bmatrix}$ (b) $D = \begin{bmatrix} 4 & -8 & 1 \\ 1 & -2 & -1 \\ 3 & -6 & -2 \end{bmatrix}$

12. Find the unique solution of the following linear system:

$$2x_1 + 6x_2 = 2$$

$$3x_1 + 6x_2 = 6$$

Use Theorem 2.8 and the fact that the coefficient matrix of this system is the matrix B of question (10a).

13. The input–output matrix A for an economy with three production sectors is given as follows:

	Inputs per $1.00 Output of		
	Agriculture (1)	Manufacturing (2)	Services (3)
Agriculture (1)	0.10	0.25	0.1333
Manufacturing (2)	0.30	0.30	0.2667
Services (3)	0.20	0.25	0.20

See Table 2.2 of frame 33 for a related input–output matrix. Here a_{ij} represents the dollars of input of sector i per dollar of output of sector j. For example, $a_{12} = 0.25$, and this means there is 0.25 dollars of input from the agriculture sector for every dollar of output in the manufacturing sector.

(a) Determine the matrix $(I_3 - A)^{-1}$.

(b) Determine the values for total outputs required from the agriculture, manufacturing, and service sectors if the consumer demands, in millions of dollars, are 24, 81, and 43, on the three production sectors, respectively.

ANSWERS TO SELF-TEST

If your answers to the test questions do not agree with the ones given here, review the frames indicated in parentheses after each answer before you go on to the next chapter.

1. (a) $A = \begin{bmatrix} 2 & -4 & 3 \\ 3 & 5 & -4 \\ 4 & -6 & 5 \end{bmatrix}$,   (b) $\mathbf{b} = \begin{bmatrix} -8 \\ 4 \\ -12 \end{bmatrix}$,   (c) $\mathbf{x} = \begin{bmatrix} x_1 \\ x_2 \\ x_3 \end{bmatrix}$,

(d) $A|\mathbf{b} = \begin{bmatrix} 2 & -4 & 3 & | & -8 \\ 3 & 5 & -4 & | & 4 \\ 4 & -6 & 5 & | & -12 \end{bmatrix}$.   (frames 5–7)

2. (a), (c), and (e) are in reduced row echelon form. For (a), $J_1 = 1$, $J_2 = 3$. For (c), $J_1 = 1$, $J_2 = 2$, $J_3 = 4$. For (e), $J_1 = 1$, $J_2 = 2$, $J_3 = 3$.   (frames 11, 12)

3. Stage 1: $\begin{bmatrix} 1 & -2 & \frac{3}{2} & | & -4 \\ 0 & 11 & -\frac{17}{2} & | & 16 \\ 0 & 2 & -1 & | & 4 \end{bmatrix}$,

Stage 2: $\begin{bmatrix} 1 & 0 & -\frac{1}{22} & | & -\frac{12}{11} \\ 0 & 1 & -\frac{17}{22} & | & \frac{16}{11} \\ 0 & 0 & \frac{6}{11} & | & \frac{12}{11} \end{bmatrix}$,

Stage 3: $\begin{bmatrix} 1 & 0 & 0 & | & -1 \\ 0 & 1 & 0 & | & 3 \\ 0 & 0 & 1 & | & 2 \end{bmatrix}$.

The unique solution, given by $x_1 = -1$, $x_2 = 3$, $x_3 = 2$, represents the point of intersection of the planes corresponding to the equations of question (1).   (frames 12–18)

4. A partial application of Method 2.1 leads to a third row in the form 0 0 0|6 after several steps; this indicates that the system has no solution. The reduced row echelon form of $A|\mathbf{b}$ is $\begin{bmatrix} 1 & 0 & 2 & | & 0 \\ 0 & 1 & -1 & | & 0 \\ 0 & 0 & 0 & | & 1 \end{bmatrix}$. Note that the coefficient matrix for the starting system is the same as for the example in frame 19. (frames 14–19)

5. The reduced row echelon form is $\begin{bmatrix} 1 & -2 & 0 & | & 4 \\ 0 & 0 & 1 & | & 2 \\ 0 & 0 & 0 & | & 0 \end{bmatrix}$, indicating infinitely

many solutions. The corresponding equations are $x_1 = 2x_2 + 4$ and $x_3 = 2$, with x_2 arbitrary, or, put differently, $x_1 = 2r + 4$, $x_2 = r$, $x_3 = 2$, with r arbitrary.   (frames 14–19)

6.   (a)   Only the trivial solution $x_1 = 0$, $x_2 = 0$.

(b)   Infinitely many solutions, given by $x_1 = 2x_2$, $x_3 = 0$, with x_2 arbitrary. *Note:* Since the coefficient matrix here is the same as in question 5, it follows that [after looking at the answer in question 5] the reduced row echelon form here is

$$\begin{bmatrix} 1 & -2 & 0 & | & 0 \\ 0 & 0 & 1 & | & 0 \\ 0 & 0 & 0 & | & 0 \end{bmatrix} \qquad \text{(frames 20–21)}$$

7.   (a)   Rank $(A|\mathbf{b})$ = Rank (A) = 3, indicating a consistent system. Also, the fact that n equals 3 indicates a unique solution.

(b).   Rank $(A|\mathbf{b})$ = 3, and Rank (A) = 2, indicating an inconsistent system (i.e., no solution).

(c)   Rank $(A|\mathbf{b})$ = Rank (A) = 2, indicating a consistent system. Also, the fact that $n = 3$ indicates infinitely many solutions.   (frames 22–24)

8.   Since Rank $(A|\mathbf{b})$ = Rank (A), the system is consistent. Since Rank $(A|\mathbf{b})$ ≤ m, in general, and, here, $m < n$ ($m = 3$ and $n = 4$), it follows that Rank $(A|\mathbf{b}) < n$, which implies that the system has infinitely many solutions. (frames 24–25)

9.   (a)   Rank $(A|\mathbf{0})$ = 2 = n, indicating that the trivial solution is the only solution.

(b)   Rank $(A|\mathbf{0})$ = 2 and $n = 3$, indicating infinitely many solutions. (frames 24–26)

10.   (a) $B^{-1} = \begin{bmatrix} -1 & 1 \\ \frac{1}{2} & -\frac{1}{3} \end{bmatrix}$. Checking, $BB^{-1} = B^{-1}B = \begin{bmatrix} 1 & 0 \\ 0 & 1 \end{bmatrix}$.

(b) $(B^t)^{-1} = (B^{-1})^t = \begin{bmatrix} -1 & \frac{1}{2} \\ 1 & -\frac{1}{3} \end{bmatrix}$, after using the result from part (a).

(c) $(AB)^{-1} = B^{-1}A^{-1} = \begin{bmatrix} 3 & -2 \\ -\frac{4}{3} & \frac{11}{12} \end{bmatrix}$.   (frames 27–30)

11.   (a) $C^{-1} = \begin{bmatrix} 0 & -2 & 1 \\ 1 & 2 & -2 \\ 0 & 1 & 0 \end{bmatrix}$;

(b) D^{-1} does not exist.   (frames 27–30)

12. $\mathbf{x} = B^{-1}\mathbf{b} = \begin{bmatrix} -1 & 1 \\ \frac{1}{2} & -\frac{1}{3} \end{bmatrix}\begin{bmatrix} 2 \\ 6 \end{bmatrix} = \begin{bmatrix} 4 \\ -1 \end{bmatrix}$, and, thus, $x_1 = 4$, $x_2 = -1$ constitutes the unique solution.   (frame 31)

13. (a)

$$(I_3 - A)^{-1} = \begin{bmatrix} 1.4425 & 0.68226 & 0.46784 \\ 0.85770 & 2.0273 & 0.81871 \\ 0.62865 & 0.80409 & 1.6228 \end{bmatrix}$$

(b) $x_1 = 110$ (agriculture), $x_2 = 220$ (manufacturing), and $x_3 = 150$ (services), where the units are in millions of dollars.   (frames 33–36)

CHAPTER THREE
Determinants

In this chapter we shall study some of the basic properties of determinants. A determinant is a function defined for an $n \times n$ matrix A, and it is a *number* pertaining to A. Determinants are useful in the study of solutions of linear systems, and historically (circa 1700), that is how the study of determinants originated. Also, the determinant provides a valuable criterion for determining whether or not a square matrix has an inverse. Our work will lead us to formulas, expressed in terms of determinants, for the solution of a linear system of n equations in n unknowns, and for the inverse of a matrix.

OBJECTIVES

When you complete this chapter, you should be able to

- Compute the determinant of an $n \times n$ matrix, either from the basic definition or by using a cofactor expansion on any row or column.
- Simplify the computation of determinants by using elementary row operations.
- Determine whether or not the inverse of a matrix exists, from the value of the determinant of the matrix.
- Calculate the inverse of a matrix (if it exists) from a formula involving determinants.
- Calculate the unique solution components $x_1, x_2, \ldots, x_n$ of a linear system of n equations in n unknowns from formulas expressed in terms of determinants (Cramer's rule).
- Determine the profit and tax relationship for a multinational corporation (application).

A. INTRODUCTION TO DETERMINANTS

1. Let A be an $n \times n$ matrix. We use the symbol $|A|$ or det (A) for the *determinant* of A. We shall use the first symbol more often in this book.

The determinant of an $n \times n$ matrix A is a number pertaining to A. It is defined as follows for 1×1 and 2×2 matrices.

Definition 3.1

If A is the 1×1 matrix $[a_{11}]$, then $|A| = a_{11}$.

If A is the 2×2 matrix $\begin{bmatrix} a_{11} & a_{12} \\ a_{21} & a_{22} \end{bmatrix}$, then $|A| = a_{11}a_{22} - a_{12}a_{21}$.

Example (a):   (i) The determinant of $A = \begin{bmatrix} 6 & 2 \\ 5 & 3 \end{bmatrix}$ is $6(3) - 2(5) = 8$. Using

the $|A|$ notation, we can write $|A| = 8$. We also can indicate this by writing

$\begin{vmatrix} 6 & 2 \\ 5 & 3 \end{vmatrix} = 8$, even though we should write $\left| \begin{bmatrix} 6 & 2 \\ 5 & 3 \end{bmatrix} \right| = 8$, to be more

precise. We will use the former type of notation extensively in this chapter.

In det (A) notation, we have det $(A) = 8$, or det $\left(\begin{bmatrix} 6 & 2 \\ 5 & 3 \end{bmatrix} \right) = 8$, or,

abbreviating, det $\begin{bmatrix} 6 & 2 \\ 5 & 3 \end{bmatrix} = 8$.

(ii) For the determinant of the matrix $[7]$, we have $|7| = 7$, and for the

determinant of the matrix $\begin{bmatrix} 3 & 2 \\ 4 & -5 \end{bmatrix}$, we have $\begin{vmatrix} 3 & 2 \\ 4 & -5 \end{vmatrix} = 3(-5) - 2(4)$

$= -23$.

Notes: (a) Henceforth, we shall occasionally refer to a determinant of a 2×2 matrix as being "a 2×2 determinant." Thus, in Example (a) we call the symbol $\begin{vmatrix} 3 & 2 \\ 4 & -5 \end{vmatrix}$ the 2×2 determinant corresponding to matrix $\begin{bmatrix} 3 & 2 \\ 4 & -5 \end{bmatrix}$. Similarly, a determinant of an $n \times n$ matrix will occasionally be called an "$n \times n$ determinant." This usage should not prove to be confusing.

(b) It is important not to confuse the symbol for a matrix, which is $\begin{bmatrix} \end{bmatrix}$, with our main symbol for a determinant, which is $\begin{vmatrix} \end{vmatrix}$. Also, we

stress again that a determinant of a matrix is a *number* associated with the matrix.

Definition 3.2 (for a 3 × 3 Matrix)

The determinant of the 3×3 matrix $A = \begin{bmatrix} a_{11} & a_{12} & a_{13} \\ a_{21} & a_{22} & a_{23} \\ a_{31} & a_{32} & a_{33} \end{bmatrix}$, which is

denoted by $\begin{vmatrix} a_{11} & a_{12} & a_{13} \\ a_{21} & a_{22} & a_{23} \\ a_{31} & a_{32} & a_{33} \end{vmatrix}$, is equal to the following expression involving

three products:

$$a_{11} \begin{vmatrix} a_{22} & a_{23} \\ a_{32} & a_{33} \end{vmatrix} - a_{12} \begin{vmatrix} a_{21} & a_{23} \\ a_{31} & a_{33} \end{vmatrix} + a_{13} \begin{vmatrix} a_{21} & a_{22} \\ a_{31} & a_{32} \end{vmatrix}$$

In a typical product we have an entry of the first row of A, say a_{1j}, multiplied by the 2×2 determinant obtained by deleting row 1 and column j of the original determinant symbol.

For example, to obtain the 2×2 determinant to multiply by a_{12}, we delete row 1 and column 2 in the original determinant symbol, as in $\begin{vmatrix} a_{11} & a_{12} & a_{13} \\ a_{21} & a_{22} & a_{23} \\ a_{31} & a_{32} & a_{33} \end{vmatrix}$, to yield the determinant $\begin{vmatrix} a_{21} & a_{23} \\ a_{31} & a_{33} \end{vmatrix}$.

Example (b): To evaluate the determinant of $A = \begin{bmatrix} 4 & 5 & 6 \\ 1 & 2 & 3 \\ 0 & 0 & 4 \end{bmatrix}$, we have

$$\begin{vmatrix} 4 & 5 & 6 \\ 1 & 2 & 3 \\ 0 & 0 & 4 \end{vmatrix} = 4 \begin{vmatrix} 2 & 3 \\ 0 & 4 \end{vmatrix} - 5 \begin{vmatrix} 1 & 3 \\ 0 & 4 \end{vmatrix} + 6 \begin{vmatrix} 1 & 2 \\ 0 & 0 \end{vmatrix}$$

$$= 4[2(4) - 3(0)] - 5[1(4) - 3(0)] + 6[1(0) - 2(0)]$$

$$= 4(8) - 5(4) + 6(0) = 12$$

Observe, for example, that, on the first line, $a_{13} = 6$ is multiplied by the 2×2 determinant $\begin{vmatrix} 1 & 2 \\ 0 & 0 \end{vmatrix}$ obtained by deleting row 1 and column 3 of the starting determinant symbol.

Example (c): Evaluate the determinant of $B = \begin{bmatrix} 7 & 8 & 10 \\ 4 & 5 & 6 \\ 1 & 2 & 3 \end{bmatrix}$.

$$|B| = 7 \begin{vmatrix} 5 & 6 \\ 2 & 3 \end{vmatrix} - 8 \begin{vmatrix} 4 & 6 \\ 1 & 3 \end{vmatrix} + 10 \begin{vmatrix} 4 & 5 \\ 1 & 2 \end{vmatrix}$$

$$= 7(15 - 12) - 8(12 - 6) + 10(8 - 5)$$

$$= 21 - 48 + 30 = 3.$$

2.   The expression in Definition 3.2 for the determinant of a 3×3 matrix is known as an expansion along the first row. Just as the determinant of a 3×3 matrix can be given by an expression involving three 2×2 determinants, the determinant of a 4×4 matrix can be given by an expression involving four 3×3 determinants. Similarly, the determinant of an $n \times n$ matrix can be given, using a similar recursive pattern, by an expression involving n $(n - 1) \times (n - 1)$ determinants. We illustrate the continuing pattern for a 4×4 matrix.

Definition 3.2 (for a 4 × 4 Matrix)

The determinant of a 4×4 matrix $A = \begin{bmatrix} a_{11} & a_{12} & a_{13} & a_{14} \\ a_{21} & a_{22} & a_{23} & a_{24} \\ a_{31} & a_{32} & a_{33} & a_{34} \\ a_{41} & a_{42} & a_{43} & a_{44} \end{bmatrix}$, which

is denoted by $\begin{vmatrix} a_{11} & a_{12} & a_{13} & a_{14} \\ a_{21} & a_{22} & a_{23} & a_{24} \\ a_{31} & a_{32} & a_{33} & a_{34} \\ a_{41} & a_{42} & a_{43} & a_{44} \end{vmatrix}$, is equal to the following expression

involving four products:

$$a_{11} \begin{vmatrix} a_{22} & a_{23} & a_{24} \\ a_{32} & a_{33} & a_{34} \\ a_{42} & a_{43} & a_{44} \end{vmatrix} - a_{12} \begin{vmatrix} a_{21} & a_{23} & a_{24} \\ a_{31} & a_{33} & a_{34} \\ a_{41} & a_{43} & a_{44} \end{vmatrix}$$

$$+ a_{13} \begin{vmatrix} a_{21} & a_{22} & a_{24} \\ a_{31} & a_{32} & a_{34} \\ a_{41} & a_{42} & a_{44} \end{vmatrix} - a_{14} \begin{vmatrix} a_{21} & a_{22} & a_{23} \\ a_{31} & a_{32} & a_{33} \\ a_{41} & a_{42} & a_{43} \end{vmatrix}$$

In a typical product, we have an entry of the first row of A, say a_{1j}, multiplied by the 3×3 determinant obtained by deleting row 1 and column j of the original 4×4 determinant symbol.

The preceding expression, involving four products, is also known as an expansion along the first row. Note that the signs of the products alternate $+, -, +, -$.

Example: Evaluate the determinant of $C = \begin{bmatrix} 3 & 0 & 0 & 2 \\ -4 & 5 & 6 & 1 \\ 8 & -1 & 2 & 7 \\ 6 & 0 & 4 & 0 \end{bmatrix}$.

Solution: From the definition we have the following first-row expansion in terms of 3×3 determinants.

$$|C| = 3 \begin{vmatrix} 5 & 6 & 1 \\ -1 & 2 & 7 \\ 0 & 4 & 0 \end{vmatrix} - 0 \begin{vmatrix} -4 & 6 & 1 \\ 8 & 2 & 7 \\ 6 & 4 & 0 \end{vmatrix}$$

$$+ 0 \begin{vmatrix} -4 & 5 & 1 \\ 8 & -1 & 7 \\ 6 & 0 & 0 \end{vmatrix} - 2 \begin{vmatrix} -4 & 5 & 6 \\ 8 & -1 & 2 \\ 6 & 0 & 4 \end{vmatrix} \tag{1}$$

Thus,

$$|C| = 3 \begin{vmatrix} 5 & 6 & 1 \\ -1 & 2 & 7 \\ 0 & 4 & 0 \end{vmatrix} - 2 \begin{vmatrix} -4 & 5 & 6 \\ 8 & -1 & 2 \\ 6 & 0 & 4 \end{vmatrix}, \tag{2}$$

since the second and third products in Eq. (1) equal zero.

Now evaluate the two 3×3 determinants in Eq. (2), and finish the problem.

- - - - - - - - - -

Using the technique of frame 1 [Examples (b) and (c)], we find

$$\begin{vmatrix} 5 & 6 & 1 \\ -1 & 2 & 7 \\ 0 & 4 & 0 \end{vmatrix} = 5 \begin{vmatrix} 2 & 7 \\ 4 & 0 \end{vmatrix} - 6 \begin{vmatrix} -1 & 7 \\ 0 & 0 \end{vmatrix} + 1 \begin{vmatrix} -1 & 2 \\ 0 & 4 \end{vmatrix}$$

$$= 5(-28) - 6(0) + 1(-4) = -144 \tag{3}$$

Similarly,

$$\begin{vmatrix} -4 & 5 & 6 \\ 8 & -1 & 2 \\ 6 & 0 & 4 \end{vmatrix} = -4 \begin{vmatrix} -1 & 2 \\ 0 & 4 \end{vmatrix} - 5 \begin{vmatrix} 8 & 2 \\ 6 & 4 \end{vmatrix} + 6 \begin{vmatrix} 8 & -1 \\ 6 & 0 \end{vmatrix}$$

$$= -4(-4) - 5(20) + 6(6) = -48 \tag{4}$$

Substituting from Eqs. (3) and (4) into Eq. (2) yields

$$|C| = 3(-144) - 2(-48) = -336 \tag{5}$$

3. In the previous two frames we indicated how to calculate determinants by means of expansions along the first row. Actually, we can use any row (or column) in similar fashion to compute the value of a determinant. Before we develop this idea in more detail, we need some more definitions.

Definition 3.3

Let A be an $n \times n$ matrix. The *minor of entry* a_{ij}, denoted by M_{ij}, is defined to be the determinant of the matrix that remains after the ith row and jth column are deleted from matrix A. The *cofactor of entry* a_{ij}, denoted by C_{ij}, is related to M_{ij} by

$$C_{ij} = (-1)^{i+j} M_{ij}$$

Example (a): Refer to Example (b) of frame 1. There,

$$A = \begin{bmatrix} 4 & 5 & 6 \\ 1 & 2 & 3 \\ 0 & 0 & 4 \end{bmatrix}$$

To obtain the minor and cofactor of a_{11}, we delete row 1 and column 1 of matrix A, obtaining

$$M_{11} = \begin{vmatrix} 2 & 3 \\ 0 & 4 \end{vmatrix} = 8 \quad \text{and} \quad C_{11} = (-1)^{1+1} M_{11} = (+1)(8) = 8$$

Observe that $(-1)^{i+j}$ is equal to either $+1$ (when i and j are both odd or both even), or -1 (if exactly one of i and j is odd).

For the minor and cofactor of a_{21}, we delete row 2 and column 1 of matrix A, obtaining

$$M_{21} = \begin{vmatrix} 5 & 6 \\ 0 & 4 \end{vmatrix} = 20 \quad \text{and}$$

$$C_{21} = (-1)^{2+1} M_{21} = (-1)(20) = -20$$

Now find minors and cofactors of a_{12}, and a_{33}.

— — — — — — — — — —

$$M_{12} = \begin{vmatrix} 1 & 3 \\ 0 & 4 \end{vmatrix} = 4 \quad \text{and}$$

$$C_{12} = (-1)^{1+2} M_{12} = (-1)^3(4) = (-1)(4) = -4$$

$$M_{33} = \begin{vmatrix} 4 & 5 \\ 1 & 2 \end{vmatrix} = 4(2) - 5(1) = 3 \quad \text{and}$$

$$C_{33} = (+1) M_{33} = 3$$

4. Note that $C_{ij} = \pm M_{ij}$. The proper sign to use is illustrated by the following checkerboard pattern that has a $+$ in the row 1, column 1 position:

$$
\begin{array}{ccccccc}
+ & - & + & - & + & \cdots \\
- & + & - & + & - & \cdots \\
+ & - & + & - & + & \cdots \\
- & + & - & + & - & \cdots \\
\vdots & \vdots & \vdots & \vdots & \vdots
\end{array}
$$

For example, $C_{24} = M_{24}$ since there is a $+$ in the row 2, column 4 position of the checkerboard pattern [also, $(-1)^{2+4} = (-1)^6 = +1$]. Also, $C_{32} = -M_{32}$ since there is a $-$ in the row 3, column 2 position [also, $(-1)^{3+2} = (-1)^5 = -1$].

The following theorem, whose proof is omitted, provides for several ways of evaluating determinants.

Theorem 3.1 (Cofactor Expansion)

Suppose A is an $n \times n$ matrix. Then the determinant $|A|$ is given by both of the following formulas:

$$|A| = a_{i1}C_{i1} + a_{i2}C_{i2} + \cdots + a_{in}C_{in} \qquad \text{(cofactor expansion along } i\text{th row)}$$

and

$$|A| = a_{1j}C_{1j} + a_{2j}C_{2j} + \cdots + a_{nj}C_{nj} \qquad \text{(cofactor expansion along } j\text{th column)}$$

Each of the preceding expressions actually represents n formulas, the first for $i = 1, 2, \ldots, n$, and the second for $j = 1, 2, \ldots, n$.

Example (a): For A, the matrix of Example (b) of frame 1, evaluate $|A|$ by a cofactor expansion along row 3.

Solution: First, note that $C_{31} = (+1) \begin{vmatrix} 5 & 6 \\ 2 & 3 \end{vmatrix}$, $\quad C_{32} = (-1) \begin{vmatrix} 4 & 6 \\ 1 & 3 \end{vmatrix}$, $\quad C_{33} =$ $(+1) \begin{vmatrix} 4 & 5 \\ 1 & 2 \end{vmatrix}$. Now the determinant equals $a_{31}C_{31} + a_{32}C_{32} + a_{33}C_{33}$, which means

$$
\begin{vmatrix} 4 & 5 & 6 \\ 1 & 2 & 3 \\ 0 & 0 & 4 \end{vmatrix} = 0 \begin{vmatrix} 5 & 6 \\ 2 & 3 \end{vmatrix} - 0 \begin{vmatrix} 4 & 6 \\ 1 & 3 \end{vmatrix} + 4 \begin{vmatrix} 4 & 5 \\ 1 & 2 \end{vmatrix}
$$

$$= 0 - 0 + 4(3) = 12.$$

This agrees with the result previously obtained.

Notes: (a) In the example it was not necessary to compute two cofactors, since each was multiplied by zero. Usually, the best approach for evaluating a determinant by cofactor expansion is to expand along a row or column having the largest number of zeros. (b) Definition 3.2 is equivalent to the special case of Theorem 3.1 pertaining to a cofactor expansion along row 1.

Example (b): For A, the matrix of Example (b) of frame 1, evaluate $|A|$ from (i) a cofactor expansion along row 1, and (ii) a cofactor expansion along column 1.

Solution: (i)

$$\begin{vmatrix} 4 & 5 & 6 \\ 1 & 2 & 3 \\ 0 & 0 & 4 \end{vmatrix} = 4 \begin{vmatrix} 2 & 3 \\ 0 & 4 \end{vmatrix} - 5 \begin{vmatrix} 1 & 3 \\ 0 & 4 \end{vmatrix} + 6 \begin{vmatrix} 1 & 2 \\ 0 & 0 \end{vmatrix}$$

$$= 4(8) - 5(4) + 6(0) = 12$$

The calculation is identical to that in Example (b) of frame 1. Now do part (ii).

— — — — — — — — — —

(ii)

$$\begin{vmatrix} 4 & 5 & 6 \\ 1 & 2 & 3 \\ 0 & 0 & 4 \end{vmatrix} = 4 \begin{vmatrix} 2 & 3 \\ 0 & 4 \end{vmatrix} - 1 \begin{vmatrix} 5 & 6 \\ 0 & 4 \end{vmatrix} + 0 \begin{vmatrix} 5 & 6 \\ 2 & 3 \end{vmatrix}$$

$$= 4(8) - 1(20) + 0 = 12$$

B. PROPERTIES OF DETERMINANTS

5. In the next few frames, we shall summarize and illustrate some of the major properties of determinants by means of statements of theorems, and examples.

Theorem 3.2

Suppose A is an $n \times n$ matrix.

(a) If A contains either a row or column of zeros, then $|A| = 0$.
(b) Recall that A^t is the symbol for the transpose of A (Chapter 1, frame 18). Now, $|A^t| = |A|$.

Example (a): For the matrix $E = \begin{bmatrix} 6 & 7 & -5 \\ 0 & 0 & 0 \\ 4 & -9 & 8 \end{bmatrix}$, $|E| = 0$. This can be seen

if one uses a cofactor expansion on row 2. This verifies part (a) of Theorem 3.2.

For the matrix $A = \begin{bmatrix} 4 & 5 & 6 \\ 1 & 2 & 3 \\ 0 & 0 & 4 \end{bmatrix}$ of the previous frame, $A^t = $

$\begin{bmatrix} 4 & 1 & 0 \\ 5 & 2 & 0 \\ 6 & 3 & 4 \end{bmatrix}$. Evaluating $|A^t|$ by a cofactor expansion along column 3 yields

$$|A^t| = 0 \begin{vmatrix} 5 & 2 \\ 6 & 3 \end{vmatrix} - 0 \begin{vmatrix} 4 & 1 \\ 6 & 3 \end{vmatrix} + 4 \begin{vmatrix} 4 & 1 \\ 5 & 2 \end{vmatrix}$$

$$= 0 - 0 + 4(3) = 12$$

Thus, $|A^t| = |A| = 12$, thus verifying part (b) of Theorem 3.2. The calculations here are essentially identical to those of Example (a) of frame 4.

The next theorem shows how an elementary row operation on a matrix affects the value of a determinant. See Definition 2.3 of Chapter Two (frame 8) for the three row operations. The three parts of the following theorem correspond directly to the effects of the three respective elementary row operations.

Theorem 3.3

Let A be any $n \times n$ matrix.

(a) If A' is the matrix that results when two rows of A are interchanged, then $|A'| = -|A|$.
(b) If A' is the matrix that results when a single row of A is multiplied by a constant k, then $|A'| = k|A|$.
(c) If A' is the matrix that results when a multiple of one row of A is added to another row A, then $|A'| = |A|$.

Note: Parts (a), (b), and (c) also hold if the word *row* is replaced by *column*.

Example (b): Consider the matrices

$$A = \begin{bmatrix} 1 & 3 & 2 \\ 2 & 4 & 1 \\ 3 & 5 & 1 \end{bmatrix}, \qquad B = \begin{bmatrix} 3 & 5 & 1 \\ 2 & 4 & 1 \\ 1 & 3 & 2 \end{bmatrix},$$

$$C = \begin{bmatrix} 3 & 9 & 6 \\ 2 & 4 & 1 \\ 3 & 5 & 1 \end{bmatrix}, \qquad D = \begin{bmatrix} 1 & 3 & 2 \\ 0 & -2 & -3 \\ 3 & 5 & 1 \end{bmatrix}$$

Evaluating the determinants by cofactor expansions, we find that

$$|A| = -2, \quad |B| = 2, \quad |C| = -6, \quad |D| = -2$$

Observe that B is obtained by interchanging the first and third rows of A; C by multiplying the first row of A by 3; and D by adding -2 times the first row to the second row of A (in the notation of Chapter Two, $r'_2 = r_2 - 2r_1$). As predicted in Theorem 3.3,

$$|B| = -|A|, \quad |C| = 3|A|, \quad |D| = |A|$$

The following two corollaries of Theorem 3.3 are easy to prove and are fairly useful.

Corollaries of Theorem 3.3

(i) If A is a matrix with two equal rows, then $|A| = 0$.
(ii) If A is a matrix where one row is a multiple of another row, then $|A| = 0$.

These results also hold if the word *row* is replaced by *column*.

Example (c): To demonstrate the first corollary, consider the following matrix A, whose first two rows are equal:

$$A = \begin{bmatrix} a & b & c \\ a & b & c \\ d & e & f \end{bmatrix}$$

Applying part (c) of Theorem 3.3 in the form $r'_2 = r_2 - r_1$, we get $|A| = |A'|$, with

$$A' = \begin{bmatrix} a & b & c \\ 0 & 0 & 0 \\ d & e & f \end{bmatrix}$$

But $|A'| = 0$ from Theorem 3.2, part (a), and thus $|A| = 0$, also.

Example (d): Demonstrate the second corollary with respect to the matrix

$$B = \begin{bmatrix} a & b & c \\ ka & kb & kc \\ d & e & f \end{bmatrix},$$

whose second row equals k times the first row. *Hint:* First, make use of Theorem 3.3, part (b).

_ _ _ _ _ _ _ _ _ _ _

Since matrix B results from matrix A of Example (c) if we multiply row 2 of A by the multiple k, we have $|B| = k|A|$ from Theorem 3.3, part (b). But, $|A| = 0$ from Example (c), and thus $|B| = k(0) = 0$.

6. Some useful notation with respect to Theorem 3.3, part (b) is obtained by considering two matrices A and A', where

$$A = \begin{bmatrix} a_{11} & a_{12} & a_{13} \\ a_{21} & a_{22} & a_{23} \\ a_{31} & a_{32} & a_{33} \end{bmatrix} \quad \text{and } A' = \begin{bmatrix} ka_{11} & ka_{12} & ka_{13} \\ a_{21} & a_{22} & a_{23} \\ a_{31} & a_{32} & a_{33} \end{bmatrix}$$

Here the first row of A' is the multiple k times the first row of A, and thus $|A'| = k|A|$. Thus, we can write

$$\begin{vmatrix} ka_{11} & ka_{12} & ka_{13} \\ a_{21} & a_{22} & a_{23} \\ a_{31} & a_{32} & a_{33} \end{vmatrix} = k \begin{vmatrix} a_{11} & a_{12} & a_{13} \\ a_{21} & a_{22} & a_{23} \\ a_{31} & a_{32} & a_{33} \end{vmatrix}$$

This "factoring out" of a constant from a single row of a determinant can be done with respect to another row or with respect to any column.

Example (a): We can express the determinant $\begin{vmatrix} 1 & 3 & 2 \\ 8 & 16 & 4 \\ 3 & 5 & 1 \end{vmatrix}$ as

$$\begin{vmatrix} 1 & 3 & 2 \\ 8 & 16 & 4 \\ 3 & 5 & 1 \end{vmatrix} = 4 \begin{vmatrix} 1 & 3 & 2 \\ 2 & 4 & 1 \\ 3 & 5 & 1 \end{vmatrix},$$

after "factoring out" the constant 4 from the second row. Since the determinant on the right of the preceding equation is equal to -2 [Example (b) of frame 5], we have

$$\begin{vmatrix} 1 & 3 & 2 \\ 8 & 16 & 4 \\ 3 & 5 & 1 \end{vmatrix} = 4(-2) = -8$$

Shortcuts in evaluating determinants are obtained by employing Theorem 3.3 [in particular, parts (b) and (c)] to convert a determinant $|A|$ to a product of a constant times a determinant $|B|$, which has a row (or column) in which all but one entry are equal to zero, and the remaining nonzero entry equals one. We illustrate in the following example.

Example (b): Evaluate the determinant $|G|$, where $G = \begin{bmatrix} 3 & 9 & 6 \\ 2 & 4 & 1 \\ 3 & 5 & 1 \end{bmatrix}$.

Solution: We shall generate a determinant whose first column consists of a one with two zeros below it. Then we shall use a cofactor expansion on the first column. The steps (a), (b), (c), etc., in the following are explained below.

$$\begin{vmatrix} 3 & 9 & 6 \\ 2 & 4 & 1 \\ 3 & 5 & 1 \end{vmatrix} \overset{(a)}{=} 3\begin{vmatrix} 1 & 3 & 2 \\ 2 & 4 & 1 \\ 3 & 5 & 1 \end{vmatrix} \overset{(b)}{=} 3\begin{vmatrix} 1 & 3 & 2 \\ 0 & -2 & -3 \\ 3 & 5 & 1 \end{vmatrix}$$

$$\overset{(c)}{=} 3\begin{vmatrix} 1 & 3 & 2 \\ 0 & -2 & -3 \\ 0 & -4 & -5 \end{vmatrix} \overset{(d)}{=} 3(1)\begin{vmatrix} -2 & -3 \\ -4 & -5 \end{vmatrix} \overset{(e)}{=} 3(10 - 12) = -6$$

Steps:

(a)   The constant 3 is "factored out" of row 1; in symbols, $r_1' = \frac{1}{3}r_1$, where the prime refers to the "new" row 1, as in Chapter Two.

(b)   Add -2 times the first row to the second row; in symbols, $r_2' = r_2 - 2r_1$. Thus, zero is generated in the row 2, column 1 position.

(c)   Add -3 times the first row to the third row; in symbols, $r_3' = r_3 - 3r_1$. Thus, zero is generated in the row 3, column 1 position.

(d)   Use cofactor expansion on column 1.

(e)   Evaluate the single 2×2 determinant.

Note that doing steps (a), (b), and (c) corresponds to the "clearing out" of column 1 in the Gauss–Jordan elimination method (Method 2.1 in frame 14 of Chapter Two).

Evaluation of determinants of matrices having more than three rows (and columns) can be done by using the same kind of strategy as in Example (b). For example, for a 4×4 determinant, first we transform to a constant times a new 4×4 determinant that has a column (or row) with one 1 and the remaining entries equal to zero. A cofactor expansion on that column (row) leads to a 3×3 determinant. Then the same type of strategy is used on the 3×3 determinant; for example, see Example (b).

Example (c): Evaluate the determinant of the 4×4 matrix C of the frame 2 example using the strategy suggested above.

Solution: The steps (a), (b), (c), etc., are explained below the sequence of determinants that appear in the following. First, our approach will generate a constant times a determinant with 0 0 0 1 in row 1. This is indicated by steps (a) and (b) below.

$$
\begin{vmatrix}
3 & 0 & 0 & 2 \\
-4 & 5 & 6 & 1 \\
8 & -1 & 2 & 7 \\
6 & 0 & 4 & 0
\end{vmatrix}
\overset{(a)}{=} 2
\begin{vmatrix}
\frac{3}{2} & 0 & 0 & 1 \\
-4 & 5 & 6 & 1 \\
8 & -1 & 2 & 7 \\
6 & 0 & 4 & 0
\end{vmatrix}
\overset{(b)}{=} 2
\begin{vmatrix}
0 & 0 & 0 & 1 \\
-\frac{11}{2} & 5 & 6 & 1 \\
-\frac{5}{2} & -1 & 2 & 7 \\
6 & 0 & 4 & 0
\end{vmatrix}
$$

$$
\overset{(c)}{=} 2(-1)
\begin{vmatrix}
-\frac{11}{2} & 5 & 6 \\
-\frac{5}{2} & -1 & 2 \\
6 & 0 & 4
\end{vmatrix}
$$

Steps:

(a)   The constant 2 is "factored out" of row 1; in symbols, $r_1' = \frac{1}{2}r_1$.

(b)   Add $-\frac{3}{2}$ times the fourth column to the first column; in symbols, $c_1' = c_1 - \frac{3}{2}c_4$, where, for example, c_1 refers to "old" column 1, and c_1' to the "new" column 1. Thus, zero is generated in the row 1, column 1 position.

(c)   Use cofactor expansion on row 1 to generate a single 3 × 3 determinant.

Now finish the computations.

———————————

Since column 2 of the 3 × 3 determinant already has a zero entry, let us transform column 2 to the desired form.

$$
(-2)
\begin{vmatrix}
-\frac{11}{2} & 5 & 6 \\
-\frac{5}{2} & -1 & 2 \\
6 & 0 & 4
\end{vmatrix}
\overset{(d)}{=} (-2)(-1)
\begin{vmatrix}
-\frac{11}{2} & 5 & 6 \\
\frac{5}{2} & 1 & -2 \\
6 & 0 & 4
\end{vmatrix}
$$

$$
\overset{(e)}{=} 2
\begin{vmatrix}
-18 & 0 & 16 \\
\frac{5}{2} & 1 & -2 \\
6 & 0 & 4
\end{vmatrix}
\overset{(f)}{=} 2
\begin{vmatrix}
-18 & 16 \\
6 & 4
\end{vmatrix}
$$

$$
\overset{(g)}{=} 2[(-18)(4) - 16(6)] = -336
$$

Steps (continued):

(d)   The constant -1 is "factored out" of row 2 $[r_2' = (-1)r_2]$.

(e)   Add -5 times the second row to the first row $[r_1' = r_1 - 5r_2]$.

(f)   Use cofactor expansion on column 2.

(g)   Evaluate the 2 × 2 determinant.

The answer, $|C| = -336$, checks with the frame 2 result.

7. Now let us consider some more theorems and examples illustrating useful properties of determinants. We shall omit the proof of the following theorem.

Theorem 3.4

If A and B are both $n \times n$ matrices, then the determinant of the matrix product AB is given by $|AB| = |A||B|$.

Note: The matrix product AB is, of course, also an $n \times n$ matrix. It is also true that the determinant of the $n \times n$ matrix BA is given by $|BA| = |B||A| = |A||B|$. (Thus, $|AB| = |BA|$.) Here we first interchange letters in Theorem 3.4 and then use the fact that for any numbers h and k, we have $hk = kh$. Recall that the determinant of a matrix is a *number* associated with the matrix.

Example (a): Let $A = \begin{bmatrix} 1 & 4 \\ 3 & 2 \end{bmatrix}$ and $B = \begin{bmatrix} 5 & 2 \\ -3 & 4 \end{bmatrix}$. Verify Theorem 3.4 for these matrices.

Solution: First, $|A| = -10$, and $|B| = 26$. Thus, $|A||B| = -260$. Also,

$$AB = \begin{bmatrix} -7 & 18 \\ 9 & 14 \end{bmatrix}$$

and, thus, $|AB| = (-7)(14) - 18(9) = -260$. Observe that $|BA|$ is also equal to -260. (The matrix $BA = \begin{bmatrix} 11 & 24 \\ 9 & -4 \end{bmatrix}$.)

It should be noted that no simple relationship exists between $|A|$, $|B|$, and $|A + B|$. In fact, $|A + B|$ is usually *not* equal to $|A| + |B|$.

Example (b): Show that $|A + B|$ is unequal to $|A| + |B|$ for the matrices of Example (a).

— — — — — — — — —

First, $A + B = \begin{bmatrix} 6 & 6 \\ 0 & 6 \end{bmatrix}$, and thus $|A + B| = 36$. Also, $|A| + |B| = 16$.

8. The next theorem indicates that the determinant of an invertible (nonsingular) matrix is unequal to zero.

Theorem 3.5

Suppose A is an invertible $n \times n$ matrix.

Then, $|A| \neq 0$.

Also, $|A^{-1}| \neq 0$, and $|A|$ and $|A^{-1}|$ are related by the equation

$$|A^{-1}| = \frac{1}{|A|}.$$

The proof, which follows from Theorem 3.4, is very instructive and will be done later. One fact needed for the proof is that for any $n \times n$ identity matrix I_n, $|I_n| = 1$.

· **Example (a):** Show that $|I_4| = 1$.

Solution: Since $I_4 = \begin{bmatrix} 1 & 0 & 0 & 0 \\ 0 & 1 & 0 & 0 \\ 0 & 0 & 1 & 0 \\ 0 & 0 & 0 & 1 \end{bmatrix}$, a cofactor expansion on row 1 of $|I_4|$,

and then on row 1 of $|I_3|$ yields

$$|I_4| = (1) \begin{vmatrix} 1 & 0 & 0 \\ 0 & 1 & 0 \\ 0 & 0 & 1 \end{vmatrix} = (1)(1) \begin{vmatrix} 1 & 0 \\ 0 & 1 \end{vmatrix}$$

$$= (1)(1)(1) = 1$$

The same pattern applies to any identity matrix, regardless of n.

Proof of Theorem 3.5

First, if matrix A is invertible, then A^{-1} exists, and

$$AA^{-1} = I_n \tag{1}$$

Taking the determinant on both sides results in

$$|AA^{-1}| = |I_n|. \tag{2}$$

From Theorem 3.4, $|AA^{-1}| = |A||A^{-1}|$, and from the preceding comment, $|I_n| = 1$. Substituting into Eq. (2) yields

$$|A||A^{-1}| = 1 \tag{3}$$

Now the left side of Eq. (3) is a product of two numbers (since determinants are numbers), and when the product of two numbers equals 1, that means that each of the numbers must be unequal to zero. That is,

$$|A| \neq 0 \qquad \text{and} \qquad |A^{-1}| \neq 0 \tag{4a), (4b}$$

Next, dividing through in Eq. (3) by $|A|$ yields

$$|A^{-1}| = \frac{1}{|A|} \tag{5}$$

Example (b): Verify Theorem 3.5 for the matrices A and A^{-1} of Example (b) of frame 29 of Chapter Two. There, $A = \begin{bmatrix} 1 & 3 \\ 2 & 4 \end{bmatrix}$ and $A^{-1} = \begin{bmatrix} -2 & \frac{3}{2} \\ 1 & -\frac{1}{2} \end{bmatrix}$.

- - - - - - - - - - -

Solution:

$$A = \begin{bmatrix} 1 & 3 \\ 2 & 4 \end{bmatrix} \qquad \text{and} \qquad |A| = (1)(4) - (3)(2) = -2$$

$$A^{-1} = \begin{bmatrix} -2 & \frac{3}{2} \\ 1 & -\frac{1}{2} \end{bmatrix} \qquad \text{and} \qquad |A^{-1}| = (-2)(\tfrac{1}{2}) - \tfrac{3}{2}(1) = -\tfrac{1}{2}$$

We see that $|A^{-1}| = \dfrac{1}{|A|}$.

C. THE INVERSE OF A MATRIX; CRAMER'S RULE

9. Refer back to Theorem 3.1 in frame 4. Suppose in the cofactor expansion along the ith row that the cofactors corresponded to entries in the kth row, where $k \neq i$. That is, suppose we wanted to calculate the quantity

$$a_{i1}C_{k1} + a_{i2}C_{k2} + \ldots + a_{in}C_{kn} \qquad \text{for } k \neq i.$$

It turns out that the value of this quantity is zero and that knowing this will be important to us in developing a new method for finding the inverse of an invertible matrix.

An illustration of the preceding comments follows for a 3×3 matrix:

Example (a): For the 3×3 matrix

$$A = \begin{bmatrix} a & b & c \\ a_{21} & a_{22} & a_{23} \\ a_{31} & a_{32} & a_{33} \end{bmatrix}, \tag{1}$$

the determinant is given by

$$|A| = aC_{11} + bC_{12} + cC_{13}, \tag{2}$$

if we use a cofactor expansion along row 1.

Now let us replace a by a_{31}, b by a_{32}, and c by a_{33} and form the matrix A':

$$A' = \begin{bmatrix} a_{31} & a_{32} & a_{33} \\ a_{21} & a_{22} & a_{23} \\ a_{31} & a_{32} & a_{33} \end{bmatrix} \tag{3}$$

Now if we evaluate the determinant $|A'|$ by using a cofactor expansion along row 1 [see (3)], we obtain

$$|A'| = a_{31}C_{11} + a_{32}C_{12} + a_{33}C_{13} \tag{4}$$

The cofactors in (2) and (4) are, respectively, equal in both expressions, since computing them (i.e., the cofactors) involves entries in rows 2 and 3. Observe that rows 2 and 3 are, respectively, identical in the preceding matrices (1) and (3).

Now from Theorem 3.3, Corollary (i),

$$|A'| = 0, \tag{5}$$

since in A' rows 1 and 3 are identical. Thus, the expression on the right side of (4) equals zero also, that is,

$$a_{31}C_{11} + a_{32}C_{12} + a_{33}C_{13} = 0 \tag{6}$$

Using an approach similar to that used in Example (a), we obtain the following theorem. In the theorem, we shall repeat the results of Theorem 3.1 for cases where the cofactors correspond directly to the row (column) entries.

Theorem 3.6 •

If A is an $n \times n$ matrix, then

$$a_{i1}C_{k1} + a_{i2}C_{k2} + \ldots + a_{in}C_{kn} = 0 \qquad \text{if } k \neq i$$

and

$$a_{i1}C_{i1} + a_{i2}C_{i2} + \ldots + a_{in}C_{in} = |A| \qquad \text{(here } k = i\text{)}$$

The preceding expressions are cofactor expansions in the row direction. Also,

$$a_{1j}C_{1k} + a_{2j}C_{2k} + \ldots + a_{nj}C_{nk} = 0 \qquad \text{if } k \neq j$$

and

$$a_{1j}C_{1j} + a_{2j}C_{2j} + \ldots + a_{nj}C_{nj} = |A| \qquad \text{(here } k = j\text{)}$$

The preceding expressions are cofactor expansions in the column direction.

Example (b): For the matrix $A = \begin{bmatrix} 2 & 3 & 4 \\ 1 & 2 & 1 \\ 1 & 2 & 3 \end{bmatrix}$, which occurs in Example (b) of frame 30 of Chapter Two,

(a)   First determine the cofactors C_{11}, C_{12}, and C_{13}.
(b)   Compute $|A|$ by using these cofactors.
(c)   Verify the rest of Theorem 3.6 with respect to these cofactors.

Solution:

(a) $C_{11} = \begin{vmatrix} 2 & 1 \\ 2 & 3 \end{vmatrix} = 4, \qquad C_{12} = (-1)\begin{vmatrix} 1 & 1 \\ 1 & 3 \end{vmatrix} = -2,$

$C_{13} = \begin{vmatrix} 1 & 2 \\ 1 & 2 \end{vmatrix} = 0$

(b)   Using a cofactor expansion on row 1, we have

$$|A| = a_{11}C_{11} + a_{12}C_{12} + a_{13}C_{13} = 2(4) + 3(-2) + 4(0) = 2.$$

(c)   Forming an expansion using entries of row 2 together with the preceding cofactors, which correspond to row 1, leads to

$$a_{21}C_{11} + a_{22}C_{12} + a_{23}C_{13} = 1(4) + 2(-2) + 1(0) = 0$$

Now proceed in similar fashion and use the entries of row 3 together with the cofactors that correspond to row 1.

- - - - - - - - - -

$$a_{31}C_{11} + a_{32}C_{12} + a_{33}C_{13} = 1(4) + 2(-2) + 3(0) = 0$$

10.   The following definition, which involves a matrix containing cofactors as its entries, will be important to us in our quest for a new method for finding the inverse of an invertible matrix.

Definition 3.4

Given the $n \times n$ matrix A. Suppose that C_{ij} denotes the cofactor of entry a_{ij}. The $n \times n$ matrix adj (A), called the *adjoint of A*, is given by

$$\text{adj } (A) = \begin{bmatrix} C_{11} & C_{21} & \cdots & C_{n1} \\ C_{12} & C_{22} & \cdots & C_{n2} \\ \cdot & \cdot & \cdots & \cdot \\ C_{1n} & C_{2n} & \cdots & C_{nn} \end{bmatrix}.$$

It is important to note the labeling of subscripts here. The row i, column j entry of adj (A) is C_{ji}, the cofactor of entry a_{ji}.

Example (a):   For matrix A in Example (b) of frame 9, determine the matrix adj (A).

Solution: We have already calculated C_{11}, C_{12}, and C_{13} (in the previous frame). Continuing, we have

$$C_{21} = (-1)\begin{vmatrix} 3 & 4 \\ 2 & 3 \end{vmatrix} = -1, \qquad C_{22} = \begin{vmatrix} 2 & 4 \\ 1 & 3 \end{vmatrix} = 2,$$

$$C_{23} = (-1)\begin{vmatrix} 2 & 3 \\ 1 & 2 \end{vmatrix} = -1$$

As a check, we have the following cofactor expansion along row 2, which yields $|A|$ again.

$$a_{21}C_{21} + a_{22}C_{22} + a_{23}C_{23} = 1(-1) + 2(2) + 1(-1) = 2 = |A|$$

Continuing, we have

$$C_{31} = \begin{vmatrix} 3 & 4 \\ 2 & 1 \end{vmatrix} = -5, \qquad C_{32} = (-1)\begin{vmatrix} 2 & 4 \\ 1 & 1 \end{vmatrix} = +2,$$

$$C_{33} = \begin{vmatrix} 2 & 3 \\ 1 & 2 \end{vmatrix} = 1$$

Thus, the adjoint matrix of A is given by

$$\text{adj}\,(A) = \begin{bmatrix} C_{11} & C_{21} & C_{31} \\ C_{12} & C_{22} & C_{32} \\ C_{13} & C_{23} & C_{33} \end{bmatrix} = \begin{bmatrix} 4 & -1 & -5 \\ -2 & 2 & 2 \\ 0 & -1 & 1 \end{bmatrix}$$

The following theorem follows directly from our definition for the adjoint matrix.

Theorem 3.7

If A is any $n \times n$ matrix with adjoint given by adj (A), then the matrix products of A with adj (A) are both equal to $|A|I_n$, where I_n is the $n \times n$ identity matrix. That is,

$$A[\text{adj}\,(A)] = |A|I_n \quad \text{and} \quad [\text{adj}\,(A)]A = |A|I_n$$

Example (b): Verify that $A[\text{adj}\,(A)] = |A|I_n$ for the matrix A of Example (b) of frame 9.

Solution: The adjoint of A is given by Example (a) of the current frame. Now multiply A on the right by adj (A), and finish the verification.

— — — — — — — — —

$$A[\text{adj}\,(A)] = \begin{bmatrix} 2 & 3 & 4 \\ 1 & 2 & 1 \\ 1 & 2 & 3 \end{bmatrix}\begin{bmatrix} 4 & -1 & -5 \\ -2 & 2 & 2 \\ 0 & -1 & 1 \end{bmatrix} = \begin{bmatrix} 2 & 0 & 0 \\ 0 & 2 & 0 \\ 0 & 0 & 2 \end{bmatrix} \quad (1)$$

Factoring the scalar (number) 2 from the right-hand matrix (Definition 1.2 in frame 8 of Chapter One) and recognizing that $|A| = 2$ leads to

$$A[\text{adj}\,(A)] = 2\begin{bmatrix} 1 & 0 & 0 \\ 0 & 1 & 0 \\ 0 & 0 & 1 \end{bmatrix} = |A|I_3 \quad (2)$$

11.   A partial proof of Theorem 3.7 for the case of a 3×3 matrix A follows:

Proof (partial) of Theorem 3.7

$$A = \begin{bmatrix} a_{11} & a_{12} & a_{13} \\ a_{21} & a_{22} & a_{23} \\ a_{31} & a_{32} & a_{33} \end{bmatrix}, \qquad \text{adj}\,(A) = \begin{bmatrix} C_{11} & C_{21} & C_{31} \\ C_{12} & C_{22} & C_{32} \\ C_{13} & C_{23} & C_{33} \end{bmatrix}$$

The row 1, column 1, entry of the matrix product $A[\text{adj}\,(A)]$ is given by

$$a_{11}C_{11} + a_{12}C_{12} + a_{13}C_{13} \tag{1}$$

From Theorem 3.6 (or Theorem 3.1), this product equals the determinant of A, that is,

$$a_{11}C_{11} + a_{12}C_{12} + a_{13}C_{13} = |A| \tag{2}$$

In similar fashion, we can show that the other main diagonal entries of $A[\text{adj}\,(A)]$ are also equal to the determinant $|A|$. For example, for the row 2, column 2 entry, we have $a_{21}C_{21} + a_{22}C_{22} + a_{23}C_{23}$, which also equals $|A|$ from Theorem 3.6.

Every entry of $A[\text{adj}\,(A)]$ which is not on the main diagonal is equal to zero. For example, the row 1, column 3 entry is equal to

$$a_{11}C_{31} + a_{12}C_{32} + a_{13}C_{33},$$

which equals zero, by Theorem 3.6 (here, $i = 1$, $k = 3$, and hence, $k \neq i$).

Thus, the matrix product $A[\text{adj}\,(A)]$ is given by

$$A[\text{adj}\,(A)] = \begin{bmatrix} |A| & 0 & 0 \\ 0 & |A| & 0 \\ 0 & 0 & |A| \end{bmatrix} = |A| \begin{bmatrix} 1 & 0 & 0 \\ 0 & 1 & 0 \\ 0 & 0 & 1 \end{bmatrix} = |A|I_3$$

The following corollary of Theorem 3.7 provides us with a new method for computing the inverse of a matrix.

Corollary of Theorem 3.7

If $|A| \neq 0$, then A^{-1} exists and is given by

$$A^{-1} = \frac{1}{|A|}[\text{adj}\,(A)]$$

Here we have the *number* $\dfrac{1}{|A|}$ times the *matrix* adj (A) because the determinant $|A|$ is itself a number.

The proof is both easy and instructive.

Proof of Corollary

From Theorem 3.7, we have

$$A[\text{adj}\,(A)] = |A|I_n \tag{1}$$

If $|A| \neq 0$, then we can divide both sides of Eq. (1) by the number $|A|$, and obtain

$$A\left[\frac{\text{adj}\,(A)}{|A|}\right] = I_n \tag{2}$$

From Note (b) of Definition 2.7 and Theorem 2.5 (frame 27 of Chapter Two), it then follows that the inverse matrix of A is given by

$$A^{-1} = \frac{\text{adj}\,(A)}{|A|}, \qquad \text{or} \qquad A^{-1} = \frac{1}{|A|}[\text{adj}\,(A)] \tag{3}$$

Example: Determine the inverse of the matrix A cited in Example (b) of frame 9 by using the corollary of Theorem 3.7.

Solution: First, we note that $|A| = 2$ (frame 9), thus indicating that the inverse A^{-1} exists. Next, make use of the adjoint matrix of A [Example (a) of frame 10], and then use the Corollary of Theorem 3.7.

— — — — — — — — — —

$$A^{-1} = \frac{1}{|A|}[\text{adj}\,(A)] = \frac{1}{2}\begin{bmatrix} 4 & -1 & -5 \\ -2 & 2 & 2 \\ 0 & -1 & 1 \end{bmatrix} = \begin{bmatrix} 2 & -\frac{1}{2} & -\frac{5}{2} \\ -1 & 1 & 1 \\ 0 & -\frac{1}{2} & \frac{1}{2} \end{bmatrix}$$

We obtained the same result for A^{-1} by using the Gauss–Jordan elimination method in frame 30 of Chapter 2.

12. From Theorem 3.5 (frame 8), the following is valid for an $n \times n$ matrix A.

If matrix A is invertible (A^{-1} exists), then $|A| \neq 0$ \hfill (I)

From the Corollary of Theorem 3.7 (frame 11), we have the following for an $n \times n$ matrix A.

If $|A| \neq 0$, then matrix A is invertible. \hfill (II)

Expressions (I) and (II) are said to be converses of one another. Now, in general, when both an expression of the form "If P, then Q" is valid and its converse. "If Q, then P" is also valid, we can summarize this compactly by writing the following "if and only if" expression relating the statements P and Q:

"P if and only if Q."

Here, statements P and Q can be interchanged.

Thus, we can combine (I) and (II) above in the following "if and only if" expression, which we call a theorem because of its importance.

Theorem 3.8

Matrix A is invertible if and only if $|A| \neq 0$.

Note: For example, "matrix A is invertible" plays the role of statement P, and "$|A| \neq 0$" plays the role of statement Q.

If an expression "If P, then Q" and its converse "If Q, then P" are both valid, we can indicate this also by saying that statements P and Q are *equivalent*. Thus, it is correct to say that the statements "matrix A is invertible" and "$|A| \neq 0$" are equivalent.

We spoke about equivalent statements in frames 26 and 32 of Chapter Two. Because of the development in the current frame, we see that we can update and extend the list of equivalent statements in Theorem 2.9 of frame 32.

Theorem 3.9 (Extension of Theorem 2.9)

If A is an $n \times n$ matrix, then the following statements are equivalent:

(a) Rank $(A) = n$.
(b) The reduced row echelon form of A is I_n.
(c) A is invertible (i.e., A^{-1} exists).
(d) The system $A\mathbf{x} = \mathbf{b}$ has a unique solution (given by $\mathbf{x} = A^{-1}\mathbf{b}$).
(e) $|A| \neq 0$.

Note: In the current frame we established that statements (c) and (e) were equivalent.

Recall that if any statement in a list of equivalent statements is true, then all the statements are true; similarly, if any statement is false, then all the statements are false.

Example: Interpret all the statements of Theorem 3.9 with respect to the

2×2 matrix $A = \begin{bmatrix} 5 & 3 \\ 2 & 4 \end{bmatrix}$. *Hint:* See Examples (a) and (b) of frame 32 of

Chapter Two for similar examples. Also, observe that it is easy to calculate the determinant $|A|$.

————————————

$$|A| = 5(4) - 3(2) = 14$$

Since $|A| \neq 0$, statement (e) of the list is true, and thus the other four statements are also true. In particular, A^{-1} exists [statement (c)], and Rank $(A) = 2$ [statement (a)].

13. There is another form of Theorem 3.5 [see frame 8 or form (I) of the preceding frame] that is very useful. It is known as the *contrapositive* of Theorem 3.5. In general, if a valid expression (or theorem) has the form "If P, then Q," then the expression "If not Q, then not P," known as the contrapositive of the former expression, is automatically valid.

Thus, referring to Theorem 3.5, if we let statement P be "matrix A is invertible" and statement Q be "$|A| \neq 0$," the following contrapositive form of Theorem 3.5 is valid for an $n \times n$ matrix A.

Contrapositive of Theorem 3.5 (or Theorem 3.5′)

If $|A| = 0$, then A is not invertible.

Note: Observe that the negation of $|A| \neq 0$ is $|A| = 0$.

Using this together with the corollary of Theorem 3.7 [frame 11 and form (II) of the preceding frame] provides us with a quick criterion, involving the determinant, for showing whether or not the inverse of a matrix exists. The corollary is repeated here.

Corollary of Theorem 3.7

If $|A| \neq 0$, then A is invertible.

Example (a): Determine if the inverse of the following matrix A exists.

$$A = \begin{bmatrix} 3 & -9 & 1 \\ -1 & 3 & 2 \\ 2 & -6 & -1 \end{bmatrix}$$

Solution: Since $|A| = 0$, it follows that the inverse of matrix A does not exist. Note that we also reached the same conclusion for this matrix in Example (a) of frame 30 of Chapter Two.

For another approach in this example, refer to the list of equivalent statements in Theorem 3.9 (frame 12). Since statement (e) is false, so are the remaining statements. The falsity of statement (c) means A^{-1} does not exist.

Suppose we consider statement (d) of Theorem 3.9 for the special case where we have a homogeneous system. (See frames 20 and 21 of Chapter Two for a review of homogeneous systems.) Thus, $\mathbf{b} = \mathbf{0}$, and the unique solution is the trivial solution $\mathbf{x} = \mathbf{0}$. Let us call this modified statement, statement (d').

(d')   The system $A\mathbf{x} = \mathbf{0}$ has only the trivial solution $\mathbf{x} = \mathbf{0}$.

If we now relate statements (d') and (e) by an "if and only if" expression, we obtain the following useful theorem for a homogeneous linear system of n equations in n unknowns.

Theorem 3.10 (for cases where coefficient matrix A is n × n)

   The homogeneous system given by $A\mathbf{x} = \mathbf{0}$ has only the trivial solution $\mathbf{x} = \mathbf{0}$ if and only if $|A| \neq 0$.

If an expression of the form "*P* if and only if *Q*" is valid, then the contrapositive form of this expression, which has the form "not *P* if and only if not *Q*," is also automatically valid. Taking the contrapositive of Theorem 3.10 yields the following variation, which is also quite useful.

Theorem 3.10' (for cases where coefficient matrix A is n × n)

   The homogeneous system given by $A\mathbf{x} = \mathbf{0}$ has nontrivial (nonzero) solutions [i.e., $\mathbf{x} \neq \mathbf{0}$] if and only if $|A| = 0$.

Note: It should be realized that when we say that a homogeneous system has nontrivial solutions, this means that it has *infinitely many* such nontrivial solutions.

Example (b): Determine all solutions of the homogeneous system given by

$$2x_1 + 3x_2 + 4x_3 = 0 \tag{1}$$

$$x_1 + 2x_2 + x_3 = 0 \tag{2}$$

$$x_1 + 2x_2 + 3x_3 = 0 \tag{3}$$

Hint: First, evaluate $|A|$ for the system.

— — — — — — — — —

Solution: We see that $|A| = \begin{vmatrix} 2 & 3 & 4 \\ 1 & 2 & 1 \\ 1 & 2 & 3 \end{vmatrix} = 2$, as was shown in Example (b)

of frame 9. Thus, from Theorem 3.10, we see that the unique solution is the trivial solution $\mathbf{x} = \mathbf{0}$. In terms of components, the unique solution is $x_1 = 0$, $x_2 = 0$, $x_3 = 0$.

14. Using Theorem 3.7 in frame 10 as a basis, one can prove the following useful theorem. (The proof is omitted here, however.)

Theorem 3.11 (Cramer's Rule)*

Given the following linear system of n equations in n unknowns:

$$a_{11}x_1 + a_{12}x_2 + \ldots + a_{1n}x_n = b_1$$
$$a_{21}x_1 + a_{22}x_2 + \ldots + a_{2n}x_n = b_2$$
$$\cdots\cdots\cdots\cdots\cdots\cdots\cdots\cdots\cdots\cdots$$
$$\cdots\cdots\cdots\cdots\cdots\cdots\cdots\cdots\cdots\cdots$$
$$a_{n1}x_1 + a_{n2}x_2 + \ldots + a_{nn}x_n = b_n$$

Using the notation of Chapter Two, the coefficient matrix A and the right-side vector $\mathbf{b}$ are given by

$$A = \begin{bmatrix} a_{11} & a_{12} & \cdots & a_{1n} \\ a_{21} & a_{22} & \cdots & a_{2n} \\ \cdots & \cdots & \cdots & \cdots \\ a_{n1} & a_{n2} & \cdots & a_{nn} \end{bmatrix}, \qquad \mathbf{b} = \begin{bmatrix} b_1 \\ b_2 \\ \vdots \\ b_n \end{bmatrix}$$

Let A_j be the matrix obtained by replacing the entries of the jth column of A by the entries of $\mathbf{b}$. Thus, for example,

$$A_2 = \begin{bmatrix} a_{11} & b_1 & \cdots & a_{1n} \\ a_{21} & b_2 & \cdots & a_{2n} \\ \cdots & \cdots & \cdots & \cdots \\ a_{n1} & b_n & \cdots & a_{nn} \end{bmatrix}$$

If $|A| \neq 0$, then the system has the unique solution given by

$$x_1 = \frac{|A_1|}{|A|}, \qquad\qquad x_2 = \frac{|A_2|}{|A|},$$

$$x_3 = \frac{|A_3|}{|A|}, \ldots, \qquad x_n = \frac{|A_n|}{|A|}$$

---

* Gabriel Cramer (1704–1752), of Swiss birth, stated the rules (basically as indicated in Theorem 3.11) for solving a system of linear equations in an appendix to a book he wrote in 1750.

Here $|A_j|$, for $j = 1, 2, 3, \ldots, n$, is the determinant corresponding to matrix A_j. We call the $|A_j|$ determinants, *numerator determinants*, since they appear in the numerators of the equations for the x_j's.

Example (a): Solve the following system for x_1, x_2, and x_3 by using Theorem 3.11 (Cramer's rule).

$$x_1 - x_2 + x_3 = 1 \tag{1}$$

$$2x_1 + x_2 + 3x_3 = 4 \tag{2}$$

$$3x_1 - x_2 - x_3 = -5 \tag{3}$$

Solution: The determinant of the matrix of coefficients, $|A|$, is given by

$$|A| = \begin{vmatrix} 1 & -1 & 1 \\ 2 & 1 & 3 \\ 3 & -1 & -1 \end{vmatrix} = -14$$

Now for the determinant $|A_1|$, we replace the column 1 entries in the expression for $|A|$ by the right-side entries 1, 4, and -5, respectively. Similarly, for $|A_2|$, we replace the column 2 entries of $|A|$ by the right-side entries, and for $|A_3|$ we replace the column 3 entries of $|A|$ by the right-side entries. Thus, we have

$$|A_1| = \begin{vmatrix} 1 & -1 & 1 \\ 4 & 1 & 3 \\ -5 & -1 & -1 \end{vmatrix} = 14,$$

$$|A_2| = \begin{vmatrix} 1 & 1 & 1 \\ 2 & 4 & 3 \\ 3 & -5 & -1 \end{vmatrix} = 0,$$

$$|A_3| = \begin{vmatrix} 1 & -1 & 1 \\ 2 & 1 & 4 \\ 3 & -1 & -5 \end{vmatrix} = -28$$

Now from Theorem 3.11 we have

$$x_1 = \frac{|A_1|}{|A|} = \frac{14}{-14} = -1, \qquad x_2 = \frac{|A_2|}{|A|} = \frac{0}{-14} = 0,$$

$$x_3 = \frac{|A_3|}{|A|} = \frac{-28}{-14} = 2$$

The answers agree with those obtained in the example of frame 18 of Chapter Two, where we employed the Gauss–Jordan elimination method (Method 2.1). In the latter we transformed the augmented matrix $A|\mathbf{b}$ to the reduced row echelon form $I_3|\mathbf{k}$. The solution vector was then given by $\mathbf{x}_s = \mathbf{k}$; for this example, $\mathbf{x}_s = [-1, \ 0, \ 2]^t$.

Note: Refer to statements (d) and (e) of Theorem 3.9 in frame 12. We know that if $|A| = 0$ [statement (e) is false], then the system does not have a unique solution [statement (d) is false]. Also if $|A| \neq 0$ [statement (e) is true], then the system $A\mathbf{x} = \mathbf{b}$ has a unique solution [statement (d) is true]. Two of the ways of finding the unique solution vector $\mathbf{x}_s$ are as follows: (1) finding the reduced row echelon form of $A|\mathbf{b}$, by using Method 2.1, for example, and (2) using Cramer's rule.

Example (b): Solve the following linear system for x_1 and x_2 by using Theorem 3.11.

$$x_1 + 3x_2 = -3$$

$$2x_1 + 4x_2 = -2$$

Solution:

$$|A| = \begin{vmatrix} 1 & 3 \\ 2 & 4 \end{vmatrix} = -2, \qquad |A_1| = \begin{vmatrix} -3 & 3 \\ -2 & 4 \end{vmatrix} = -6,$$

$$|A_2| = \begin{vmatrix} 1 & -3 \\ 2 & -2 \end{vmatrix} = 4,$$

$$x_1 = \frac{|A_1|}{|A|} = \frac{-6}{-2} = 3,$$

$$x_2 = \frac{|A_2|}{|A|} = \frac{4}{-2} = -2$$

This solution was previously obtained in the example of frame 31 of Chapter Two. There we used the following approach. First, we calculated the inverse A^{-1} by transforming (using Method 2.1, for example) the augmented matrix $A|I_2$ to the reduced row echelon form $I_2|K$. The inverse was given by $A^{-1} = K$. Then we used $\mathbf{x} = A^{-1}\mathbf{b}$ (Theorem 2.8) to calculate the solution vector $\mathbf{x}$.

15. It is useful to review and compare at this point the main computational methods covered thus far in this book. To make things simple let us focus on problems for which the matrix A is $n \times n$. Thus, we might say that two of our main problems thus far have been:

(I) Find the solution vector $\mathbf{x}$ for a system of n equations in n unknowns (variables) represented by $A\mathbf{x} = \mathbf{b}$.

(II) Find the inverse (if it exists) of matrix A.

For both problems we have used two fairly different methods. In Chapter Two we employed the Gauss–Jordan elimination method (Method 2.1 in frame 14) to find a reduced row echelon form for the particular problem. Then from the latter the answer to the problem was read off. In the current chapter we have used determinant methods. We summarize in the following table (again, assume A is $n \times n$), where the main methods are listed in the body of the table.

TABLE 3.1   Major Calculational Methods in Chapters Two and Three

Problem / Chapter	(I) Solve $A\mathbf{x} = \mathbf{b}$ for $\mathbf{x}$.	(II) Find A^{-1} if given A.
Chapter Two	Use Method 2.1 to transform $A|\mathbf{b}$ to $I_n|\mathbf{k}$. Then, $\mathbf{x}_c = \mathbf{k}$.	Use Method 2.1 to transform $A|I_n$ to $I_n|K$. Then $A^{-1} = K$.
Chapter Three	Use Theorem 3.11 (Cramer's Rule), and compute the x_j's from $$x_j = \frac{|A_j|}{|A|} \text{ for } j = 1, 2, \dots, n.$$	Use Corollary of Theorem 3.7 in frame 11. Compute A^{-1} from $$A^{-1} = \frac{\text{adj }(A)}{|A|}$$

For the table it is assumed that A^{-1} exists or, equivalently, that $|A| \neq 0$.

Example: For the system of equations

$$4x_1 - 2x_2 - 6x_3 = 14$$

$$-3x_1 + 4x_2 + 9x_3 = -9$$

$$2x_1 + 3x_2 + 3x_3 = 7,$$

find the solution by using (a) Method 2.1, and (b) Cramer's rule.

Solution: (a) Here $A|\mathbf{b} = \begin{bmatrix} 4 & -2 & -6 & 14 \\ -3 & 4 & 9 & -9 \\ 2 & 3 & 3 & 7 \end{bmatrix}$, and we find that the reduced

row echelon form is $\begin{bmatrix} 1 & 0 & 0 & 5 \\ 0 & 1 & 0 & -3 \\ 0 & 0 & 1 & 2 \end{bmatrix}$. Thus, the solution is $x_1 = 5$, $x_2 = -3$,

$x_3 = 2$.

Now do part (b).

_ _ _ _ _ _ _ _ _ _

(b)   We find that

$$|A| = \begin{vmatrix} 4 & -2 & -6 \\ -3 & 4 & 9 \\ 2 & 3 & 3 \end{vmatrix} = -12,$$

$$|A_1| = \begin{vmatrix} 14 & -2 & -6 \\ -9 & 4 & 9 \\ 7 & 3 & 3 \end{vmatrix} = -60, \qquad |A_2| = 36, \qquad |A_3| = -24$$

Thus,

$$x_1 = \frac{|A_1|}{|A|} = \frac{-60}{-12} = 5,$$

$$x_2 = \frac{|A_2|}{|A|} = \frac{36}{-12} = -3,$$

$$x_3 = \frac{|A_3|}{|A|} = \frac{-24}{-12} = 2$$

D. AN APPLICATION

16.   The application of this section involves a linear system with the same number of equations as unknowns. Thus, the methods of Chapter Two can be used in the analysis of such a system. However, to gain practice using the techniques of the current chapter, our method of analysis will involve determinants.

The following application deals with corporate taxes for a multinational corporation.

Example: The Apex Corporation does business in the United States and in Canada. Henceforth, we shall denote the United States and Canada by the labels 1 and 2, respectively. The tax rates set by the United States for profits earned in the U.S. and Canada are, respectively, 5% (or 0.05) and 0.08. Likewise, Canada taxes the Apex Corporation on its profits. The Canadian tax rate on profits earned in the U.S. is 0.07, and its tax rate for profits earned in Canada is 0.04.

If we let r_{ij} be the tax rate by country i on profits earned in country j, we have the following tax rate matrix R, both in general, and for the example at hand:

$$R = \begin{bmatrix} r_{11} & r_{12} \\ r_{21} & r_{22} \end{bmatrix} = \begin{bmatrix} 0.05 & 0.08 \\ 0.07 & 0.04 \end{bmatrix} \tag{1}$$

For example, $r_{21} = 0.07$ is the tax rate by country 2 (Canada) on profits earned in country 1 (the U.S.).

Suppose p_1 and p_2 are the total profits per year, earned in countries 1 and 2, respectively, and that t_1 and t_2 are the total taxes per year, paid in countries 1 and 2, respectively. Furthermore, let

$$\mathbf{p} = \begin{bmatrix} p_1 \\ p_2 \end{bmatrix} \qquad \text{and} \qquad \mathbf{t} = \begin{bmatrix} t_1 \\ t_2 \end{bmatrix} \qquad\qquad (2), \quad (3)$$

be symbols for the profit and tax vectors, respectively.

(a)   Write down general equations relating p_1, p_2, t_1 and t_2, and the r_{ij}'s.
(b)   Write down a general matrix–vector equation relating $\mathbf{p}$, $\mathbf{t}$, and the matrix R.

Solution: (a)   The general equation relating taxes t_1 paid to country 1 (U.S.) to the profits earned in both countries is

$$t_1 = r_{11}p_1 + r_{12}p_2 \qquad \text{(taxes to country 1)} \qquad\qquad (4a)$$

Similarly, for t_2, the taxes paid to country 2, we have

$$t_2 = r_{21}p_1 + r_{22}p_2 \qquad \text{(taxes to country 2)} \qquad\qquad (4b)$$

Now do part (b).

— — — — — — — — — —

(b)   We can combine Eqs. (4a) and (4b) in the following column vector equation:

$$\begin{bmatrix} t_1 \\ t_2 \end{bmatrix} = \begin{bmatrix} r_{11}p_1 + r_{12}p_2 \\ r_{21}p_1 + r_{22}p_2 \end{bmatrix} \qquad\qquad (5)$$

The right-side vector of Eq. (5) is equal to the product of matrix R times vector $\mathbf{p}$, and the left side is the vector $\mathbf{t}$. Thus, we have

$$\mathbf{t} = R\mathbf{p} \qquad\qquad (6)$$

17.

Example (a): Suppose in 1980 that the profits for the Apex company are $800,000 in the United States (p_1) and p_2 = $350,000 in Canada. Compute the taxes paid to both countries. Use the r_{ij} tax rate data given in frame 16.

Solution: From Eqs. (4a) and (4b) of frame 16, we have

$$t_1 = 0.05(800,000) + 0.08(350,000) = \$68,000 \quad \text{(taxes paid to U.S.)}$$

$$t_2 = 0.07(800,000) + 0.04(350,000) = \$70,000 \quad \text{(taxes paid to Canada)}$$

Very often a problem that is the reverse of that solved in Example (a) is posed. That is, a meaningful problem is to calculate the profits in both countries, if given figures for the total taxes paid by the corporation to both countries in a given year. One approach to solving this problem involves using the following equation for the profit vector $\mathbf{p}$, which is derived by premultiplying both sides of Eq. (6) of frame 16 by the inverse matrix R^{-1}:

$$\mathbf{p} = R^{-1}\mathbf{t} \qquad \text{(here, we assume } R^{-1} \text{ exists)} \tag{7}$$

Example (b): (i) Calculate R^{-1} for the matrix R given in Eq. (1) of frame 16. Use the corollary of Theorem 3.7 here (frame 11). (ii) Then use R^{-1} and Eq. (7) to determine profits in both countries if the total taxes paid to both countries in 1981 are $t_1 = \$70,000$ and $t_2 = \$62,000$, respectively.

Solution: (i) First,

$$R = \begin{bmatrix} 0.05 & 0.08 \\ 0.07 & 0.04 \end{bmatrix} = \frac{1}{100} \begin{bmatrix} 5 & 8 \\ 7 & 4 \end{bmatrix} \tag{8}$$

Then, letting

$$B = \begin{bmatrix} 5 & 8 \\ 7 & 4 \end{bmatrix}, \tag{9}$$

we have

$$R^{-1} = 100 B^{-1}, \tag{10}$$

after using Theorem 2.6, part (c)—here, $1/k = 1/(1/100) = 100$. Now let us find B^{-1} by using the Corollary of Theorem 3.7.
    First,

$$|B| = 20 - 56 = -36 \tag{11}$$

Then the cofactors for matrix B are as follows:

$$C_{11} = 4, \qquad C_{12} = -7, \qquad C_{21} = -8, \qquad C_{22} = 5 \tag{12}$$

For example, for C_{12}, we have a (-1) times the entry 7 ($= b_{21}$), which remains when we delete row 1 and column 2 in matrix B. Recall that the determinant of a 1×1 matrix $[a]$ is merely the number a. Then, from (12), we have

$$\text{adj } (B) = \begin{bmatrix} 4 & -8 \\ -7 & 5 \end{bmatrix} \tag{13}$$

Now we make use of $|B|$, adj (B), and the corollary of Theorem 3.7 to obtain

$$B^{-1} = \frac{\text{adj }(B)}{|B|} = \left(-\frac{1}{36}\right)\begin{bmatrix} 4 & -8 \\ -7 & 5 \end{bmatrix} \tag{14}$$

Substituting from Eq. (14) into Eq. (10) yields

$$R^{-1} = \left(\frac{-100}{36}\right)\begin{bmatrix} 4 & -8 \\ -7 & 5 \end{bmatrix} = \left(\frac{100}{36}\right)\begin{bmatrix} -4 & 8 \\ 7 & -5 \end{bmatrix} \tag{15}$$

Now do part (ii).

— — — — — — — — — —

(ii)   Substituting R^{-1} and $\mathbf{t} = \begin{bmatrix} 70,000 \\ 62,000 \end{bmatrix}$ into Eq. (7) yields

$$\begin{aligned}
\mathbf{p} &= \left(\frac{100}{36}\right)\begin{bmatrix} -4 & 8 \\ 7 & -5 \end{bmatrix}\begin{bmatrix} 70,000 \\ 62,000 \end{bmatrix} \\
&= \left(\frac{100}{36}\right)\begin{bmatrix} 216,000 \\ 180,000 \end{bmatrix} = \begin{bmatrix} 600,000 \\ 500,000 \end{bmatrix}
\end{aligned} \tag{16}$$

Thus, the profits are $600,000 in the United States and $500,000 in Canada.

The reader now has a fairly good acquaintance with some of the major aspects of the matrix algebra part of linear algebra. In fact, at this point, the reader is well prepared for some of the material in Chapters Eight and Nine, and might wish to jump ahead to either of these chapters (Chapter Nine is especially of interest since it deals with computer and programmable calculator usage). Some of the material in Chapter Eight involves concepts from Chapters Four through Seven. Thus, should the reader jump ahead, he or she should anticipate the occasional need to refer back to the intermediate chapters whenever unfamiliar symbols and concepts are encountered.

SELF-TEST

This Self-Test will help you determine whether or not you have mastered the chapter objectives and are ready to go on to another chapter. Correct answers are given at the end of the test.

1. Use the preliminary definitions (frames 1, 2) to compute the determinants of the following matrices:

(a) $\begin{bmatrix} 2 & -5 \\ 6 & 4 \end{bmatrix}$,　(b) $\begin{bmatrix} 3 & 5 & 7 \\ 2 & 4 & 6 \\ 8 & -2 & 9 \end{bmatrix}$,　(c) $\begin{bmatrix} 7 & 0 & -4 \\ 6 & 5 & 8 \\ -3 & 6 & 12 \end{bmatrix}$,

(d) $\begin{bmatrix} 2 & 0 & 2 & 0 \\ 7 & 1 & -6 & 3 \\ 8 & 2 & 4 & 9 \\ 3 & 5 & 6 & 1 \end{bmatrix}$

2. (a) Do question (1b) by using a cofactor expansion along the second row.

(b) Do question (1c) by doing a cofactor expansion along the second column.

In the next two questions use the time saving approach used in frame 6. Show intermediate determinant expressions.

3. Do question (1b).

4. Do question (1d).

5. (a) If A is $n \times n$, then $|kA| = k^n|A|$. Demonstrate this for the case of a 2×2 determinant. *Hint:* Use Theorem 3.3, part (b).

(b) Use the part (a) result to find the determinant of $\frac{1}{4}\begin{bmatrix} 32 & 2 \\ 4 & 1 \end{bmatrix}$.

6. (a) Using a method from Chapter Two, find the inverse of

$$F = \begin{bmatrix} 3 & 2 \\ 5 & 6 \end{bmatrix}.$$

(b) Then verify that $|F^{-1}| = 1/|F|$.

7. For the matrix $A = \begin{bmatrix} -4 & 2 & 0 \\ 6 & 1 & 3 \\ 2 & 3 & -1 \end{bmatrix}$:

(a) find all cofactors, and then find the matrix adj (A).

(b) Find A^{-1} by using the corollary of Theorem 3.7 (frame 11).

8. (a) Solve the following linear system for x_1, x_2, and x_3 by using Cramer's rule (Theorem 3.11 in frame 14).

$$2x_1 + 3x_2 + 4x_3 = 47$$
$$x_1 + 2x_2 + x_3 = 23$$
$$x_1 + 2x_2 + 3x_3 = 31$$

List $|A|$ and the three numerator determinants.

(b) Determine Rank (A) and the reduced row echelon form of A (see Theorem 3.9).

9. Suppose the Aglo Corporation does business in the United States and Japan. We label the two countries 1 and 2, respectively. Each country taxes the corporation profits according to the following tax rate matrix:

$$
\begin{array}{cc}
 & \begin{array}{cc} \text{U.S. (1)} & \text{Japan (2)} \end{array} \\
\begin{array}{c} \text{U.S.} \quad (1) \\ \text{Japan} \quad (2) \end{array} &
\left[\begin{array}{cc} 0.08 & 0.10 \\ 0.12 & 0.06 \end{array} \right]
\end{array}
\qquad
\begin{array}{l} \text{The tax rate} \\ \text{matrix } R \end{array}
$$

Each entry r_{ij} is the tax rate by country i on profits earned in country j. For example, $r_{21} = 0.12$ is the tax rate by country 2 (Japan) on profits earned in country 1 (U.S.) by the Aglo Corporation.

(a) Determine R^{-1} by using the corollary of Theorem 3.7.

(b) Make use of R^{-1} to determine profits in 1980 if the corporation paid taxes of \$26,400 to the United States and \$28,800 to Japan.

ANSWERS TO SELF-TEST

If your answers to the test questions do not agree with the ones given here, review the frames indicated in parentheses after each answer before you go on to another chapter.

1. (a) 38,   (b) 42,   (c) -120,   (d) -1024.   (frames 1–2)

2. (a) 42,   (b) -120.   (frames 3–4)

3.
$$
\begin{vmatrix} 3 & 5 & 7 \\ 2 & 4 & 6 \\ 8 & -2 & 9 \end{vmatrix}
\overset{(a)}{=} 2\begin{vmatrix} 3 & 5 & 7 \\ 1 & 2 & 3 \\ 8 & -2 & 9 \end{vmatrix}
\overset{(b)}{=} 2\begin{vmatrix} 0 & -1 & -2 \\ 1 & 2 & 3 \\ 0 & -18 & -15 \end{vmatrix}
$$

$$
\overset{(c)}{=} (-2)\begin{vmatrix} -1 & -2 \\ -18 & -15 \end{vmatrix}
\overset{(d)}{=} (-2)[15 - 36] = 42
$$

Steps: (a) $r_2' = r_2/2$, (b) $r_1' = r_1 - 3r_2$ and $r_3' = r_3 - 8r_2$, (c) cofactor expansion along column 1, (d) evaluate 2×2 determinant.   (frames 5–6)

4.
$$
\begin{vmatrix} 2 & 0 & 2 & 0 \\ 7 & 1 & -6 & 3 \\ 8 & 2 & 4 & 9 \\ 3 & 5 & 6 & 1 \end{vmatrix}
\overset{(a)}{=} 2\begin{vmatrix} 2 & 0 & 1 & 0 \\ 7 & 1 & -3 & 3 \\ 8 & 2 & 2 & 9 \\ 3 & 5 & 3 & 1 \end{vmatrix}
\overset{(b)}{=} 2\begin{vmatrix} 0 & 0 & 1 & 0 \\ 13 & 1 & -3 & 3 \\ 4 & 2 & 2 & 9 \\ -3 & 5 & 3 & 1 \end{vmatrix}
$$

$$
\overset{(c)}{=} 2\begin{vmatrix} 13 & 1 & 3 \\ 4 & 2 & 9 \\ -3 & 5 & 1 \end{vmatrix}
\overset{(d)}{=} 2\begin{vmatrix} 13 & 1 & 3 \\ -22 & 0 & 3 \\ -68 & 0 & -14 \end{vmatrix}
\overset{(e)}{=} (-2)\begin{vmatrix} -22 & 3 \\ -68 & -14 \end{vmatrix}
$$

$$
\overset{(f)}{=} (-2)[308 + 204] = -1024
$$

Steps: (c stands for column in the following.) (a) $c_3' = c_3/2$, (b) $c_1' = c_1 - 2c_3$, (c) cofactor expansion on row 1, (d) $r_2' = r_2 - 2r_1$, $r_3' = r_3 - 5r_1$,

(e) cofactor expansion on column 2, (f) evaluate 2×2 determinant. (frames 5–6)

5. (a) If $A = \begin{bmatrix} a_{11} & a_{12} \\ a_{21} & a_{22} \end{bmatrix}$, then $kA = \begin{bmatrix} ka_{11} & ka_{12} \\ ka_{21} & ka_{22} \end{bmatrix}$ from Definition 1.2 (frame 8 of Chapter One). Thus, $|kA| = \begin{vmatrix} ka_{11} & ka_{12} \\ ka_{21} & ka_{22} \end{vmatrix} = k \begin{vmatrix} a_{11} & a_{12} \\ ka_{21} & ka_{22} \end{vmatrix} = k^2 \begin{vmatrix} a_{11} & a_{12} \\ a_{21} & a_{22} \end{vmatrix} = k^2|A|$ after using part (b) of Theorem 3.3 twice in succession.

    (b) Since $k = \frac{1}{4}$ here, we get $\left(\frac{1}{4}\right)^2 \begin{vmatrix} 32 & 2 \\ 4 & 1 \end{vmatrix} = \frac{1}{16}(32 - 8) = \frac{3}{2}$.   (frame 5)

6. (a) $F^{-1} = \frac{1}{8} \begin{bmatrix} 6 & -2 \\ -5 & 3 \end{bmatrix}$

    (b) $|F| = 8, |F^{-1}| = \frac{8}{64} = \frac{1}{8}$.   (frames 7–8)

7. (a) $\text{adj}(A) = \begin{bmatrix} -10 & 2 & 6 \\ 12 & 4 & 12 \\ 16 & 16 & -16 \end{bmatrix}$.

    (b) First, $|A| = 64$. Then $A^{-1} = \frac{1}{64} \begin{bmatrix} -10 & 2 & 6 \\ 12 & 4 & 12 \\ 16 & 16 & -16 \end{bmatrix}$.   (frames 9–11)

8. (a) $|A| = 2, |A_1| = 10, |A_2| = 14, |A_3| = 8$. Thus, $x_1 = \frac{10}{2} = 5; x_2 = 7;$ $x_3 = 4$.

    (b) Rank $(A) = 3$; reduced row echelon form of A is I_3.   (frames 12–15)

9. (a) $R^{-1} = \frac{100}{72} \begin{bmatrix} -6 & 10 \\ 12 & -8 \end{bmatrix}$.

    (b) $p_1 = \$180{,}000, p_2 = \$120{,}000$.   (frames 11, 16–17)

CHAPTER FOUR

Geometrical Properties of the Vector Space R^n

In this chapter we shall give a geometrical interpretation to vectors and shall learn about some of the properties of the n-dimensional vector space, R^n (Euclidean n-space). Our discussion will lead us to methods for characterizing geometrical structures in Euclidean n-space. Two such structures to be studied in this chapter are the hyperplane and the straight line.

OBJECTIVES

When you complete this chapter you should be able to

- Determine geometrical representations of column vectors, and geometrically illustrate the addition, subtraction, and scalar multiplication of such vectors.
- Determine the properties of the n-dimensional vector space, R^n.
- Do computations involving the dot product and lengths (norms) of vectors.
- Use and illustrate the Cauchy–Schwarz and triangle inequalities.
- Do computations and diagrams involving the hyperplane and straight line in R^n.

A. GEOMETRICAL REPRESENTATIONS OF VECTORS

1. It is very useful to give vectors a geometrical interpretation. This is particularly simple if we are dealing with two- or three-dimensional spaces. In Figure 4.1 we represent the two-component column vector $\mathbf{b} = [b_1, b_2]^t$ as a directed line segment (arrow) from an initial point at the origin to a terminal point at (b_1, b_2). That is, the x_1 coordinate of the terminal point is b_1 and the x_2 coordinate is b_2.

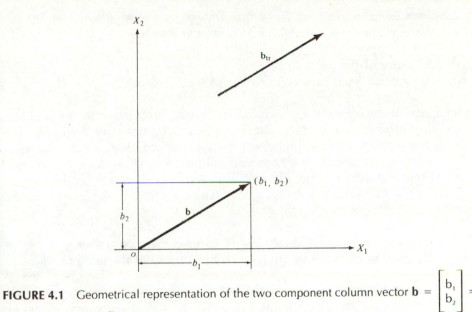

FIGURE 4.1   Geometrical representation of the two component column vector $\mathbf{b} = \begin{bmatrix} b_1 \\ b_2 \end{bmatrix} = [b_1, b_2]^t$.

The axes shown in Figure 4.1 are the mutually perpendicular X_1 and X_2 axes. These axes are also traditionally labeled, respectively, as the X and Y axes. We often shall refer to two-dimensional space as "the plane."

Geometrical representations of typical numerical two-component column vectors are presented in Figure 4.2. In two-dimensional space (and, similarly, for three and higher dimensions), the directed line segment drawn from the initial point $(0, 0)$ to the terminal point (b_1, b_2), with an arrowhead at the latter point, is a geometrical representation of $\mathbf{b} = [b_1, b_2]^t$. In summary, this geo-

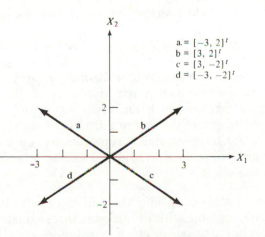

$$\begin{aligned} a. &= [-3, \ 2]^t \\ b &= [3, \ 2]^t \\ c &= [3, \ -2]^t \\ d &= [-3, \ -2]^t \end{aligned}$$

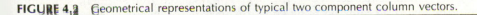

FIGURE 4.2   Geometrical representations of typical two component column vectors.

metrical representation allows for two possible geometrical interpretations (or correspondences, or associations) for a column vector $\mathbf{b} = [b_1, b_2]^t$:

(i)　A *locater* of the point (b_1, b_2).
(ii)　A *direction* from the origin to the point (b_1, b_2).

Both interpretations will be useful to us in our further development of the geometrical properties of vectors. These two interpretations are useful also for three-dimensional space and higher-dimensional space (more on this later).

For example, for the three-component column vector $[b_1, b_2, b_3]^t$, our geometrical representation is the directed line segment drawn from the origin, $(0, 0, 0)$, to the terminal point, (b_1, b_2, b_3). Then, for $[b_1, b_2, b_3]^t$, we have the following geometrical interpretations:

(i)　A *locater* of the point (b_1, b_2, b_3) in three-dimensional space.
(ii)　A *direction* from the origin [i.e., from $(0, 0, 0)$] to the point (b_1, b_2, b_3).

Let us tentatively visualize n-dimensional space as being a collection of all possible points such that each point has n coordinates. For example, a point P in n-dimensional space has n coordinates indicated by $(p_1, p_2, \ldots, p_n)$, say. Now for a geometrical representation of the n-component column vector $\mathbf{b} = [b_1, b_2, \ldots, b_n]^t$, we first try to conceive of a directed line segment drawn from the origin [i.e., from $(0, 0, \ldots, 0)$] to the terminal point with coordinates $(b_1, b_2, \ldots, b_n)$. Thus, for the n-component column vector $\mathbf{b}$, we have the following two possible interpretations:

(i)　A *locater* of the point $(b_1, b_2, \ldots, b_n)$ in n-dimensional space.
(ii)　A *direction* from the origin to the point $(b_1, b_2, \ldots, b_n)$.

By the way, the n-component zero vector, $\mathbf{0} = [0, 0, \ldots, 0]^t$, is geometrically represented by the origin point $(0, 0, \ldots, 0)$. Thus, in two- and three-dimensional spaces, the zero vector is represented by the origin points $(0, 0)$ and $(0, 0, 0)$, respectively.

The directed line segment labeled $\mathbf{b}_{tr}$ in Figure 4.1 is called a "translated vector" corresponding to $\mathbf{b}$. It is *not* a geometrical representation of a vector since it does not emanate from the origin. Its initial point can be anywhere in the plane, except at the origin. It is parallel to (or on the same line as) the directed line segment representing $\mathbf{b}$, and its direction and length are the same as the direction and length of the directed line segment representing $\mathbf{b}$.

Thus, in a plane, for the geometrical representation of a vector, the initial point is at the origin, while for a "translated vector," the initial point can be anywhere else in the plane.

Notes:　(a)　Some authors consider the (infinite) collection of all possible translated vectors, together with the directed line segment with initial point at the origin (the latter being *our* geometrical representation for the vector), as representing the vector $\mathbf{b} = [b_1, b_2]^t$. The directed line segment starting from the origin is referred to by such authors as the *position representation of the vector.* (b)   In dealing with geometrical interpretations of two-component column vectors, we could just as well make our interpretations with respect to row vectors (such as $\mathbf{b} = [b_1, b_2]$), or, for that matter, with respect to an "abstract vector," which is just an ordered pair $\langle b_1, b_2 \rangle$. Similar comments apply to n-component vectors. In many applications of linear algebra (e.g., linear programming), there are definite advantages to making geometrical interpretations with respect to column vectors. In this book our geometrical interpretations will usually be for column vectors.

As indicated previously, the situation is similar for three-component column vectors. The geometrical representation of $[b_1, b_2, b_3]^t$ is a directed line segment from initial point $(0, 0, 0)$ to the terminal point at (b_1, b_2, b_3). The terminal point (b_1, b_2, b_3) in three-dimensional space is the point whose x_1 coordinate is b_1, x_2 coordinate is b_2, and x_3 coordinate is b_3. The X_1, X_2, and X_3 axes are mutually perpendicular. In Figure 4.3 the X_2 and X_3 axes are in the plane of the paper, while the X_1 axis is "supposed to" look as if it is perpendicular to both the X_2 and X_3 axes and "coming out" of the paper. Clearly, we have some distortion here, since our diagram is on a two-dimensional page.

Example: Give geometrical representations of the column vectors (a) $\mathbf{a} = [3, 4, 5]^t$, and (b) $\mathbf{b} = [3, -4, 7]^t$.

Solution: (a) For $\mathbf{a} = [3, 4, 5]^t$, first we locate the point $(3, 4, 5)$ in three-dimensional space. See Figure 4.3. We can do this as follows: Go out three units on the positive X_1 axis, then go out four units in the positive x_2 direction along a line parallel to the X_2 axis, and then go out five units in the positive x_3 direction along a line parallel to the X_3 axis. For the geometrical representation of the vector $\mathbf{a}$, we then connect a directed line segment from the origin to the point $(3, 4, 5)$.

Now do part (b).

— — — — — — — — — —

(b)   First, the point $(3, -4, 7)$ is located. For the second coordinate we go four units in the *negative* x_2 direction along a line parallel to the X_2 axis. (See Figure 4.3.) For the geometrical representation of the vector $\mathbf{b}$, we connect a directed line segment from the origin to the point $(3, -4, 7)$.

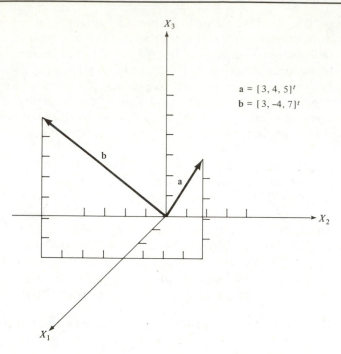

$$a = [3, 4, 5]^t$$
$$b = [3, -4, 7]^t$$

FIGURE 4.3   Geometric representations of three component column vectors.

2.   Just as the column vector $\mathbf{b} = [b_1, b_2]^t$ can be interpreted as locating a point in two-dimensional space (and similarly for $[b_1, b_2, b_3]^t$ with respect to three-dimensional space), the n-component column vector $\mathbf{b} = [b_1, b_2, \ldots, b_n]^t$ can be interpreted as locating a point in n-dimensional space. We shall have more to say shortly about this abstract concept known as n-dimensional space.

Observe that in Figure 4.1, the (terminal) point that characterizes the column vector $\mathbf{b}$ is labeled (b_1, b_2). The reason for this is that the convention for representing points in analytic geometry is to use parentheses, that is, ( ). On occasion we shall label the point with either $\begin{bmatrix} b_1 \\ b_2 \end{bmatrix}$ or $[b_1, b_2]^t$.

We recall that a k-component column vector is the same as a $k \times 1$ matrix (frame 5 of Chapter One). The two operations, addition and multiplication by a scalar, are thus defined for column vectors in the same way as for general matrices (frames 6–8 of Chapter One).

Thus, for example, in the two-component case, we have

$$\mathbf{a} + \mathbf{b} = \begin{bmatrix} a_1 \\ a_2 \end{bmatrix} + \begin{bmatrix} b_1 \\ b_2 \end{bmatrix} = \begin{bmatrix} a_1 + a_2 \\ b_1 + b_2 \end{bmatrix} = [a_1 + a_2, \quad b_1 + b_2]^t$$

Example (a): If $\mathbf{a} = \begin{bmatrix} 5 \\ 2 \end{bmatrix} = [5,\ 2]^t$ and $\mathbf{b} = [1,\ 6]^t$, then the vector

$$\mathbf{a} + \mathbf{b} = \begin{bmatrix} 5 \\ 2 \end{bmatrix} + \begin{bmatrix} 1 \\ 6 \end{bmatrix} = \begin{bmatrix} 5+1 \\ 2+6 \end{bmatrix} = \begin{bmatrix} 6 \\ 8 \end{bmatrix}$$

It is useful to represent the process of addition of vectors geometrically. We shall do this for the two-dimensional case, but it should be noted that a similar process applies to three-dimensional space. In the following geometrical discussion we shall often refer to the geometrical representation of a column vector $\mathbf{a}$ as *being* the column vector $\mathbf{a}$.

To geometrically find the vector sum $\mathbf{a} + \mathbf{b}$, which we shall denote as $\mathbf{c}$, we first construct the translated vector $\mathbf{b}_{tr}$ such that its initial point coincides with the terminal point (arrowhead) of $\mathbf{a}$ (see Figure 4.4a). The directed line segment from the origin to the terminal point of $\mathbf{b}_{tr}$ is then the representation of $\mathbf{a} + \mathbf{b}$. The coordinates of the terminal point are $(a_1 + b_1,\ a_2 + b_2)$ as expected. This process is known as the triangle rule for adding vectors.

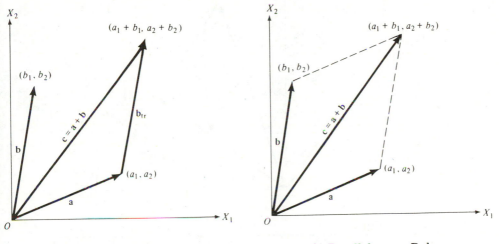

(a) Triangle Rule.   (b) Parallelogram Rule.

FIGURE 4.4   Vector Addition.

Another device for geometrically determining the sum $\mathbf{a} + \mathbf{b}$ is the parallelogram rule for adding vectors. Here we construct a parallelogram with sides parallel to the geometrical representations of vectors $\mathbf{a}$ and $\mathbf{b}$ (see Figure 4.4b). Thus, in the diagram the two new sides are shown as broken line segments parallel to $\mathbf{a}$ and $\mathbf{b}$, respectively. The diagonal arrow that starts at the origin is then the geometrical representation of $\mathbf{a} + \mathbf{b}$.

Note: The diagrams in Figure 4.4 for vectors **a**, **b**, and **a** + **b** are consistent with the numerical vectors cited in Example (a).

The geometrical process of vector subtraction can also be demonstrated by the triangle rule. Refer to Figure 4.5. Suppose we are given vectors **a** and **b** and we wish to represent **b** − **a** geometrically. Now **b** − **a** is the vector that when added to **a** will yield **b**. That is,

a + (**b** − **a**) = **b**

Thus, we construct the translated vector (**b** − **a**)$_{tr}$, which extends from the terminal point of **a** to the terminal point of **b**. Now the geometrical representation for the vector (**b** − **a**) is obtained by constructing a directed line segment from the origin, which is parallel to and of the same length and direction as (**b** − **a**)$_{tr}$. See Figure 4.5.

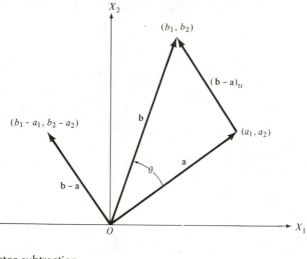

FIGURE 4.5   Vector subtraction.

The process of scalar multiplication (see frame 8 of Chapter One) of a column vector is illustrated geometrically in Figure 4.6. There the vector 3**a** is three times as long as the vector **a**, and both vectors have the same direction. The vector − 2**a** is twice as long as the vector **a**, and vector − 2**a** has a direction opposite from **a**.

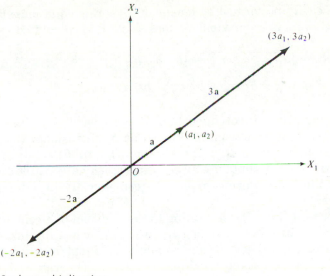

FIGURE 4:6   Scalar multiplication.

In short, if k is a positive real number (scalar), then **a** and k**a** have the same direction, while if k is a negative scalar, then **a** and k**a** have opposite directions. We refer to k**a** as a (scalar) multiple of vector **a**.

Example (b):   Given that $\mathbf{a} = [4, 3]^t$ and $\mathbf{b} = [2, 6]^t$, compute $\mathbf{b} - \mathbf{a}$, 3**a**, and $-2\mathbf{a}$, and represent them geometrically.

Solution:

$$\mathbf{b} - \mathbf{a} = \begin{bmatrix} 2 \\ 6 \end{bmatrix} - \begin{bmatrix} 4 \\ 3 \end{bmatrix} = \begin{bmatrix} 2 - 4 \\ 6 - 3 \end{bmatrix} = \begin{bmatrix} -2 \\ 3 \end{bmatrix}$$

This result is consistent with the diagram of Figure 4.5.
Now complete the problem.

Performing the usual scalar multiplication,

$$3\mathbf{a} = 3\begin{bmatrix} 4 \\ 3 \end{bmatrix} = \begin{bmatrix} 3(4) \\ 3(3) \end{bmatrix} = \begin{bmatrix} 12 \\ 9 \end{bmatrix}, \qquad -2\mathbf{a} = -2\begin{bmatrix} 4 \\ 3 \end{bmatrix} = \begin{bmatrix} -8 \\ -6 \end{bmatrix}$$

These results conform to the diagram of Figure 4.6.

3.   Calculations such as those preceding lend themselves to a ready geometrical interpretation in the case of two and three dimensions. In higher dimensions we lose the geometric interpretation but the processes of column vector addition, column vector subtraction, and multiplication of a column vector by a scalar are easily performed by means of algebraic equations. For example, the sum **c** of two four-component vectors **a** and **b** is calculated as follows:

$$\mathbf{c} = \mathbf{a} + \mathbf{b} = [a_1, a_2, a_3, a_4]^t + [b_1, b_2, b_3, b_4]^t$$

$$= [a_1 + b_1, a_2 + b_2, a_3 + b_3, a_4 + b_4]^t$$

Such manipulations involving column vectors with any number of components lead to the abstract concept known as n-dimensional Euclidean space (indicated symbolically as R^n). The structure for this Euclidean space is algebraic. However, we shall use two-, and occasionally three-, dimensional diagrams to illustrate various processes geometrically in n-dimensional Euclidean space.

Remember that the rules pertaining to the addition of n-component column vectors and the multiplication of a column vector by a scalar are special cases of the definitions and theorems for the addition of matrices and the multiplication of a matrix by a scalar (Definitions 1.1, 1.2, Theorems 1.1, 1.2). This is so because an n-component column vector is the same as an $n \times 1$ matrix.

Because of these rules pertaining to the addition of n-component column vectors and the multiplication of a column vector by a scalar, it can be shown that n-component column vectors satisfy the 10 properties listed in Theorem 4.1. By the way, recall that the n-component zero column vector is given by $\mathbf{0} = [0, 0, \ldots, 0]^t$; that is, each of its components is zero. Also, if $\mathbf{a} = [a_1, a_2, \ldots, a_n]^t$, then $-\mathbf{a} = (-1)\mathbf{a} = [-a_1, -a_2, \ldots, -a_n]^t$.

Theorem 4.1

For any n-component column vectors **a**, **b**, **c**, and any scalars h and k, the following properties hold:

(i)   The sum of any two n-component column vectors is an n-component column vector.

(ii) $(\mathbf{a} + \mathbf{b}) + \mathbf{c} = \mathbf{a} + (\mathbf{b} + \mathbf{c})$.

(iii) $\mathbf{a} + \mathbf{0} = \mathbf{a}$.

(iv) $\mathbf{a} + (-\mathbf{a}) = \mathbf{0}$.

(v) $\mathbf{a} + \mathbf{b} = \mathbf{b} + \mathbf{a}$.

(vi)   The product of any scalar times any n-component column vector is an n-component column vector.

(vii) $h(\mathbf{a} + \mathbf{b}) = h\mathbf{a} + h\mathbf{b}$.

(viii) $(h + k)\mathbf{a} = h\mathbf{a} + k\mathbf{a}$.

(ix) $(hk)\mathbf{a} = h(k\mathbf{a})$.

(x) $1\mathbf{a} = \mathbf{a}$.   (Here 1 is the scalar 1.)

Properties (i) and (vi), the so-called *closure* properties, follow from Definitions 1.1 and 1.2. The rest follow from Theorems 1.1, 1.2, and other definitions in Chapter One (such as for the negative of a matrix, after Definition 1.2 in frame 8, and the zero matrix in frame 8).

Since the preceding 10 properties are satisfied, the set of all n-component column vectors (with real components) is said to *constitute* what is known as a *vector space*. In Chapters Five, Six, and Seven we shall study vector space concepts extensively. That vector space we shall focus on the most is the set of n-component column vectors discussed above.

Note: Several of the properties cited above follow directly from the corresponding properties for matrices. For example, refer to Theorems 1.1 and 1.2.

Definition 4.1 (Euclidean n-space, R^n)

Consider two typical column vectors **a** and **b** that have n components:

$$\mathbf{a} = [a_1, a_2, \ldots, a_n]^t, \qquad \mathbf{b} = [b_1, b_2, \ldots, b_n]^t$$

As discussed previously we have the following operations of column vector addition and scalar multiplication:

$$\mathbf{a} + \mathbf{b} = [a_1 + b_1, a_2 + b_2, \ldots, a_n + b_n]^t$$
$$k\mathbf{a} = [ka_1, ka_2, \ldots, ka_n]^t$$

The set of all such column vectors with real number components, with these two operations, is called n-dimensional space or n-dimensional Euclidean space, and will be denoted by R^n. The "R" here refers to the requirement that the components be real numbers, and the "n" refers to the number of components. Other expressions for this are Euclidean n-space or just plain n-space.

Notes: (a)   Here, Definition 4.1 is algebraic in nature. For example, we would refer to two-dimensional "space" R^2 as consisting of all possible two-component column vectors of the form $\mathbf{x} = [x_1, x_2]^t$, for which x_1 and x_2 range over all real number values. Previously, however, we thought of two-dimensional space as being the geometrical structure consisting of all points with coordinates (x_1, x_2), where x_1 is the X_1 coordinate and x_2 is the X_2 coordinate. Here, too, x_1 and x_2 range over all real number values. It is clear that each point indicated geometrically by (x_1, x_2) corresponds to the column vector $[x_1, x_2]^t$, an algebraic quantity, and conversely. In this book our point of view on the vector space R^2 will be to think of it as being either (i) a geometrical structure consisting of points with coordinates (x_1, x_2) or (ii) an algebraic structure consisting of column vectors of the form $[x_1, x_2]^t$. Similarly, our attitude will be the same with respect to R^n, regardless of what n is.

Refer to Figures 4.1, 4.2, and 4.3 of frame 1 for geometrical illustrations for R^2 and R^3, respectively.

(b)   To emphasize the correspondence further between the algebraic and geometric structures for R^n, recall our convention for the geometrical representation of the column vector $\mathbf{b} = [b_1, b_2, \ldots, b_n]^t$. We thought of this as being a directed line segment from the origin point $(0, 0, \ldots, 0)$ to a terminal point with coordinates $(b_1, b_2, \ldots, b_n)$. See frame 1.

The commentary in the preceding paragraph leads us to still another geometrical interpretation of R^n, namely, that R^n consists of all possible directed line segments coming out from the origin.

(c)   Because the 10 properties of Theorem 4.1 are satisfied by n-component column vectors, the Euclidean n-space R^n is a vector space (in fact, the most important vector space to be covered in this book).

(d)   The word *Euclidean* stems from Euclid (330?–275? B.C.), who was a famous Greek mathematician.

A third operation pertaining to column vectors that will be useful to us is that of dot product (also called scalar product or Euclidean inner product).

Definition 4.2 (Dot Product or Euclidean Inner Product)

If $\mathbf{a}$ and $\mathbf{b}$ are any two column vectors in R^n (see Definition 4.1), then the dot product, denoted as $\mathbf{a} \cdot \mathbf{b}$, is defined by the following equation:

$$\mathbf{a} \cdot \mathbf{b} = a_1 b_1 + a_2 b_2 + \cdots + a_n b_n$$

Note that the dot product of two column vectors is a number (not a column vector). Also, for $\mathbf{a} \cdot \mathbf{b}$ to have meaning, both $\mathbf{a}$ and $\mathbf{b}$ have to have the same number of components.

Note: Observe that the *ordinary* multiplication between numbers, sometimes also indicated by a dot, does not apply to vectors.

Example: Compute the following dot products: (a) $\mathbf{a} \cdot \mathbf{b}$ if $\mathbf{a} = [1, 4]^t$ and $\mathbf{b} = [-3, 2]^t$, (b) $\mathbf{c} \cdot \mathbf{d}$ if $\mathbf{c} = \mathbf{0} = [0, 0, 0]^t$ and $\mathbf{d} = [4, -2, 7]^t$, (c) $\mathbf{e} \cdot \mathbf{f}$ if $\mathbf{e} = [1, 5, -2, -1]^t$ and $\mathbf{f} = [-3, 2, 0, 6]^t$, (d) $\mathbf{g} \cdot \mathbf{h}$ if $\mathbf{g} = [1, 3, 4, -1]^t$ and $\mathbf{h} = [-4, -2, 1, -6]^t$.

Solution:

(a) $\mathbf{a} \cdot \mathbf{b} = 1(-3) + 4(2) = 5.$

Now finish the problem.

— — — — — — — — — — —

(b) $\mathbf{c \cdot d} = \mathbf{0 \cdot d} = 0(4) + 0(-2) + 0(7) = 0.$

This illustrates the fact that $\mathbf{0 \cdot v} = 0$ for any n-component column vector $\mathbf{v}$.

(c) $\mathbf{e \cdot f} = 1(-3) + 5(2) + (-2)(0) + (-1)(6) = 1.$

(d) $\mathbf{g \cdot h} = 1(-4) + 3(-2) + 4(1) + (-1)(-6) = 0.$

B. PROPERTIES OF THE
DOT PRODUCT AND LENGTH

4.   An equivalent expression for the dot product of two n-component column vectors is

$$\mathbf{a \cdot b} = \mathbf{a}^t \mathbf{b}$$

as we see by following the rule for matrix multiplication. Observe that $\mathbf{a}^t$ is $1 \times n$ and $\mathbf{b}$ is $n \times 1$, and thus $\mathbf{a}^t\mathbf{b}$ is a 1×1 matrix. Now, 1×1 matrices are equivalent to numbers (scalars). Illustrating for three-component column vectors, we have

$$\mathbf{a}^t\mathbf{b} = [a_1, a_2, a_3]\begin{bmatrix} b_1 \\ b_2 \\ b_3 \end{bmatrix} = a_1b_1 + a_2b_2 + a_3b_3 = \mathbf{a \cdot b}$$

The main arithmetic properties of the dot product are given in Theorem 4.2.

Theorem 4.2

If $\mathbf{u}$, $\mathbf{v}$, and $\mathbf{w}$ are column vectors in R^n and k is any scalar, then

(a) $\mathbf{u \cdot v} = \mathbf{v \cdot u}$ (commutative property).

(b) $(\mathbf{u + v}) \cdot \mathbf{w} = \mathbf{u \cdot w} + \mathbf{v \cdot w}$ (distributive property).

(c) $(k\mathbf{u}) \cdot \mathbf{v} = k(\mathbf{u \cdot v}).$

(d)   In general, $\mathbf{v \cdot v} \geqslant 0$. Also, $\mathbf{v \cdot v} = 0$ if and only if $\mathbf{v} = \mathbf{0}$.

Note the "if and only if" wording in part (d). For a review see frame 12 of Chapter 3.

Example: Demonstrate the validity of the "if and only if" part of part (d) of Theorem 4.2 for the $n = 3$ case.

Solution: If $\mathbf{v} = \mathbf{0}$, we have $\mathbf{v} = [0, 0, 0]^t$, and thus $\mathbf{v} \cdot \mathbf{v} = 0^2 + 0^2 + 0^2 = 0$, which proves the "if" part.

To prove the "only if" part, first we set $\mathbf{v} \cdot \mathbf{v} = 0$ with $\mathbf{v} = [v_1, v_2, v_3]^t$. (Our goal then is to show that $\mathbf{v} = \mathbf{0}$.) This leads to

$$(v_1)^2 + (v_2)^2 + (v_3)^2 = 0 \tag{1}$$

Now each v_i is a real number, and a real number squared cannot be negative. That is, each $(v_i)^2 \geq 0$.
Now finish the proof.

— — — — — — — — — — —

Thus, the only way that Eq. (1) can hold is if each $(v_i)^2$ term equals zero [if any $(v_i)^2$ term were positive, then the left side of Eq. (1) would be positive]. But $(v_i)^2 = 0$ implies $v_i = 0$, and thus $v_1 = v_2 = v_3 = 0$, and thus,

$$\mathbf{v} = [0, 0, 0]^t = \mathbf{0} \tag{2}$$

5. *Henceforth, we shall often refer to a "column vector" merely as a "vector."* Let us now consider the concept of "length" of a vector. Here we let what we know to be true about two and three dimensions to guide us to a general definition for n dimensions. In particular, refer to the two-dimensional diagram of Figure 4.5 in frame 2. By "length" of the column vector $\mathbf{a} = [a_1, a_2]^t$, for example, we mean the length from the origin to the arrowhead (or terminal point) of the geometrical representation of vector $\mathbf{a}$. Let us denote the length of the column vector $\mathbf{a}$ by the symbol $\|\mathbf{a}\|$. In typical mathematical jargon $\|\mathbf{a}\|$ is referred to as the *norm* of vector $\mathbf{a}$. The Pythagorean theorem of geometry leads to the equations:

$$\|\mathbf{a}\| = \sqrt{a_1^2 + a_2^2} \qquad \text{(two dimensions, i.e., for } R^2\text{)}$$

$$\|\mathbf{a}\| = \sqrt{a_1^2 + a_2^2 + a_3^2} \qquad \text{(three dimensions, i.e., for } R^3\text{)}$$

By distance between vectors $\mathbf{a}$ and $\mathbf{b}$, we mean the straight-line distance between the points (a_1, a_2) and (b_1, b_2), respectively, if we are working in two-dimensional space (i.e., in R^2). See Figure 4.5, for example. Again following the Pythagorean theorem, we see that the distance between points (a_1, a_2) and (b_1, b_2) is $\sqrt{(b_1 - a_1)^2 + (b_2 - a_2)^2}$. Observe that this is equal to $\|\mathbf{b} - \mathbf{a}\|$, since $\mathbf{b} - \mathbf{a} = [b_1 - a_1, b_2 - a_2]^t$. A similar discussion applies to R^3.
Let us now extend these ideas to Euclidean n-space (i.e., R^n).

Definition 4.3

The *length* or *magnitude* or *norm* or *Euclidean norm* of the vector $\mathbf{a} = [a_1, a_2, \ldots, a_n]^t$ in R^n is the nonnegative number defined by

$$\|\mathbf{a}\| = \sqrt{(a_1)^2 + (a_2)^2 + \cdots + (a_n)^2}$$

The *distance between two vectors* $\mathbf{a}$ *and* $\mathbf{b}$ (where $\mathbf{b} = [b_1, b_2, \ldots, b_n]^t$) is defined to be the length of the vector $\mathbf{b} - \mathbf{a}$. That is, the distance is given by

$$\|\mathbf{b} - \mathbf{a}\| = \sqrt{(b_1 - a_1)^2 + (b_2 - a_2)^2 + \cdots + (b_n - a_n)^2}$$

Notes: (a) We see that the length of $\mathbf{a}$ can be related to the dot product by

$$\|\mathbf{a}\|^2 = \mathbf{a} \cdot \mathbf{a} \quad \text{or} \quad \|\mathbf{a}\| = \sqrt{\mathbf{a} \cdot \mathbf{a}}$$

(b) The distance between vectors $\mathbf{a}$ and $\mathbf{b}$ is also equal to $\|\mathbf{a} - \mathbf{b}\|$, since $(a_i - b_i)^2 = (b_i - a_i)^2$. Also, the distance between any two vectors is a non-negative number.

Example: Consider the four-component vectors: $\mathbf{a} = [3, 5, -2, 0]^t$ and $\mathbf{b} = [6, -3, 5, 1]^t$. Compute $\|\mathbf{a}\|$, $\mathbf{b} \cdot \mathbf{b}$, $\|\mathbf{b}\|$, $\mathbf{a} \cdot \mathbf{b}$, and the distance between vectors $\mathbf{a}$ and $\mathbf{b}$ (i.e., $\|\mathbf{b} - \mathbf{a}\|$).

Solution:

$$\|\mathbf{a}\| = \sqrt{3^2 + 5^2 + (-2)^2 + 0^2} = \sqrt{9 + 25 + 4 + 0} = \sqrt{38}$$

$$\mathbf{b} \cdot \mathbf{b} = 6^2 + (-3)^2 + 5^2 + 1^2 = 71$$

Now finish the calculations.

— — — — — — — — — —

$$\|\mathbf{b}\| = \sqrt{6^2 + (-3)^2 + 5^2 + 1^2}$$
$$= \sqrt{71}. \quad \text{(Note that this equals } \sqrt{\mathbf{b} \cdot \mathbf{b}}.\text{)}$$

$$\mathbf{a} \cdot \mathbf{b} = 3(6) + 5(-3) + (-2)(5) + 0(1) = -7$$

$$\mathbf{b} - \mathbf{a} = [6-3, -3-5, 5-(-2), 1-0]^t = [3, -8, 7, 1]^t$$

Thus, the distance between $\mathbf{a}$ and $\mathbf{b}$ is given by

$$\|\mathbf{b} - \mathbf{a}\| = \sqrt{3^2 + (-8)^2 + 7^2 + 1^2} = \sqrt{123}$$

6. It is useful to discuss the concept of an angle between two vectors. In the following discussion we shall often refer to a geometrical representation of a column vector **a** as *being* the vector **a**.

For example, for the two vectors **a** and **b** represented in Figure 4.5, the angle between them is labeled as θ (theta). (We take the angle to be the *smaller* of the two possible angles that can be formed between the vectors; in Figure 4.5 this other possible angle would be 360 − θ, in degrees.) The limiting values for θ are thus 0 degrees and 180 degrees (or 0 and π radians; note that π = 3.1416 to five significant digits). That is,

$0° \leqslant θ \leqslant 180°,$ where ° means degrees.

An angle of 0 degrees occurs when the two vectors point in the same direction (as with the vectors **a** and 3**a** in Figure 4.6), that is, when one vector equals a positive multiple of the other, say **b** = k**a** where k > 0. An angle of 180° occurs when the vectors point in opposite directions (e.g., **a** and − 2**a** in Figure 4.6), that is, when one vector equals a negative multiple of the other, say **b** = k**a** where k < 0.

An important theorem, which states an equation relating lengths of vectors, dot product, and the angle between vectors in R^2 or R^3 is the following, which can be proved by using the law of cosines of trigonometry.

Theorem 4.3 (for R^2 or R^3)

If **a** and **b** are two nonzero vectors (i.e., **a** ≠ **0** and **b** ≠ **0**), then

$\mathbf{a} \cdot \mathbf{b} = \|\mathbf{a}\| \, \|\mathbf{b}\| \cos θ$

Here cos θ means the cosine of the angle θ between the vectors. Also,

$0° \leqslant θ \leqslant 180°.$

Note: If either **a** or **b** were equal to the zero vector, then the angle θ would be undefined. In fact, in such a case, if cos θ were solved for from Theorem 4.3, we would get cos θ = 0/0; the expression 0/0 is undefined.

A proof of Theorem 4.3 for R^2 is found on p. 124 of Kolman (1980).*

Example: Determine the angle between the vectors $\mathbf{a} = [4, 3]^t$ and $\mathbf{b} = [2, 6]^t$. The relative positions of **a** and **b**, and also the angle θ, are illustrated in Figure 4.5. of frame 2. *Hint:* Use Theorem 4.3.

— — — — — — — — — —

---

* Recall that our method of referring to references (see back of book for References section) is to list the name(s) of the author(s), followed by the date of publication in parentheses.

Solution:

$$\|\mathbf{a}\| = \sqrt{4^2 + 3^2} = \sqrt{25} = 5$$

$$\|\mathbf{b}\| = \sqrt{2^2 + 6^2} = \sqrt{40} = 6.324$$

$$\mathbf{a} \cdot \mathbf{b} = 4(2) + 3(6) = 26$$

$$\cos \theta = \frac{\mathbf{a} \cdot \mathbf{b}}{\|\mathbf{a}\| \|\mathbf{b}\|} = \frac{26}{5(6.324)} = 0.822$$

Using a trigonometry table or a calculator, we find that $\theta = 34.7°$ (or 0.605 radians).

7.   If $\cos \theta$ were negative, that would mean that θ would lie between $90°$ and $180°$ (but not including $90°$, since $\cos 90° = 0$). Recall that $\cos 180° = -1$. Note that the limiting cases for θ can be characterized in the following ways for two vectors $\mathbf{a}$ and $\mathbf{b}$ (where $\mathbf{a} \neq \mathbf{0}$ and $\mathbf{b} \neq \mathbf{0}$).

θ	$\cos \theta$	Relationship Between $\mathbf{a}$ and $\mathbf{b}$
$0°$	$+1$	$\mathbf{a} = k\mathbf{b}$ where $k > 0$ ($\mathbf{a}$ and $\mathbf{b}$ have same direction)
$180°$	-1	$\mathbf{a} = k\mathbf{b}$ where $k < 0$ ($\mathbf{a}$ and $\mathbf{b}$ have opposite directions)

Thus, both limiting cases hold when $\mathbf{a}$ is a multiple of $\mathbf{b}$.

An important case of relative directions of two vectors occurs when they are mutually perpendicular. That is, the angle between them (actually between the directed line segments corresponding to them) is $90°$ or $\pi/2 = 1.571$ radians. Since $\cos 90° = 0$, we see from Theorem 4.3 that we have the following way of characterizing perpendicularity.

Theorem 4.4

Suppose, in the following, that $\mathbf{a} \neq \mathbf{0}$ and $\mathbf{b} \neq \mathbf{0}$. Two vectors $\mathbf{a}$ and $\mathbf{b}$ are perpendicular if and only if $\mathbf{a} \cdot \mathbf{b} = 0$.

Note (i):   Recalling the comments about the "if and only if" wording (frame 12 of Chapter 3), we see that Theorem 4.4 is actually the following two theorems:

Theorem 4.4a

If $\mathbf{a} \cdot \mathbf{b} = 0$, then $\mathbf{a}$ and $\mathbf{b}$ are perpendicular.

Theorem 4.4b

If $\mathbf{a}$ and $\mathbf{b}$ are perpendicular, then $\mathbf{a} \cdot \mathbf{b} = 0$.

Note (ii):   Another name for perpendicular is orthogonal.

Example: Determine the angle between the two-component vectors $\mathbf{a} = \begin{bmatrix} 2 \\ 1 \end{bmatrix}$ and $\mathbf{b} = \begin{bmatrix} -2 \\ 4 \end{bmatrix}$. Draw a sketch.

— — — — — — — — —

Solution: Here, we have $\mathbf{a} \cdot \mathbf{b} = 2(-2) + 1(4) = 0$. Thus, $\mathbf{a}$ and $\mathbf{b}$ are mutually perpendicular ($\theta = 90°$). An appropriate sketch is Figure 4.7.

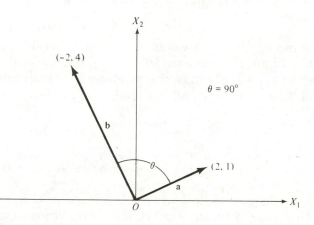

FIGURE 4.7   Perpendicular vectors: $\mathbf{a} \cdot \mathbf{b} = 0$. Here, $\mathbf{a} = [2, 1]'$ and $\mathbf{b} = [-2, 4]'$.

8.   The following very important theorem presents an inequality involving the lengths of two vectors and the dot product between the vectors.

Theorem 4.5 (Cauchy–Schwarz Inequality*)

   If $\mathbf{a}$ and $\mathbf{b}$ are both vectors in R^n, where $\mathbf{a} \neq \mathbf{0}$ and $\mathbf{b} \neq \mathbf{0}$, then

$$|\mathbf{a} \cdot \mathbf{b}| \leq \|\mathbf{a}\| \|\mathbf{b}\|$$

---

* Augustin-Louis Cauchy (1789–1857), one of the greatest mathematicians, made significant contributions in calculus, complex variable theory, differential equations, and the theory of determinants.
   Herman A. Schwarz (1843–1921) made contributions in differential equations and in geometrical aspects of calculus.

Notes: (i) To say a vector **a** is in R^n means that when **a** is expressed as a column vector, **a** has n real components (see Definition 4.1 in frame 3). Also, recall that R^n is the symbol for Euclidean n-space. (ii) Proofs of this theorem appear in Kolman (1980; p. 142), and Bloch and Michaels (1977; p. 281), where the proof in the latter book has a nice geometrical motivation. (iii) The | | on the left side of the inequality stands for the absolute value of a real number; ‖ ‖ on the right side stands for the length of a vector. (iv) The inequality in Theorem 4.5 holds as a pure equality if either **a** or **b** is equal to **0**. For example, if **a** = **0**, then **a·b** = 0 (left side) and ‖**a**‖ = 0, and thus ‖**a**‖‖**b**‖ = 0 (right side).

The $\leq$ inequality in Theorem 4.5 is actually a strict inequality (i.e., $\leq$ becomes $<$) for all possible pairs of vectors **a** and **b**, where **a** $\neq$ **0** and **b** $\neq$ **0**, except for the case where one vector is a multiple of the other. In the latter case we have the following corollary.

Corollary to Theorem 4.5

In the following, **a** and **b** are both nonzero vectors in R^n.

|**a·b**| = ‖**a**‖‖**b**‖   if and only if **a** = k**b**, for some nonzero scalar k.

Theorem 4.5 allows us to define an angle θ between nonzero vectors in R^n, regardless of what n is. In fact, our definition for θ (actually for $\cos\theta$) will transform the inequality in Theorem 4.5 into the equation that appears in Theorem 4.3. To show this, first observe that an inequality of the form $|u| \leq h$, where u is a real number and h is some positive number, is equivalent to

$$-h \leq u \leq h \tag{1}$$

For example, $|u| \leq 5$ is equivalent to $-5 \leq u \leq 5$. Thus, from Theorem 4.5, if we regard u as **a·b** and h as ‖**a**‖‖**b**‖, we have

$$-‖\mathbf{a}‖‖\mathbf{b}‖ \leq \mathbf{a·b} \leq ‖\mathbf{a}‖‖\mathbf{b}‖ \tag{2}$$

Now if we divide all three parts in (2) by ‖**a**‖‖**b**‖, we obtain

$$-1 \leq \frac{\mathbf{a·b}}{‖\mathbf{a}‖‖\mathbf{b}‖} \leq 1 \tag{3}$$

An inequality like that in (3) holds for $\cos\theta$. That is, we know from trigonometry that, for any angle θ,

$$-1 \leq \cos\theta \leq 1 \tag{4}$$

Thus, from (3) and (4), we make the following definition.

Definition 4.4

Given any nonzero vectors **a** and **b** in R^n. The quantity $\cos \theta$, where θ is the *angle between* the vectors **a** and **b**, is defined by

$$\cos \theta = \frac{\mathbf{a} \cdot \mathbf{b}}{\|\mathbf{a}\| \|\mathbf{b}\|}$$

where $0° \le \theta \le 180°$.

Notes: (i)   If we multiply the equation of Definition 4.4 by $\|\mathbf{a}\| \|\mathbf{b}\|$, we obtain the equation that appears in Theorem 4.3 (frame 6). (ii)   The decision to confine θ to be between 0° and 180° inclusive is based on a decision to make the result in Definition 4.4 compatible with the result of Theorem 4.3. Also, observe that $\cos \theta$ assumes every value between $+1$ and -1, inclusive, as θ varies between 0° and 180°, inclusive. (iii)   Our prior results from frames 6 and 7 now apply also for R^n, regardless of n. For example, the prior results for the special cases $\theta = 0°$, 90°, and 180° apply also for R^n (see frame 7), and not just for R^2 and R^3.

Example: Verify Theorem 4.5 and calculate $\cos \theta$ and then θ for the following pairs of vectors: (a) $\mathbf{a} = [1, 5, -2, 2]^t$ and $\mathbf{b} = [2, 1, 2, 3]^t$; (b) $\mathbf{c} = [2, 6, -1, 3]^t$ and $\mathbf{d} = [-1, 2, 8, -2]^t$; (c) $\mathbf{e} = [2, -4, 3, 5, 0]^t$ and $\mathbf{f} = [-4, 8, -6, -10, 0]^t$; (d) $\mathbf{g} = [3, 2, -1, -5]^t$ and $\mathbf{h} = [2, 1, -2, 2]^t$.

Solution:
   (a) $\mathbf{a} \cdot \mathbf{b} = 2 + 5 - 4 + 6 = 9$ and $|\mathbf{a} \cdot \mathbf{b}| = 9$. $\|\mathbf{a}\| = \sqrt{1 + 25 + 4 + 4} = \sqrt{34}$, $\|\mathbf{b}\| = \sqrt{4 + 1 + 4 + 9} = \sqrt{18}$. Thus, $\|\mathbf{a}\| \|\mathbf{b}\| = \sqrt{34(18)} = \sqrt{612} = 24.74$. Thus, Theorem 4.5 holds as a strict inequality.

$$\cos \theta = \frac{\mathbf{a} \cdot \mathbf{b}}{\|\mathbf{a}\| \|\mathbf{b}\|} = \frac{9}{24.74} = 0.3638,$$

and $\theta = 68.67° = 1.198$ radians.
   Now, do parts (b), (c), and (d).

— — — — — — — — — —

   (b) $\mathbf{c} \cdot \mathbf{d} = -2 + 12 - 8 - 6 = -4$, and $|\mathbf{c} \cdot \mathbf{d}| = 4$. $\|\mathbf{c}\| = \sqrt{4 + 36 + 1 + 9} = \sqrt{50}$, $\|\mathbf{d}\| = \sqrt{1 + 4 + 64 + 4} = \sqrt{73}$. Thus, $\|\mathbf{c}\| \|\mathbf{d}\| = \sqrt{50(73)} = \sqrt{3650} = 60.415$, and Theorem 4.5 holds as a strict inequality.

$$\cos \theta = \frac{\mathbf{c} \cdot \mathbf{d}}{\|\mathbf{c}\| \|\mathbf{d}\|} = \frac{-4}{60.415} = -0.06621, \qquad \text{and}$$

   $\theta = 93.80° = 1.637$ radians.

(c) $\mathbf{e \cdot f} = -8 - 32 - 18 - 50 = -108$ and $|\mathbf{e \cdot f}| = 108$. $\|\mathbf{e}\| = \sqrt{4 + 16 + 9 + 25 + 0} = \sqrt{54}$, $\|\mathbf{f}\| = \sqrt{16 + 64 + 36 + 100} = \sqrt{216}$. Thus, $\|\mathbf{e}\| \|\mathbf{f}\| = \sqrt{54(216)} = \sqrt{11664} = 108$, and Theorem 4.5 holds as an equation.

$$\cos \theta = \frac{\mathbf{e \cdot f}}{\|\mathbf{e}\| \|\mathbf{f}\|} = \frac{-108}{108} = -1 \qquad \text{and}$$

$\theta = 180° = \pi$ radians.

Also, we observe that $\mathbf{f} = -2\mathbf{e}$ and, hence, the Corollary to Theorem 4.5 applies. Here $\theta = 180°$ since $\mathbf{f} = k\mathbf{e}$ with $k < 0$ (frame 7, limiting case).

(d) $\mathbf{g \cdot h} = 6 + 2 + 2 - 10 = 0$, and thus Theorem 4.5 holds as a strict inequality since $\|\mathbf{g}\| \|\mathbf{h}\|$ is positive. Also, $\cos \theta = 0$, thus indicating $\theta = 90° = \pi/2$ radians. The vectors $\mathbf{g}$ and $\mathbf{h}$ are perpendicular.

9. It is useful to summarize some of the key properties of the length (Euclidean norm); refer to Definition 4.3 of frame 5.

Theorem 4.6 (Properties of Length)

(a) For any vector $\mathbf{v}$ in R^n, $\|\mathbf{v}\| \geq 0$. The equality case, $\|\mathbf{v}\| = 0$, holds if and only if $\mathbf{v} = \mathbf{0}$.

(b)
$$\|k\mathbf{v}\| = \begin{cases} k\|\mathbf{v}\| & \text{if } k \geq 0 \\ -k\|\mathbf{v}\| & \text{if } k < 0 \end{cases}$$
In short, $\|k\mathbf{v}\| = |k| \|\mathbf{v}\|$

Part (a) follows from part (d) of Theorem 4.2 (frame 4) since $\mathbf{v \cdot v} = \|\mathbf{v}\|^2$, and $\|\mathbf{v}\|^2 \geq 0$ implies $\|\mathbf{v}\| \geq 0$. Similarly, $\mathbf{v \cdot v} = 0$ means $\|\mathbf{v}\|^2 = 0$, which means $\|\mathbf{v}\| = 0$.

Example (a): Illustrate part (a) of Theorem 4.6 for the vector $\mathbf{v} = \begin{bmatrix} -4 \\ 3 \end{bmatrix}$.

Illustrate part (b) of Theorem 4.6 with respect to $\mathbf{v}$ if $k = -2$, say.

Solution: (a) Here $\|\mathbf{v}\| = \sqrt{\mathbf{v \cdot v}} = \sqrt{(-4)^2 + 3^2} = \sqrt{25} = 5$, and 5 is greater than zero. (b) If $k = -2$, then $k\mathbf{v} = \begin{bmatrix} 8 \\ -6 \end{bmatrix}$, and thus, $\|k\mathbf{v}\| = \sqrt{8^2 + (-6)^2} = \sqrt{100} = 10$. Thus, we see that $\|k\mathbf{v}\| = 2\|\mathbf{v}\|$ here. This verifies part (b) since $-k = |k| = 2$, here.

The following "triangle inequality" theorem is a consequence of the Cauchy–Schwarz inequality. A proof is found on p. 145 of Kolman (1980).

Theorem 4.7 (Triangle Inequality)

If **a** and **b** are both vectors in R^n, where $\mathbf{a} \neq \mathbf{0}$ and $\mathbf{b} \neq \mathbf{0}$, then

$$\|\mathbf{a} + \mathbf{b}\| \leq \|\mathbf{a}\| + \|\mathbf{b}\|$$

Actually, the only case for which $\leq$ becomes $=$, that is, for which $\|\mathbf{a} + \mathbf{b}\| = \|\mathbf{a}\| + \|\mathbf{b}\|$, is when $\mathbf{a} = k\mathbf{b}$, where k is positive. In all other cases (including $\mathbf{a} = k\mathbf{b}$ for $k < 0$), $\|\mathbf{a} + \mathbf{b}\| < \|\mathbf{a}\| + \|\mathbf{b}\|$.

Note: For the case where either **a** or **b** is equal to the zero vector (**0**), the expressions in both Theorems 4.5 (frame 8) and 4.7 would hold as equalities (recall that $\|\mathbf{0}\| = 0$). Such a case was excluded in the hypotheses of these theorems.

The triangle inequality in R^2 or R^3 confirms the fact that the sum of the lengths of two sides of a triangle is greater than the length of the third side. See Figure 4.8 for a typical case in R^2. Note that **b** is not a multiple of **a** (i.e., $\mathbf{b} \neq k\mathbf{a}$), and, hence, $\|\mathbf{a} + \mathbf{b}\| < \|\mathbf{a}\| + \|\mathbf{b}\|$. Also, the triangle shown is formed by **a**, the translated vector $\mathbf{b}_{tr}$, and $\mathbf{a} + \mathbf{b}$. Observe that the length of $\mathbf{b}_{tr}$ is equal to $\|\mathbf{b}\|$, which, of course, is the length of **b**.

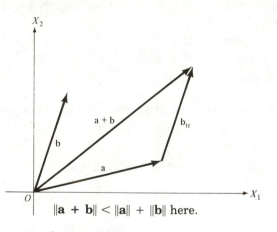

$$\|\mathbf{a} + \mathbf{b}\| < \|\mathbf{a}\| + \|\mathbf{b}\| \text{ here.}$$

FIGURE 4.8   Demonstrating the triangle inequality.

Example (b): Verify Theorem 4.7 for the following pairs of vectors: (i) $\mathbf{a} = \begin{bmatrix} 4 \\ 1 \end{bmatrix}$ and $\mathbf{b} = \begin{bmatrix} 1 \\ 3 \end{bmatrix}$, (ii) $\mathbf{c} = \begin{bmatrix} 3 \\ 4 \end{bmatrix}$ and $\mathbf{d} = \begin{bmatrix} 9 \\ 12 \end{bmatrix}$, and (iii) $\mathbf{v} = \begin{bmatrix} 8 \\ -6 \end{bmatrix}$ and $\mathbf{w} = \begin{bmatrix} -4 \\ 3 \end{bmatrix}$.

Solution: (i) $\|\mathbf{a}\| = \sqrt{16 + 1} = \sqrt{17} = 4.123$, $\|\mathbf{b}\| = \sqrt{1 + 9} = \sqrt{10} = 3.162$, and thus, $\|\mathbf{a}\| + \|\mathbf{b}\| = 7.285$. Now $\mathbf{a} + \mathbf{b} = \begin{bmatrix} 5 \\ 4 \end{bmatrix}$ and $\|\mathbf{a} + \mathbf{b}\| = \sqrt{25 + 16} = \sqrt{41} = 6.403$, thus verifying Theorem 4.7 as a strict inequality. The diagram in Figure 4.8 is consistent with the numerical data here. (ii) $\|\mathbf{c}\| = \sqrt{9 + 16} = \sqrt{25} = 5$, $\|\mathbf{d}\| = \sqrt{225} = 15$, $\mathbf{c} + \mathbf{d} = [12, 16]^t$, and $\|\mathbf{c} + \mathbf{d}\| = \sqrt{400} = 20$. Thus, Theorem 4.7 is verified as an equality, which is no surprise, since $\mathbf{d} = 3\mathbf{c}$. Now do part (iii).

— — — — — — — — — —

(iii) $\|\mathbf{v}\| = \sqrt{100} = 10$, $\|\mathbf{w}\| = \sqrt{25} = 5$, $\mathbf{v} + \mathbf{w} = [4, -3]^t$, and $\|\mathbf{v} + \mathbf{w}\| = \sqrt{25} = 5$. Thus, Theorem 4.7 is verified as a strict inequality ($5 < 15$). Observe that $\mathbf{v} = -2\mathbf{w}$ here.

10. A *unit vector* $\mathbf{u}$ in R^n is a vector of unit length; that is, $\|\mathbf{u}\| = 1$. For example, the vector $\mathbf{b} = [1/\sqrt{5}, 2/\sqrt{5}]^t$ is a unit vector, since $\|\mathbf{b}\| = \sqrt{(1/\sqrt{5})^2 + (2/\sqrt{5})^2} = \sqrt{\frac{1}{5} + \frac{4}{5}} = \sqrt{1} = 1$. Given any nonzero vector $\mathbf{v}$, the unit vector having the *same direction* as $\mathbf{v}$, which is called $\mathbf{u}_v$, is defined by the equation

$$\mathbf{u}_v = \frac{\mathbf{v}}{\|\mathbf{v}\|} = \left(\frac{1}{\|\mathbf{v}\|} \right) \mathbf{v} \tag{I}$$

Refer to Figure 4.9 for a sketch in R^2 illustrating a typical pair, $\mathbf{v}$ and $\mathbf{u}_v$. Clearly, $\mathbf{v}$ has the same direction as $\mathbf{u}_v$; the multiple k, in the equation $\mathbf{v} = k\mathbf{u}_v$, is the positive number $\|\mathbf{v}\|$.

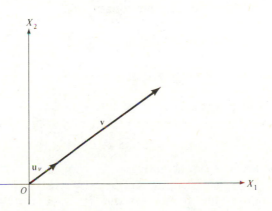

FIGURE 4.9   The unit vector $\mathbf{u}_v$ has the same direction as the vector $\mathbf{v}$, but has a length (magnitude) of one.

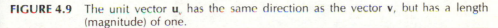

Example: Given that $\mathbf{v} = [4, 3]^t$, determine the unit vector $\mathbf{u}_v$. Show both $\mathbf{v}$ and $\mathbf{u}_v$ on the same diagram.

– – – – – – – – – –

Solution: Since $\|\mathbf{v}\| = \sqrt{16 + 9} = \sqrt{25} = 5$,

$$\mathbf{u}_v = \left(\frac{1}{5}\right)\begin{bmatrix} 4 \\ 3 \end{bmatrix} = \begin{bmatrix} \frac{4}{5} \\ \frac{3}{5} \end{bmatrix}$$

Checking, we see that $\|\mathbf{u}_v\| = \sqrt{\frac{16}{25} + \frac{9}{25}} = \sqrt{\frac{25}{25}} = 1$. The diagram in Figure 4.9 is consistent with the numerical data here.

C. HYPERPLANES AND STRAIGHT LINES IN R^n

11.   In the following discussion we shall continue our practice of (sometimes) referring to the geometrical representation of a column vector as *being* the vector.

It is important to recall that we have assigned a dual geometrical interpretation to the algebraic entity known as a column vector. Sometimes, we interpret the n-component column vector $[b_1, b_2, \ldots, b_n]^t$ as locating the point $(b_1, b_2, \ldots, b_n)$ in n-dimensional space. This point is also the terminal point in the geometrical representation of the vector. Also, we interpret the column vector as a *direction* that goes from the origin to the terminal point $(b_1, b_2, \ldots, b_n)$. An arrowhead located at the terminal point reinforces the idea of direction.

Both interpretations are relevant to our upcoming discussion of hyperplanes, lines, and line segments.

Consider the following linear equation in n variables $x_1, x_2, \ldots, x_n$, where $c_1, c_2, \ldots, c_n$, and b are real number constants.

$$c_1 x_1 + c_2 x_2 + \ldots + c_n x_n = b \tag{1}$$

If we define the column vectors $\mathbf{c}$ and $\mathbf{x}$ from

$$\mathbf{c} = [c_1, c_2, \ldots, c_n]^t \qquad \text{and}$$

$$\mathbf{x} = [x_1, x_2, \ldots, x_n]^t,$$

then equivalent forms of Eq. (1) are

$$\mathbf{c} \cdot \mathbf{x} = b \qquad \text{(dot product form)} \qquad \text{and} \tag{1a}$$

$$\mathbf{c}^t \mathbf{x} = b \qquad \text{(matrix product form)} \tag{1b}$$

By the way, we assume that $\mathbf{c} \neq \mathbf{0}$; that is, at least one of the c_i's in vector $\mathbf{c}$ is unequal to zero.

Recall that we denote a point in R^n by $(x_1, x_2, \ldots, x_n)$. The column vector that corresponds to this point is $\mathbf{x} = [x_1, x_2, \ldots, x_n]^t$, and vice versa.

Definition 4.5 (Hyperplane)

The set of all points $(x_1, x_2, \ldots, x_n)$ that satisfy Eq. (1) above is called a hyperplane in R^n. We will sometimes refer to Eq. (1) as the *scalar equation* for a hyperplane.

Alternately, the set of all points in R^n such that their *corresponding column vectors* satisfy Eq. (1a) or Eq. (1b) is called a hyperplane in R^n. The vector $\mathbf{c}$ is a constant vector since its components are constants.

In two- and three-dimensional spaces we have the following equations for hyperplanes:

$$c_1 x_1 + c_2 x_2 = b \qquad \text{(2-space)}$$
$$c_1 x_1 + c_2 x_2 + c_3 x_3 = b \qquad \text{(3-space)}$$

Note: We can define a hyperplane in terms of column vectors as follows. A hyperplane in R^n is the set of column vectors $\mathbf{x} = [x_1, x_2, \ldots, x_n]^t$ for which Eq. (1) or (1a) or (1b) is satisfied.

In R^2 (i.e., 2-space) we see that hyperplanes are straight lines, and in R^3 the hyperplanes are the usual planes. We see from these examples that the dimension of the hyperplane is 1 less then the dimension of the containing space. For example, the line $c_1 x_1 + c_2 x_2 = b$ contained in 2-space has dimension 1. In general, the dimension of a hyperplane lying in n-dimensional space is $(n-1)$, that is, 1 less than the dimension of the space. The proof that this last statement is true is difficult relative to the level of this book.

Example (a): Find the equation of the hyperplane in 2-space (R^2) for which $\mathbf{c} = \begin{bmatrix} 2 \\ 1 \end{bmatrix}$ and $b = 20$. Draw a sketch.

Solution: Since the constant vector $\mathbf{c} = [2, 1]^t$ and $\mathbf{x} = [x_1, x_2]^t$, we have $\mathbf{c} \cdot \mathbf{x} = c_1 x_1 + c_2 x_2$, and the equation for the hyperplane [see Eq. (1a) above] is

$$2x_1 + x_2 = 20$$

This is the equation for a straight line. See Figure 4.10. Let us locate the intercept points and the column vectors that correspond to them:

When $x_2 = 0$, $x_1 = 10$, and when $x_1 = 0$, $x_2 = 20$

Labeling the x_1 intercept point (i.e., the intercept point on the X_1 axis) as Q and the x_2 intercept point as P, the coordinates for P and Q are indicated as follows:

$$P(0, 20) \quad \text{and} \quad Q(10, 0)$$

The corresponding column vectors are $\mathbf{p} = [0, 20]^t$ and $\mathbf{q} = [10, 0]^t$. In Figure 4.10 we locate the intercept points with the geometrical vectors $\mathbf{p}$ and $\mathbf{q}$. The reason for this will be indicated in the discussion that follows.

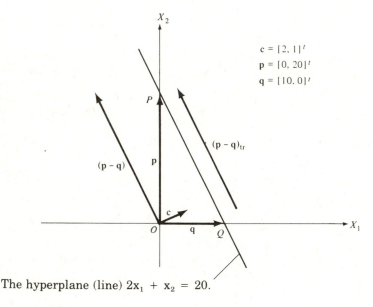

The hyperplane (line) $2x_1 + x_2 = 20$.

FIGURE 4.10   A hyperplane in 2-dimensional space.

We now introduce the useful concept of a *direction vector* from one vector to another vector. The direction vector from vector $\mathbf{a}$ to vector $\mathbf{b}$ is defined to be the vector $\mathbf{b} - \mathbf{a}$. Refer to Figure 4.5 (frame 2) for a diagram illustrating $\mathbf{a}$, $\mathbf{b}$, and $\mathbf{b} - \mathbf{a}$ for a typical case in R^2. We also note that the direction vector from $\mathbf{b}$ to $\mathbf{a}$ is $\mathbf{a} - \mathbf{b}$; this vector, of course, has a direction opposite that of $\mathbf{b} - \mathbf{a}$, since

$$\mathbf{a} - \mathbf{b} = -(\mathbf{b} - \mathbf{a}); \quad \text{that is, } \mathbf{a} - \mathbf{b} = (-1)(\mathbf{b} - \mathbf{a})$$

Note: In Definition 4.3 of frame 5, we spoke of the distance between two vectors $\mathbf{b}$ and $\mathbf{a}$. This was equal to $\|\mathbf{b} - \mathbf{a}\|$ or $\|\mathbf{a} - \mathbf{b}\|$.

Theorem 4.8

Suppose that **c** is a constant vector for a hyperplane with equation $\mathbf{c} \cdot \mathbf{x} = b$, and that **p** and **q** are two distinct vectors corresponding, respectively, to any two distinct points P and Q that lie on the hyperplane. Then **c** is perpendicular to $(\mathbf{p} - \mathbf{q})$, the direction vector from **q** to **p**.

Note: Because of Theorem 4.8 (to be proved shortly), we say that **c** is perpendicular to the hyperplane with equation $\mathbf{c} \cdot \mathbf{x} = b$. In fact, the geometrical representation of **c** is perpendicular to the line $c_1 x_1 + c_2 x_2 = b$ to which the hyperplane reduces in two dimensions, and to the plane $c_1 x_1 + c_2 x_2 + c_3 x_3 = b$ to which the hyperplane reduces in three dimensions.

The constant vector **c** (or any multiple s**c**, where $s \neq 0$) is called a *normal vector* for the hyperplane $\mathbf{c} \cdot \mathbf{x} = b$. Normal is still another word that means perpendicular.

Refer to Figure 4.10 for an illustration of the two-dimensional case. Observe that the geometrical representation of the vector **c** appears to be perpendicular to the straight line (hyperplane) $2x_1 + x_2 = 20$.

Example (b): Refer to Example (a). Show that **c** is perpendicular to $(\mathbf{p} - \mathbf{q})$, where **p** and **q** are the specific vectors cited in Example (a). *Hint:* Recall from Theorem 4.4 (frame 7) how we characterize perpendicularity in terms of dot product.

— — — — — — — — — —

Solution: Here, $\mathbf{p} - \mathbf{q} = \begin{bmatrix} 0 \\ 20 \end{bmatrix} - \begin{bmatrix} 10 \\ 0 \end{bmatrix} = \begin{bmatrix} -10 \\ 20 \end{bmatrix}$, and $\mathbf{c} = \begin{bmatrix} 2 \\ 1 \end{bmatrix}$. Thus,

$$\mathbf{c} \cdot (\mathbf{p} - \mathbf{q}) = 2(-10) + 1(20) = 0$$

Since the dot product of **c** and $(\mathbf{p} - \mathbf{q})$ equals zero, this means that **c** and $(\mathbf{p} - \mathbf{q})$ are perpendicular. Observe, in Figure 4.10, that the geometrical representation of **c** appears to be perpendicular to the geometrical representation of $(\mathbf{p} - \mathbf{q})$.

12. Another way of showing that **c** and $(\mathbf{p} - \mathbf{q})$ are perpendicular in 2-space is by computing the slopes of the corresponding directed line segments (i.e., the geometrical representations of the vectors) and showing that the product equals -1. Let us use the letter m for slope. Recalling that the slope of a line segment is equal to the change in x_2 divided by the corresponding change in x_1, we have

$$m_{\mathbf{c}} = \frac{1 - 0}{2 - 0} = \frac{1}{2} \qquad \text{and} \qquad m_{(\mathbf{p}-\mathbf{q})} = \frac{20 - 0}{-10 - 0} = -2$$

Thus, $m_c \cdot m_{(p-q)} = \frac{1}{2}(-2) = -1$, as expected.

Example (a):   Find an equation of the hyperplane in 3-space (R^3) for which $\mathbf{c} = [2, 2, 3]^t$ and $b = 60$. Draw a sketch of the portion of the hyperplane in the first octant.

Solution: Referring to the general equation of a hyperplane in 3-space (Definition 4.5 in frame 11), we have $c_1x_1 + c_2x_2 + c_3x_3 = b$. Thus, here we have

$$2x_1 + 2x_2 + 3x_3 = 60 \tag{1}$$

Equation (1) is an equation of an *actual plane* in 3-space. The intercept points are now calculated from Eq. (1):

When $x_1 = x_2 = 0$, $x_3 = 20$ (point K).
When $x_1 = x_3 = 0$, $x_2 = 30$ (point L).
When $x_2 = x_3 = 0$, $x_1 = 20$ (point M).

The sketch for the first octant is shown in Figure 4.11. The normal vector $\mathbf{c} = [2, 2, 3]^t$, which is perpendicular to the hyperplane, is not shown in the figure. Again we stress that the hyperplane is an actual plane here.

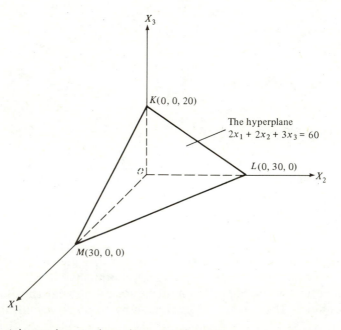

FIGURE 4.11   A hyperplane in three dimensional space.

Example (b):   Determine an equation of the (hyper)plane that is parallel to the (hyper)plane of Example (a) if the origin $(0, 0, 0)$ lies on the new plane. *Hint:* Since the two planes are parallel, an acceptable normal vector for the new plane is $\mathbf{c}' = \mathbf{c} = [2, 2, 3]^t$. Actually, $\mathbf{c}' = k\mathbf{c}$, for any $k \neq 0$, will serve just as well as a normal vector for the new plane.

— — — — — — — — —

Solution: A tentative equation for the new plane is

$$2x_1 + 2x_2 + 3x_3 = b \tag{2}$$

Since $(0, 0, 0)$ lies on this plane, we have

$$2(0) + 2(0) + 3(0) = 0 = b \tag{3}$$

That is, $b = 0$, and our equation (no longer tentative) becomes

$$2x_1 + 2x_2 + 3x_3 = 0 \tag{4}$$

In general, if a hyperplane has a normal vector $\mathbf{c} = [c_1, c_2, \ldots, c_n]^t$, and it passes through the origin, then an equation for the hyperplane is

$$c_1 x_1 + c_2 x_2 + \ldots + c_n x_n = 0 \tag{5}$$

In 2-space, this reduces to

$$c_1 x_1 + c_2 x_2 = 0, \tag{6}$$

an equation for a straight line that passes through the origin.

13.   It is instructive to prove Theorem 4.8 (frame 11). Refer to Figure 4.12 for the appropriate diagram in 2-space.

Proof of Theorem 4.8

Given a hyperplane with equation

$$\mathbf{c} \cdot \mathbf{x} = b$$

Let $\mathbf{p}$ and $\mathbf{q}$ be two column vectors whose corresponding points P and Q lie on the hyperplane (e.g., P is the terminal point on the geometrical representation of $\mathbf{p}$). The direction vector from $\mathbf{q}$ to $\mathbf{p}$ is $\mathbf{p} - \mathbf{q}$; refer to Figure 4.12. What we have to show is that $\mathbf{c}$ is perpendicular to $\mathbf{p} - \mathbf{q}$. Now

$$\mathbf{c} \cdot (\mathbf{p} - \mathbf{q}) = \mathbf{c} \cdot \mathbf{p} - \mathbf{c} \cdot \mathbf{q}, \tag{1}$$

as can be seen from Theorem 4.2 in frame 4. Now, since P and Q are both points on the hyperplane,

$$\mathbf{c} \cdot \mathbf{p} = b \qquad \text{and} \qquad \mathbf{c} \cdot \mathbf{q} = b \qquad\qquad (2), \quad (3)$$

Thus, Eq. (1) becomes

$$\mathbf{c} \cdot (\mathbf{p} - \mathbf{q}) = b - b = 0$$

which indicates that $\mathbf{c}$ is perpendicular to $(\mathbf{p} - \mathbf{q})$.

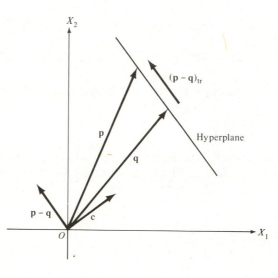

FIGURE 4.12   Diagram for proof of Theorem 4.8.

We shall now turn to the mathematical definition of a straight line through two distinct points in space. We let the situation in two- and three-dimensional space guide us. Consider the 2-space diagram of Figure 4.13. We wish to find an equation (involving vectors) for the straight line that goes through points P and Q. Let $\mathbf{p}$ and $\mathbf{q}$ be the two column vectors corresponding to the points P and Q. (The points P and Q are the terminal points of the geometrical representations of $\mathbf{p}$ and $\mathbf{q}$, respectively.) The vector $(\mathbf{p} - \mathbf{q})$, which determines *direction* for the line, is shown in Figure 4.13. Let $\mathbf{x} = [x_1, x_2]^t$ denote the column vector that corresponds to a typical point (x_1, x_2), which lies on the line through P and Q. A geometrical representation of $\mathbf{x}$, for the case where the typical point lies *strictly between* P and Q, is shown in Figure 4.13. Using the triangle rule for vector addition (frame 2), we see that $\mathbf{x}$ equals the vector sum of $\mathbf{q}$ plus a vector whose corresponding *translated* vector extends from the terminal point of $\mathbf{q}$ to the terminal point of $\mathbf{x}$. Thus, the vector added to $\mathbf{q}$ has the same direction as $(\mathbf{p} - \mathbf{q})$, but has a smaller length. We express this vector as the scalar multiple t times $(\mathbf{p} - \mathbf{q})$. For Figure 4.13 t is a real number strictly between 0 and 1 (i.e., $0 < t < 1$). Thus, from the triangle rule for vector addition, we have

$$\mathbf{x} = \mathbf{q} + t(\mathbf{p} - \mathbf{q}) \tag{I}$$

We can also write this as

$$\mathbf{x} = t\mathbf{p} + (1 - t)\mathbf{q} \tag{Ia}$$

The vector $\mathbf{x}$ corresponding to the varying point on the line reduces to $\mathbf{q}$ when $t = 0$ and to $\mathbf{p}$ when $t = 1$. Thus, the point Q is located when $t = 0$, and point P is located when $t = 1$. For a point strictly between P and Q, we have $0 < t < 1$.

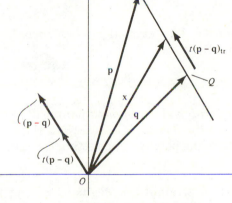

$$x = q + t(p - q) = tp + (1 - t)q$$

FIGURE 4.13 A straight line in 2-space through two points.

In Figure 4.13, for a point on the straight line below and to the right of Q, we have $t < 0$, and for a point on the straight line above and to the left of P, we have $t > 1$. The preceding discussion is also valid for 3-space. We use the preceding ideas to define the straight line (in terms of vectors) in n-dimensional space that passes through two points P and Q. For the following, suppose the distinct points P and Q are defined as follows:

P has coordinates $(p_1, p_2, \ldots, p_n)$.
Q has coordinates $(q_1, q_2, \ldots, q_n)$.

The corresponding column vectors are $\mathbf{p} = [p_1, p_2, \ldots, p_n]^t$ and $\mathbf{q} = [q_1, q_2, \ldots, q_n]^t$.

Also, $\mathbf{x} = [x_1, x_2, \ldots, x_n]^t$ denotes the column vector that corresponds to a typical point X with coordinates $(x_1, x_2, \ldots, x_n)$, which lies on the line through P and Q.

Definition 4.6 (Straight Line in R^n)

   Given two distinct points P and Q in n-dimensional space, to which correspond the column vectors $\mathbf{p}$ and $\mathbf{q}$. The *straight line* through the points P and Q consists of all points X such that the corresponding column vector $\mathbf{x}$ satisfies the following vector equation:

$$\mathbf{x} = t\mathbf{p} + (1 - t)\mathbf{q}, \qquad \text{where } t \text{ is any real number.} \tag{Ia}$$

If t is restricted to lie between 0 and 1 inclusive, then the corresponding set of points that lie on the straight line is called the straight-line segment joining P and Q.

Definition 4.6a (Straight-Line Segment in R^n)

   Refer to Definition 4.6. The set of all points X such that the corresponding column vector $\mathbf{x}$ satisfies

$$\mathbf{x} = t\mathbf{p} + (1 - t)\mathbf{q}, \qquad \text{where } 0 \leqslant t \leqslant 1,$$

is defined to be the *straight-line segment* joining P and Q. We denote this straight-line segment by the symbol $\overline{PQ}$. Note that setting $t = 0$ causes $\mathbf{x} = \mathbf{q}$, thereby generating point Q, and setting $t = 1$ causes $\mathbf{x} = \mathbf{p}$, thereby generating point P.

Note: The concept of a straight-line segment joining two points in n-dimensional space is very important in linear programming and convex geometry.

Example: Given two points $P(4, 12)$ and $Q(10, 4)$ in 2-space to which correspond the column vectors $\mathbf{p} = [4, 12]^t$ and $\mathbf{q} = [10, 4]^t$. Determine points on the line for which $t = -\frac{1}{4}, 0, \frac{1}{4}, \frac{1}{2}, \frac{3}{4}, 1, \frac{5}{4}$. Sketch the line, indicating the points corresponding to the preceding values of t.

Solution: If X is a typical point on the straight line, then the column vector $\mathbf{x}$ corresponding to X satisfies the following equation:

$$\mathbf{x} = t\mathbf{p} + (1 - t)\mathbf{q} = t\begin{bmatrix} 4 \\ 12 \end{bmatrix} + (1 - t)\begin{bmatrix} 10 \\ 4 \end{bmatrix} = \begin{bmatrix} 10 - 6t \\ 4 + 8t \end{bmatrix} \tag{1}$$

   For $t = 0$, $\mathbf{x} = \mathbf{q} = [10, 4]^t$, thereby generating $Q(10, 4)$.
   For $t = 1$, $\mathbf{x} = \mathbf{p} = [4, 12]^t$, thereby generating $P(4, 12)$.

Setting $t = \frac{1}{4}, \frac{1}{2}$, and $\frac{3}{4}$ will generate three more points on the line segment $\overline{PQ}$.

For $t = \frac{1}{4}$, $\mathbf{x} = \begin{bmatrix} 10 - 6(\frac{1}{4}) \\ 4 + 8(\frac{1}{4}) \end{bmatrix} = \begin{bmatrix} 8.5 \\ 6 \end{bmatrix}$, thereby generating point (8.5, 6).

For $t = \frac{1}{2}$, $\mathbf{x} = \begin{bmatrix} 7 \\ 8 \end{bmatrix}$, thereby generating point (7, 8).

For $t = \frac{3}{4}$, $\mathbf{x} = [5.5, 10]^t$, thereby generating the point (5.5, 10).

Do the calculations for $t = -\frac{1}{4}$ and $t = \frac{5}{4}$.

— — — — — — — — — —

Solution: For $t = -\frac{1}{4}$, $\mathbf{x} = [10 + 6(\frac{1}{4}), \ 4 + 8(-\frac{1}{4})]^t = [11.5, 2]^t$, thereby generating point (11.5, 2). For $t = \frac{5}{4}$, $\mathbf{x} = [2.5, 14]^t$, thereby generating point (2.5, 14).

Since neither $-\frac{1}{4}$ nor $\frac{5}{4}$ is between 0 and 1, the last two points, though on the straight line, are not on the straight line segment $\overline{PQ}$. The line through P and Q, featuring the preceding seven points, is shown on Figure 4.14.

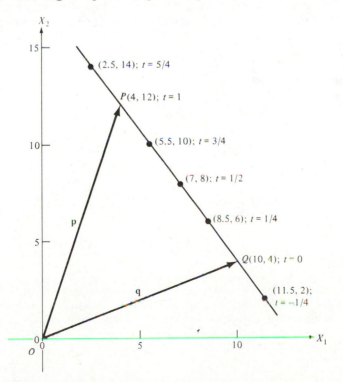

FIGURE 4.14 Points on a straight line.

14.   From intermediate algebra we already have equations for a straight line in two-dimensional space. For example, an equation for a straight line with slope m, which passes through a given point (x_0, y_0) is given by the following equation in (x, y) coordinates:

$$y - y_0 = m(x - x_0) \qquad \text{(point-slope form)} \tag{1}$$

If we replace the coordinates (x, y) by (x_1, x_2) and the coordinates (x_0, y_0) of the given point by (q_1, q_2), this becomes

$$x_2 - q_2 = m(x_1 - q_1), \tag{2}$$

in symbols compatible to those we have been using in this chapter.

Example (a):   Show that Eq. (Ia) in Definition 4.6 can be transformed to Eq. (2) immediately above in the two-dimensional case.

Solution: Equation (Ia), for the case of R^2, is

$$\begin{bmatrix} x_1 \\ x_2 \end{bmatrix} = t \begin{bmatrix} p_1 \\ p_2 \end{bmatrix} + (1 - t) \begin{bmatrix} q_1 \\ q_2 \end{bmatrix} = \begin{bmatrix} tp_1 + (1 - t)q_1 \\ tp_2 + (1 - t)q_2 \end{bmatrix} \tag{3}$$

Here we have a two-component vector on both sides of the equation. This leads to the following scalar equations, one for each component:

$$x_1 = tp_1 + (1 - t)q_1 = q_1 + t(p_1 - q_1) \tag{4a}$$

$$x_2 = tp_2 + (1 - t)q_2 = q_2 + t(p_2 - q_2) \tag{4b}$$

Solving for t from both Eqs. (4a) and (4b) leads to

$$\frac{x_2 - q_2}{p_2 - q_2} = \frac{x_1 - q_1}{p_1 - q_1} = t \tag{5}$$

[Here we assume that $(p_2 - q_2) \neq 0$ and $(p_1 - q_1) \neq 0$.] From the first two parts of Eq. (5), we obtain

$$x_2 - q_2 = \left(\frac{p_2 - q_2}{p_1 - q_1} \right)(x_1 - q_1) \tag{6}$$

We recognize that the slope m of the line through points $P(p_1, p_2)$ and $Q(q_1, q_2)$ is given by

$$m = \frac{p_2 - q_2}{p_1 - q_1} \tag{7}$$

and thus Eq. (6) is identical to Eq. (2).

Example (b): Refer to the example of frame 13. Determine the point-slope form [see Eq. (2) above] for the straight line passing through the points P and Q of that example.

— — — — — — — — — —

Solution: For the points $Q(10, 4)$ and $P(4, 12)$, we have

$$m = \frac{p_2 - q_2}{p_1 - q_1} = \frac{12 - 4}{4 - 10} = -\tfrac{4}{3} \tag{8}$$

Thus, the point-slope form is

$$x_2 - 4 = \left(-\tfrac{4}{3}\right)(x_1 - 10) \tag{9}$$

If we let $x_1 = 7$, then we find $x_2 = 8$. This checks out since the point $(7, 8)$ occurs when $t = \tfrac{1}{2}$ in the example of frame 13.

15. A straight line in R^n is also determined by specifying a direction vector $\mathbf{v}$ for the line as well as a single point $Q(q_1, q_2, \ldots, q_n)$, which lies on the line. (The column vector $\mathbf{q} = [q_1, q_2, \ldots, q_n]^t$ corresponds to the point Q.) For example, refer to Figure 4.15 for a diagram that pertains to 2-space. Let $X(x_1, x_2, \ldots, x_n)$ be a typical point on the line and let $\mathbf{x} = [x_1, x_2, \ldots, x_n]^t$ be the corresponding column vector. The translated vector $(\mathbf{x} - \mathbf{q})_{tr}$ and the vector $(\mathbf{x} - \mathbf{q})$ are obtained by using the rule for vector subtraction (Figure 4.5 in frame 2). Since $(\mathbf{x} - \mathbf{q})$ is also a direction vector for the straight line, $(\mathbf{x} - \mathbf{q})$ must be a multiple of $\mathbf{v}$; that is, $\mathbf{x} - \mathbf{q} = t\mathbf{v}$, where t is any real number.

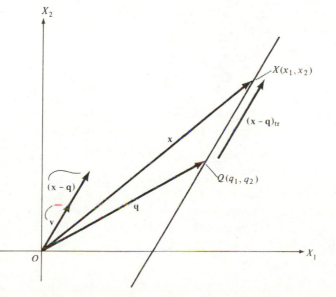

FIGURE 4.15   Here, $\mathbf{x} = \mathbf{q} + t\mathbf{v}$, where $\mathbf{v}$ is a direction vector of the straight line.

Definition 4.7

Suppose that a straight line in R^n contains a point Q (to which corresponds column vector $\mathbf{q}$) and that a direction vector of the line is given to be the column vector $\mathbf{v}$. Then a column vector equation for the line is

$\mathbf{x} = \mathbf{q} + t\mathbf{v}$, where t is any real number

Here $\mathbf{x}$ is the column vector corresponding to a point X on the straight line.

Notes: (i) In 2-space, where $\mathbf{v} = [v_1, v_2]^t$, it should be clear that the slope of the straight line is given by $m = v_2/v_1$. For example, if $\mathbf{v} = \begin{bmatrix} 2 \\ 6 \end{bmatrix}$, then $m = \frac{6}{2} = 3$. (ii) If $\mathbf{v}$ is a direction vector for a straight line, then so is $k\mathbf{v}$, where $k \neq 0$.

Example: (a) Find a vector equation of a line in R^3 if a direction vector is $\mathbf{v} = [2, 3, -4]^t$ and a point on the line is $(1, -2, 5)$. (b) Find equations for the components x_1, x_2, and x_3 in terms of t. (c) Determine points on the line for $t = -1$ and $t = 2$.

Solution: (a) Here $\mathbf{v} = [2, 3, -4]^t$, Q is $(1, -2, 5)$, and thus $\mathbf{q} = [1, -2, 5]^t$. Then, the vector equation $\mathbf{x} = \mathbf{q} + t\mathbf{v}$ becomes

$$\mathbf{x} = \begin{bmatrix} 1 \\ -2 \\ 5 \end{bmatrix} + t \begin{bmatrix} 2 \\ 3 \\ -4 \end{bmatrix} = \begin{bmatrix} 1 + 2t \\ -2 + 3t \\ 5 - 4t \end{bmatrix} \tag{1}$$

(b) Since $\mathbf{x} = [x_1, x_2, x_3]^t$, Eq. (1) leads to the following three equations for the components of $\mathbf{x}$.

$$x_1 = 1 + 2t, \qquad x_2 = -2 + 3t, \qquad x_3 = 5 - 4t \qquad \text{(2a), (2b), (2c)}$$

These are known as *parametric equations* for the straight line. If we solve for t from Eqs. (2a), (2b), and (2c) and equate the corresponding expressions, we obtain the following so-called *symmetric equations* for the straight line:

$$\frac{x_1 - 1}{2} = \frac{x_2 + 2}{3} = \frac{x_3 - 5}{(-4)} \tag{3}$$

Parametric and symmetric equations such as these are studied in the three-dimensional analytic geometry part of a usual calculus course sequence.
Now do part (c).

— — — — — — — — —

(c)   Substituting $t = -1$ in Eqs. (2a), (2b), and (2c), we obtain

$$x_1 = -1, \qquad x_2 = -5, \qquad \text{and} \qquad x_3 = 9$$

Similarly, for $t = 2$, we obtain $x_1 = 5$, $x_2 = 4$, and $x_3 = -3$.

16.   A very important special case for a straight line in R^n occurs wnen the line goes through the origin. Referring to Definition 4.7 in frame 15, let us set $\mathbf{q} = \mathbf{0}$. Thus, we have

$$\mathbf{x} = t\mathbf{v} \tag{I}$$

as the vector equation for a straight line passing through the origin, which has $\mathbf{v}$ as a direction vector.

We shall encounter the vector equation for a straight line through the origin again in Chapter Five.

Example: (a) Find a vector equation of a straight line in R^4 that passes through the origin if a direction vector is $\mathbf{v} = [1, 3, -2, 4]^t$. (b) Determine equations for vectors of the line, and corresponding points on the line, for $t = -1, 0, 1, 2$, and 2.5.

Solution: (a)   A vector equation of the line is $\mathbf{x} = t\mathbf{v}$, which becomes

$$\mathbf{x} = t[1, 3, -2, 4]^t$$

(b)   For $t = -1$, $\mathbf{x} = [-1, -3, 2, -4]^t$ and the corresponding point is $(-1, -3, 2, -4)$. For $t = 0$, $\mathbf{x} = \mathbf{0} = [0, 0, 0, 0]^t$, thus reaffirming that the line contains the origin $(0, 0, 0, 0)$.

Now do the rest of the calculations.

— — — — — — — — — — — —

For $t = 1$, $\mathbf{x} = [1, 3, -2, 4]^t$, and the corresponding point is $(1, 3, -2, 4)$. For $t = 2$, $\mathbf{x} = [2, 6, -4, 8]^t$, and the corresponding point is $(2, 6, -4, 8)$. For $t = 2.5$, $\mathbf{x} = [2.5, 7.5, -5, 10]^t$, and the corresponding point is $(2.5, 7.5, -5, 10)$.

The study of hyperplanes and straight lines provides an introduction to some of the geometrical shapes encountered in n-dimensional space. Also, hyperplanes, straight lines, and straight-line segments figure prominently in the subjects convex geometry and mathematical programming, both of which have been found to have tremendous practical application. In particular, geometrical shapes such as these are encountered in the field linear programming, which itself is a branch of the larger field mathematical programming. [See, for example, the books by Kolman and Beck (1980) and Rothenberg (1979).]

It is hoped that the current chapter will provide the reader with geometrical insights that will be useful in the study of the next three chapters, which themselves form the heart of a linear algebra course.

SELF-TEST

This Self-Test will help you determine whether or not you have mastered the chapter objectives and are ready to go on to another chapter. Correct answers are given at the end of the test.

1. Given that $\mathbf{a} = [6, 2]^t$, $\mathbf{b} = [-3, 2]^t$, and $\mathbf{c} = [4, -3]^t$, compute the following, and illustrate with a diagram when appropriate:

   (a) $\mathbf{a} + \mathbf{b}$.
   (b) $\mathbf{a} + \mathbf{c}$.
   (c) $\mathbf{a} - \mathbf{b}$.
   (d) $2\mathbf{a} + 3\mathbf{b}$.
   (e) $\mathbf{c} - \mathbf{a}$.
   (f) $\mathbf{a} - \mathbf{c}$.
   (g) $-2\mathbf{b}$.
   (h) $\mathbf{a} \cdot \mathbf{b}$.
   (i) $\mathbf{b} \cdot \mathbf{a}$.
   (j) $\mathbf{a} \cdot (\mathbf{b} + \mathbf{c})$.

2. Compute the following (when defined):

   (a) $[3, 5, 7, -6]^t + [-7, 2, 4, 3]^t$.
   (b) $[3, 5, 7, -6]^t + 3[-7, 2, 4, 3]^t$.
   (c) $[7, 2, -4, 6]^t + [8, -4, 5]^t$.
   (d) $3[2, 4, -3]^t + 2[6, -1, -5]^t - 4[3, 8, 2]^t$.
   (e) $[-2, 4, -3]^t \cdot [1, 5, 7]^t$. This is a dot product.

3. Given that $\mathbf{a} = [4, 3, -2]^t$ and $\mathbf{b} = [-2, 1, -4]^t$, compute the following:

   (a) $\|\mathbf{a}\|$.
   (b) $\|\mathbf{b}\|$.
   (c) $\|\mathbf{b} - \mathbf{a}\|$.
   (d) $\mathbf{a} \cdot \mathbf{b}$.
   (e) $\cos \theta$.
   (f) θ.

4. Given the vectors $\mathbf{a} = [5, -2, 3]^t$, $\mathbf{b} = [2, 1, -4]^t$, $\mathbf{c} = [1, 4, 1]^t$, $\mathbf{d} = [3, 4.5, -2]^t$, $\mathbf{e} = [-6, -3, 12]^t$. Which of these pairs of vectors are (a) perpendicular, (b) multiples.

5. For the vectors $\mathbf{a}$ and $\mathbf{b}$ in question 3, verify

   (a) The Cauchy–Schwarz inequality, and
   (b) The triangle inequality.

6. (a) Find an equation of the hyperplane in 2-space for which $\mathbf{c} = \begin{bmatrix} 4 \\ -3 \end{bmatrix}$ and $b = 15$.

   (b) Determine a unit vector in the direction of $\mathbf{c}$.
   (c) Find an equation of the hyperplane in 3-space for which $\mathbf{c} = [-2, 5, 3]^t$ and $b = -6$.

7. Given a straight line in 2-space that passes through the two points $P(-2, -3)$ and $Q(5, 1)$.

   (a) Find a vector equation for the straight line.
   (b) Determine points on the line for $t = -\frac{1}{2}, 0, \frac{1}{4}, \frac{1}{2}, \frac{3}{4}, 1$, and $\frac{3}{2}$. Sketch the straight line, indicating the points corresponding to the preceding values of t.
   (c) Which of the previous points lie on the line segment $\overline{PQ}$?
   (d) Determine the usual point-slope form for the straight line.
   (e) Determine a direction vector for the straight line.

8. (a) Find a vector equation for a straight line that is parallel to the line of question 7 and that passes through the origin.
    (b) Determine vectors of the line for several values of t.

ANSWERS TO SELF-TEST

If your answers to the test questions do not agree with the ones given here, review the frames indicated in parentheses after each answer before you go on to another chapter.

1. Diagrams have been omitted. (a) $[3, 4]^t$, (b) $[10, -1]^t$, (c) $[9, 0]^t$, (d) $[3, 10]^t$, (e) $[-2, -5]^t$, (f) $[2, 5]^t$, (g) $[6, -4]^t$, (h) -14, (i) -14, (j) 4.   (frames 1–4)

2. (a) $[-4, 7, 11, -3]^t$, (b) $[-18, 11, 19, 3]^t$, (c) undefined, (d) $[6, -22, -27]^t$, (e) -3.   (frames 1–4)

3. (a) $\sqrt{29} = 5.385$, (b) $\sqrt{21} = 4.583$, (c) $\sqrt{44} = 6.633$, (d) 3, (e) 0.12157, (f) $83.02° = 1.449$ radians.   (frames 4–6)

4. (a) Perpendicular pairs are **a** and **c**; **a** and **d**. (b) **e** is a multiple of **b** (**e** $= -3$**b**).   (frames 5–7)

5. (a) $|\mathbf{a} \cdot \mathbf{b}| < \|\mathbf{a}\|\|\mathbf{b}\|$ since $|\mathbf{a} \cdot \mathbf{b}| = |3| = 3$, and $\|\mathbf{a}\|\|\mathbf{b}\| = \sqrt{29}\sqrt{21} = \sqrt{609}$ $= 24.68$. Note that the Cauchy–Schwarz inequality is equivalent to $|\cos \theta| < 1$; here, $\cos \theta = 0.1216$. (b) $\|\mathbf{a} + \mathbf{b}\| < \|\mathbf{a}\| + \|\mathbf{b}\|$ since $\|\mathbf{a} + \mathbf{b}\| = \sqrt{56} = 7.483$ and $\|\mathbf{a}\| + \|\mathbf{b}\| = \sqrt{29} + \sqrt{21} = 9.968$.   (frames 6–9)

6. (a) $4x_1 - 3x_2 = 15$. The hyperplane is a straight line. (b) $\mathbf{u}_c = \begin{bmatrix} \frac{4}{5} \\ -\frac{3}{5} \end{bmatrix}$.

    (c) $-2x_1 + 5x_2 + 3x_3 = -6$. The hyperplane is an actual plane. (frames 10–13)

7. (a) $\mathbf{x} = t\mathbf{p} + (1 - t)\mathbf{q}$, where $\mathbf{p} = [-2, -3]^t$ and $\mathbf{q} = [5, 1]^t$.
    (b) The coordinates of the points are given in the following table:

t	$-\frac{1}{2}$	0	$\frac{1}{4}$	$\frac{1}{2}$	$\frac{3}{4}$	1	$\frac{3}{2}$
x_1	8.5	5	3.25	1.5	$-.25$	-2	-5.5
x_2	3	1	0	-1	$.-2$	-3	-5

    The sketch has been omitted.

    (c) The points for which $t = 0, \frac{1}{4}, \frac{1}{2}, \frac{3}{4}$ and 1. (d) $x_2 - 1 = (4/7)(x_1 - 5)$; (e) $\mathbf{p} - \mathbf{q} = [-7, -4]^t$ is a direction vector, as is any multiple $s[-7, -4]^t$, where $s \neq 0$.   (frames 13–15)

8. A direction vector for both lines is $s(\mathbf{p} - \mathbf{q})$ for any scalar $s \neq 0$. Thus, using $s = -1$, $\mathbf{v} = \mathbf{q} - \mathbf{p} = \begin{bmatrix} 7 \\ 4 \end{bmatrix}$, and a vector equation for the new line is $\mathbf{x} = \hat{t}\mathbf{v}$. For $\hat{t} = 0$, $\mathbf{x} = 0 = \begin{bmatrix} 0 \\ 0 \end{bmatrix}$; for $\hat{t} = -1$, $\mathbf{x} = \begin{bmatrix} -7 \\ -4 \end{bmatrix}$; for $\hat{t} = 1$, $\mathbf{x} = \begin{bmatrix} 7 \\ 4 \end{bmatrix}$; for $\hat{t} = 2$, $\mathbf{x} = \begin{bmatrix} 14 \\ 8 \end{bmatrix}$.   (frames 15, 16)

CHAPTER FIVE
Vector Spaces

In Chapter Four we studied and illustrated the properties of the sets R^2 (vectors in the plane) and R^3 (vectors in three-dimensional space). Furthermore, we algebraically extended our ideas about R^2 and R^3 to apply to R^n, Euclidean n-dimensional space, where n could be any positive integer.

These sets (i.e., R^2, R^3, and R^n, for any positive integer n) are called *vector spaces*. In this chapter we shall give a general definition for a vector space and then present models of this abstract concept. Then we shall further explore ideas and structural properties that hold for a general vector space. For the most part, our illustrations and examples will be from R^n, and in particular from R^2 and R^3.

Perhaps the most important concept to be studied in this chapter is that of a set of *basis* vectors for a vector space. If a basis set is available for a vector space, then any vector in the vector space can be expressed as a unique linear combination of the basis vectors. This fact is very important for the further development of linear algebra and for practical applications.

OBJECTIVES

When you complete this chapter you should be able to

- Determine whether a set of elements, together with rules for addition of elements and multiplication of an element by a scalar, satisfies the properties of a vector space.
- Determine whether a subset of a vector space is itself a vector space (subspace).
- Do calculations with an expression known as a linear combination of vectors from a vector space.
- Determine whether a set of vectors S from a vector space V spans the vector space and is linearly independent or dependent. (A set S that both spans V and is linearly independent is known as a basis for the vector space.)
- Do calculations with respect to both nonorthogonal and orthogonal bases.

- Determine the orthogonal projection of a vector $\mathbf{x}$ from vector space R^n onto a subspace W of vector space R^n.
- Use the Gram–Schmidt process to transform any basis of a vector space (where the vector space is R^n or a subspace of R^n) into an orthogonal basis for the vector space.

A.   VECTOR SPACES AND SUBSPACES

1.   In frame 3 of Chapter Four we indicated that the set of all n-component column vectors (with real components) was a vector space. This vector space, whose symbol is R^n, is often referred to as Euclidean n-space. We could just as well have worked with n-component row vectors, such as $[b_1, b_2, \ldots, b_n]$, or ordered sets of n numbers, such as $\langle b_1, b_2, \ldots, b_n \rangle$, (such ordered sets of n numbers are also called n-tuples) in our development of the structure of R^n. It should be understood, though, that addition and scalar multiplication for such objects would be the usual matrix addition and multiplication of a matrix by a scalar.

In this section we shall generalize the concept of a vector and shall refer to a collection of such generalized vectors as being a *vector space*. In many areas of application, such as in engineering and the natural sciences, these general vector spaces occur.

Definition 5.1

Let V be a set of elements (or objects) on which the operations of addition and multiplication by a scalar (real number) are defined. By addition, we mean a rule for associating with each pair of elements $\mathbf{x}$ and $\mathbf{y}$ in V an element $\mathbf{x} + \mathbf{y}$, called the *sum* of $\mathbf{x}$ and $\mathbf{y}$. By scalar multiplication we mean a rule for associating with each scalar k and each element $\mathbf{x}$ in V an element $k\mathbf{x}$, called a *scalar multiple* of $\mathbf{x}$.

In the following, $\mathbf{x}$, $\mathbf{y}$, and $\mathbf{z}$ denote any elements of V, and h and k denote any scalars. The set V is called a real *vector space* and the elements of V are called *vectors* if the following properties (axioms) are satisfied:

(i)	The sum $\mathbf{x} + \mathbf{y}$ of any two elements $\mathbf{x}$ and $\mathbf{y}$ of V is itself an element of V (closure under addition property).
(ii)	$(\mathbf{x} + \mathbf{y}) + \mathbf{z} = \mathbf{x} + (\mathbf{y} + \mathbf{z})$.
(iii)	There is a unique element $\mathbf{0}$ in V (the *zero vector* for V) such that $\mathbf{x} + \mathbf{0} = \mathbf{x}$ for any element $\mathbf{x}$ in V.
(iv)	For each $\mathbf{x}$ in V there is a unique element $-\mathbf{x}$ in V (called the *negative*, or *additive inverse*, of $\mathbf{x}$) such that $\mathbf{x} + (-\mathbf{x}) = \mathbf{0}$.
(v)	$\mathbf{x} + \mathbf{y} = \mathbf{y} + \mathbf{x}$.
(vi)	If $\mathbf{x}$ is any element of V and k is any real number, then $k\mathbf{x}$ is also an element of V (closure under scalar multiplication property).

(vii) $k(\mathbf{x} + \mathbf{y}) = k\mathbf{x} + k\mathbf{y}$.

(viii) $(h + k)\mathbf{x} = h\mathbf{x} + k\mathbf{x}$.

(ix) $(hk)\mathbf{x} = h(k\mathbf{x})$.

(x) $1\mathbf{x} = \mathbf{x}$ (Here 1 is the real number 1.)

Example (a): From Theorem 4.1 (frame 3 of Chapter Four), we see that the set R^n of n-component column vectors, with the usual operations of vector addition and scalar multiplication, is a vector space. The addition and scalar multiplication of vectors here are special cases of matrix addition and multiplication of a matrix by a scalar.

For example, in R^2, if $\mathbf{x} = [x_1, x_2]^t$, then $-\mathbf{x} = (-1)\mathbf{x} = [-x_1, -x_2]^t$, and the zero vector is $[0, 0]^t$. Similarly, for R^n.

Example (b): Consider the solutions of the homogeneous equation

$$x_1 - 2x_2 + 4x_3 = 0. \tag{1}$$

Here it is easy to "solve" this equation as

$$x_1 = 2x_2 - 4x_3, \tag{2}$$

or, putting $x_2 = r$ and $x_3 = s$, with r and s arbitrary scalars, we have

$$x_1 = 2r - 4s, \qquad x_2 = r, \qquad x_3 = s \tag{3}$$

In column vector form, with $\mathbf{x} = [x_1, x_2, x_3]^t$, we have

$$\mathbf{x} = \begin{bmatrix} 2r - 4s \\ r \\ s \end{bmatrix} = r\begin{bmatrix} 2 \\ 1 \\ 0 \end{bmatrix} + s\begin{bmatrix} -4 \\ 0 \\ 1 \end{bmatrix} \tag{4}$$

It is not hard to show that the collection of column vectors determined by Eq. (4), with r and s arbitrary, comprise a vector space.

For example, say $\mathbf{u}$ is determined by $r = 1$ and $s = 0$ and $\mathbf{v}$ is determined by $r = 2$ and $s = 3$, that is,

$$\mathbf{u} = [2, 1, 0]^t \qquad \text{and} \qquad \mathbf{v} = [-8, 2, 3]^t$$

Let us now demonstrate the closure properties (i) and (vi) of Definition 5.1. For example, $\mathbf{u} + \mathbf{v} = [-6, 3, 3]^t$ and the scalar multiple $3\mathbf{v} = [-24, 6, 9]^t$ can be expressed in the form given by Eq. (4) above; for $\mathbf{u} + \mathbf{v}$, $r = 3$ and $s = 3$, and for $3\mathbf{v}$, $r = 6$ and $s = 9$.

Geometrically, we know that Eq. (1) represents a plane that passes through the origin of three-dimensional space [the coordinates of the origin $(0, 0, 0)$ satisfy the equation]. Thus, the set of points corresponding to Eq. (4), namely, the points with coordinates $(2r - 4s, r, s)$, with r and s arbitrary, are the points that lie on this plane. The origin itself occurs when $r = s = 0$.

The two points on the plane corresponding to **u** and **v** cited above are $(2, 1, 0)$ and $(-8, 2, 3)$, respectively.

Example (c):   Consider the solutions of the system of two homogeneous equations:

$$x_1 + x_2 + x_3 = 0 \tag{1}$$

$$x_1 + 2x_2 - x_3 = 0 \tag{2}$$

Employing Method 2.1 on the augmented coefficient matrix $\begin{bmatrix} 1 & 1 & 1 & 0 \\ 1 & 2 & -1 & 0 \end{bmatrix}$, we obtain the reduced row echelon form

$$\begin{bmatrix} 1 & 0 & 3 & 0 \\ 0 & 1 & -2 & 0 \end{bmatrix} \tag{3}$$

From this we obtain $x_1 + 3x_3 = 0$, and $x_2 - 2x_3 = 0$, which leads to

$$x_1 = -3r, \qquad x_2 = 2r, \qquad x_3 = r \tag{4}$$

with r arbitrary. In column vector form, with $\mathbf{x} = [x_1, x_2, x_3]^t$, we have

$$\mathbf{x} = \begin{bmatrix} -3r \\ 2r \\ r \end{bmatrix} = r \begin{bmatrix} -3 \\ 2 \\ 1 \end{bmatrix} \tag{5}$$

It is not hard to show that the collection of column vectors determined by Eq. (5), with r arbitrary, comprises a vector space.

Interpret the set of solutions of Eqs. (1) and (2) geometrically.

— — — — — — — — — —

Since Eqs. (1) and (2) are both equations for planes that pass through the origin of 3-space, the set of solutions to both equations represents a line that passes through the origin (the line of intersection of the planes). From Eqs. (4) or (5), the points on this line have coordinates $(-3r, 2r, r)$, with r arbitrary.

From frame 16 of Chapter Four we know that $\mathbf{x} = r\mathbf{v}$ (with $\mathbf{v} = [-3, 2, 1]^t$ here) is a vector equation of a straight line that passes through the origin.

2.   We continue to present some typical vector spaces. Recall that a polynomial in t, $p(t)$, is a function that can be expressed as

$$p(t) = a_n t^n + a_{n-1} t^{n-1} + \cdots + a_1 t + a_0$$

where n is a nonnegative integer and $a_0, a_1, \ldots, a_{n-1}$, and a_n are real numbers. If $a_n \neq 0$, then $p(t)$ is said to have *degree n*. Thus, the degree of a polynomial is the exponent of the highest power having a nonzero coefficient. For example, $3t^2 - 2t + 1$ has degree 2, $4t + 3$ has degree 1, and the constant polynomial 7 has degree 0. The zero polynomial is defined as $0t^n + 0t^{n-1} + \cdots + 0t + 0$ and is symbolized by $0(t)$; we note that the zero polynomial is equal to the number 0 for all t. Also, for the zero polynomial, the degree *does not exist*.

We define P_n as the set of all polynomials of degree $\leq n$ (read $\leq n$ as "less than or equal to" n), together with the zero polynomial. Let

$$p(t) = a_n t^n + a_{n-1} t^{n-1} + \cdots + a_1 t + a_0,$$

and $$q(t) = b_n t^n + b_{n-1} t^{n-1} + \cdots + b_1 t + b_0$$

denote any two polynomials in P_n. For the definition of addition we have the usual addition for functions, that is,

$$p(t) + q(t) = (a_n + b_n)t^n + (a_{n-1} + b_{n-1})t^{n-1}$$
$$+ \cdots + (a_1 + b_1)t + (a_0 + b_0)$$

For multiplication by a scalar we have the usual multiplication of a function by a scalar, that is,

$$kp(t) = (ka_n)t^n + (ka_{n-1})t^{n-1} + \cdots + (ka_1)t + (ka_0)$$

It can be shown that P_n is a vector space. The zero polynomial is the zero vector. The negative of $p(t)$ above is given by

$$-p(t) = (-a_n)t^n + (-a_{n-1})t^{n-1} + \cdots + (-a_1)t + (-a_0),$$

since this causes $p(t) + (-p(t))$ to equal zero for all t. Thus property (iii) in Definition 5.1 is verified. The other properties in the definition follow directly from the properties of real numbers.

Example (a):   Let M_{22} be the set of all 2×2 matrices under the usual operations of matrix addition and scalar multiplication (Definitions 1.1 and 1.2). In Chapter One we established that all the properties in Definition 5.1 hold. Thus, M_{22} is a vector space.

Similarly, the set of all $m \times n$ matrices under the usual operations of matrix addition and scalar multiplication is a vector space. We denote this vector space by M_{mn}.

Example (b):   Consider the solutions of the equation

$$x_1 - 2x_2 + 4x_3 = 8. \tag{1}$$

In Example (b) of frame 1 we have the same left-side expression, but the right side was zero. Show that the set of solutions of Eq. (1), when expressed in column vector form, is not a vector space.

Solution: We solve Eq. (1) to obtain

$$x_1 = 2x_2 - 4x_3 + 8 \tag{2}$$

or putting $x_2 = r$ and $x_3 = s$, with r and s arbitrary, we have

$$x_1 = 2r - 4s + 8, \qquad x_2 = r, \qquad x_3 = s \tag{3}$$

In column vector form, with $\mathbf{x} = [x_1, x_2, x_3]^t$, we have

$$\mathbf{x} = \begin{bmatrix} 2r - 4s + 8 \\ r \\ s \end{bmatrix} = r\begin{bmatrix} 2 \\ 1 \\ 0 \end{bmatrix} + s\begin{bmatrix} -4 \\ 0 \\ 1 \end{bmatrix} + \begin{bmatrix} 8 \\ 0 \\ 0 \end{bmatrix} \tag{4}$$

Now finish the problem. *Hint:* Pick two solution vectors from Eq. (4) by letting r and s take on definite values and show that a condition of Definition 5.1 does not hold.

— — — — — — — — — —

From Eq. (4), let $\mathbf{u}$ be the vector for which $r = 1$ and $s = 0$ and let $\mathbf{v}$ be the vector for which $r = 0$ and $s = 1$. Thus,

$$\mathbf{u} = \begin{bmatrix} 10 \\ 1 \\ 0 \end{bmatrix} \quad \text{and} \quad \mathbf{v} = \begin{bmatrix} 4 \\ 0 \\ 1 \end{bmatrix}$$

Now $\mathbf{u} + \mathbf{v} = [14, 1, 1]^t$ *cannot* be put in the form of Eq. (4), and thus we do not have a vector space. An easy way of seeing this is by showing that Eq. (1) is not satisfied by the components of $\mathbf{u} + \mathbf{v}$. Substituting $x_1 = 14$, $x_2 = x_3 = 1$ into the left side of Eq. (1) yields 16, which is unequal to 8 (the right side).

Another way of showing that we do not have a vector space is by considering the scalar multiple $3\mathbf{u} = [30, 3, 0]^t$. The coordinates $x_1 = 30$, $x_2 = 3$, $x_3 = 0$ fail to satisfy Eq. (1).

Geometrically, Eq. (1) represents a plane in three-dimensional space that does not pass through the origin.

3. **Example (a):** Determine whether the set of all two-component row vectors of the form $[a, 0]$, with the usual rules for matrix addition and scalar multiplication of a matrix, i.e., $[a, 0] + [b, 0] = [a + b, 0]$ and $k[a, 0] = [ka, 0]$, is a vector space. (Here a, b, and k symbolize real numbers.)

Solution: Since all 10 properties of Definition 5.1 hold, we have a vector space.

Example (b): Determine whether the set of all two-component row vectors of the form $[a, 1]$, with the usual rules for matrix addition and scalar multiplication of a matrix is a vector space.

— — — — — — — — — —

Solution: The set is not a vector space. For example,

$$[a, 1] + [b, 1] = [a + b, 2],$$

and since the second component of the row vector on the right is not 1, property (i) of Definition 5.1 fails. Also, properties (iii), (iv), and (vi) fail.

4. The following theorem lists several properties that hold for a general vector space.

Theorem 5.1

Let V be a vector space. Then

(a) $0\mathbf{x} = \mathbf{0}$ for every vector $\mathbf{x}$ in V. (On the left is the scalar 0, and on the right is the vector $\mathbf{0}$.)

(b) $k\mathbf{0} = \mathbf{0}$ for every scalar k.

(c) If $k\mathbf{x} = \mathbf{0}$, then either $k = 0$ or $\mathbf{x} = \mathbf{0}$.

(d) $(-1)\mathbf{x} = -\mathbf{x}$, for every $\mathbf{x}$ in V. This says the scalar (-1) times $\mathbf{x}$ is equal to the negative (or additive inverse) of $\mathbf{x}$.

Suppose V is a vector space. Then certain subsets of V themselves form vector spaces for the vector addition and scalar multiplication defined on V. Such subsets are called subspaces.

Notes: (i) In order to understand the concept of a subspace, we first have to know what a subset is. We say that the set of elements A is a subset of a set B if every element in A is also an element in B. For example, suppose that $A = \{2, 4, 6\}$, which means that set A contains elements 2, 4, and 6. (Traditionally, one uses the curled brackets { } to enclose the elements of a set.) Also, say $B = \{1, 2, 3, 4, 5, 6, 7, 8\}$ and $C = \{1, 2, 3, 4, 6, 9\}$. Thus, set A is a subset of set B and of set C. Set C is not a subset of set B since 9 is in C but not in B.

(ii) Given any set A that contains at least one element. Then two subsets of A are the set A itself and the empty (or null) set $\emptyset$. The latter set is the set that contains no elements.

Definition 5.2

Let V be a vector space and W be a nonempty subset of V. If W is a vector space with respect to the operations of addition and scalar multiplication defined on V, then W is called a *subspace* of V.

Example (a): Let W be the subset of R^3 consisting of all column vectors of the form $[x_1, x_2, 0]^t$, where x_1 and x_2 denote any real numbers, which can vary freely. Show that W is a subspace of R^3.

Solution: Let $\hat{\mathbf{x}} = [x_1, x_2, 0]^t$ and $\hat{\mathbf{y}} = [y_1, y_2, 0]^t$ be any two vectors in W. We see that properties (i) and (vi) of Definition 5.1 hold, since $\hat{\mathbf{x}} + \hat{\mathbf{y}} = [x_1 + y_1, x_2 + y_2, 0]^t$ is in W, as is $k\hat{\mathbf{x}} = [kx_1, kx_2, 0]^t$. (Both have their third component equal to zero.) The other properties are easily verified too. For example, $\hat{\mathbf{x}} + \hat{\mathbf{y}} = \hat{\mathbf{y}} + \hat{\mathbf{x}}$ [property (v)] since this holds for *all* vectors in R^3, and the vector $[0, 0, 0]^t$, which is in W (set $x_1 = x_2 = 0$), is the zero vector of W. The latter, of course, is also the zero vector of R^3.

Geometrically, the points $(x_1, x_2, 0)$ corresponding to the vectors of W are the totality of the points on the X_1X_2 plane of three-dimensional space; recall that for such points $x_3 = 0$.

We observe that to verify that a subset W of a vector space is a subspace, we have to show that all the properties of Definition 5.1 (frame 1) hold. Actually, matters are simpler than this, as the following theorem indicates.

Theorem 5.2

Let W be a nonempty subset of the vector space V. Then W is a subspace of V if and only if both of the following conditions hold:

    (a)   If $\mathbf{x}$ and $\mathbf{y}$ are vectors in W, then $\mathbf{x} + \mathbf{y}$ is in W.
    (b)   If k is any scalar and $\mathbf{x}$ is any vector in W, then $k\mathbf{x}$ is in W.

The theorem says that to verify if W is a subspace of V it is sufficient to check only properties (i) and (vi) of Definition 5.1 (the so-called closure properties). Remember that the addition and scalar multiplication operations in (a) and (b) above are those that hold for the containing vector space V.

For any vector space V the subset $\{\mathbf{0}\}$ consisting of the zero vector alone is a subspace, since $\mathbf{0} + \mathbf{0} = \mathbf{0}$ and $k\mathbf{0} = \mathbf{0}$ for every scalar k. It is called the *zero* or *trivial subspace*.

Also, V itself is a subspace of the vector space V. In the examples of this and the following frames, we will present illustrations of more meaningful subspaces.

Example (b): Refer to Example (b) of frame 1, where we had the equation $x_1 - 2x_2 + 4x_3 = 0$ of a plane through the origin of three-dimensional space. The set of vector solutions given by $\mathbf{x} = [2r - 4s, r, s]^t$ is a subspace of R^3.

Example (c): Refer to Example (c) of frame 1, where we had the system of equations $x_1 + x_2 + x_3 = 0$ and $x_1 + 2x_2 - x_3 = 0$ for a straight line passing through the origin of 3-space. The set of vector solutions given by

$$\mathbf{x} = [-3r,\ 2r,\ r]^t$$

for the system is a subspace of R^3.

Let us think for a while of the geometrical structure for R^3. (The geometrical structure of R^n is discussed in the Notes following Definition 4.1 in frame 3 of Chapter Four.) The previous two examples illustrate, from a geometrical point of view, that the following are subspaces of R^3.

1.   Any plane through the origin. Such a plane can be specified by a single homogeneous equation of the form

$$a_{11}x_1 + a_{12}x_2 + a_{13}x_3 = 0$$

where the coefficients (the a's) are constants and at least one a_{1j} is unequal to zero.

2.   Any straight line through the origin. Such a straight line can be specified by a system of two distinct homogeneous equations of the form

$$a_{11}x_1 + a_{12}x_2 + a_{13}x_3 = 0$$
$$a_{21}x_1 + a_{22}x_2 + a_{23}x_3 = 0$$

Each of these equations is the equation of a plane through the origin. The straight line through the origin is the line of intersection of the two planes.

The only other subspaces of R^3 are R^3 itself and the origin $(0, 0, 0)$. Here the origin corresponds to the zero subspace $\{\mathbf{0}\}$, since $\mathbf{0} = [0, 0, 0]^t$, the zero column vector.

If we look at R^2 (the "plane") from a geometrical point of view, its subspaces are R^2 itself, the origin (algebraically, this is $\{\mathbf{0}\}$, where $\mathbf{0} = [0, 0]^t$), and all straight lines through the origin. Such a straight line can be specified by the homogeneous equation $a_{11}x_1 + a_{12}x_2 = 0$, where a_{11} and a_{12} are constants and at least one of them is unequal to zero.

Example (d): Refer to Example (a) of frame 2. Show that the set W of all 2×2 matrices having a zero in the row 1, column 2, location is a subspace of the vector space M_{22} of all 2×2 matrices. *Hint:* Let

$$A = \begin{bmatrix} a_{11} & 0 \\ a_{21} & a_{22} \end{bmatrix} \quad \text{and} \quad B = \begin{bmatrix} b_{11} & 0 \\ b_{21} & b_{22} \end{bmatrix}$$

be any two matrices in W and let k be any scalar. Show that conditions (a) and (b) of Theorem 5.2 are satisfied.

— — — — — — — — —

$$A + B = \begin{bmatrix} a_{11} + b_{11} & 0 \\ a_{21} + b_{21} & a_{22} + b_{22} \end{bmatrix} \quad \text{and} \quad kA = \begin{bmatrix} ka_{11} & 0 \\ ka_{21} & ka_{22} \end{bmatrix}$$

Since $A + B$ and kA have a 0 in the row 1, column 2, location, they are in the set W. Thus, W is a subspace of M_{22}.

5.   Our main attention in this chapter will be focused on the vector space R^n and subspaces of it. From time to time, however, we shall give illustrations pertaining to the vector space M_{mn} of all $m \times n$ matrices and the vector space P_n of all polynomials of degree $\leq n$, together with the zero polynomial (refer to the beginning of frame 2).

Example (a):   The vector space P_2 is a subspace of the vector space P_3. (In fact, P_2 is a subspace of vector space P_n if $n \geq 2$.) First, we observe that P_2 is certainly a subset of P_3, since a typical polynomial $a_2 t^2 + a_1 t + a_0$ in P_2 is the polynomial $a_3 t^3 + a_2 t^2 + a_1 t + a_0$ in P_3 for which $a_3 = 0$. Now consider polynomials $p(t)$ and $q(t)$ in P_2,

$$p(t) = a_2 t^2 + a_1 t + a_0 \quad \text{and} \quad q(t) = b_2 t^2 + b_1 t + b_0$$

and let k be a scalar. Then

$$p(t) + q(t) = (a_2 + b_2)t^2 + (a_1 + b_1)t + (a_0 + b_0),$$

$$\text{and} \quad kp(t) = (ka_2)t^2 + (ka_1)t + (ka_0)$$

are both polynomials in P_2, thereby verifying conditions (a) and (b) of Theorem 5.2.

Using the same type of approach as in Example (a), we can show that P_m is a subspace of P_n if $m \leq n$. Several subspaces of the vector space P_3 are P_2, P_1, P_0, and $\{0\}$, where 0 here is the zero polynomial (frame 2). Symbols for typical members of P_1 and P_0 are $a_1 t + a_0$ and a_0, respectively. The latter "polynomials" of degree zero are real number constants.

Example (b):   Let W be the subset of R^3 consisting of all column vectors of the form $[x_1, x_2, 2]^t$, where x_1 and x_2 are any real numbers. Determine whether W is a subspace. *Hint:* See if conditions (a) and (b) of Theorem 5.2 are satisfied.

— — — — — — — —

Solution: Let $\mathbf{x} = [x_1, x_2, 2]^t$ and $\mathbf{y} = [y_1, y_2, 2]^t$. Then $\mathbf{x} + \mathbf{y} = [x_1 + y_1, x_2 + y_2, 4]^t$, which is not in W, since the third component is 4 and not 2. Since condition (a) does not hold, W is not a subspace of R^3. It is not necessary to check condition (b), which, by the way, fails also.

It is useful to compare this example with Example (a) of frame 4. There the set of vectors $[x_1, x_2, 0]^t$ did constitute a subspace of R^3. Geometrically, the points $(x_1, x_2, 0)$ lie on a plane passing through the origin, the so-called X_1X_2 plane whose equation is $x_3 = 0$. The points $(x_1, x_2, 2)$, which correspond to the vectors $[x_1, x_2, 2]^t$, lie on a plane that is everywhere two units above the X_1X_2 plane; its equation is $x_3 = 2$. The latter plane does not contain the origin.

6. We are now in a position to tie together material from Chapter Two with concepts from the current chapter. Consider a general *homogeneous* linear system of m equations in n unknowns (Definition 2.5 of frame 20 of Chapter Two):

$$a_{11}x_1 + a_{12}x_2 + \ldots + a_{1n}x_n = 0 \tag{1.1}$$

$$a_{21}x_1 + a_{22}x_2 + \ldots + a_{2n}x_n = 0 \tag{1.2}$$

$$\cdots\cdots\cdots\cdots\cdots\cdots\cdots$$

$$\cdots\cdots\cdots\cdots\cdots\cdots\cdots$$

$$a_{m1}x_1 + a_{m2}x_2 + \ldots + a_{mn}x_n = 0 \tag{1.m}$$

The matrix–vector form is

$$A\mathbf{x} = \mathbf{0} \tag{I}$$

where A is the $m \times n$ matrix for which row i consists of $a_{i1}, a_{i2}, \ldots, a_{in}$, for $i = 1, 2, \ldots, m$, and $\mathbf{x}$ is an n-component column vector, indicated by $\mathbf{x} = [x_1, x_2, \ldots, x_n]^t$. Also, $\mathbf{0}$ denotes the zero column vector $[0, 0, \ldots, 0]^t$ with m components. It is important to note here that $\mathbf{x}$ is a vector of R^n, since it has n components.

In frame 1, Example (b), we have a homogeneous linear system where $m = 1$ (one equation) and $n = 3$ ($x_1, x_2,$ and x_3 occur), and in the homogeneous system of frame 1, Example (c), $m = 2$ and $n = 3$.

One of the most important ways to characterize subspaces of R^n is indicated in the following theorem.

Theorem 5.3

Given the homogeneous linear system $A\mathbf{x} = \mathbf{0}$, where A is an $m \times n$ matrix. A solution consists of an n-component column vector $\hat{\mathbf{x}}$ in R^n for which $A\hat{\mathbf{x}} = \mathbf{0}$. Let W be the subset of R^n consisting of all solutions to the homogeneous system.

Then W is a subspace of R^n.

Proof of Theorem 5.3

Let **u** and **v** be solutions of $A\mathbf{x} = \mathbf{0}$. This means $A\mathbf{u} = \mathbf{0}$ and $A\mathbf{v} = \mathbf{0}$. Thus,

$$A(\mathbf{u} + \mathbf{v}) = A\mathbf{u} + A\mathbf{v} = \mathbf{0} + \mathbf{0} = \mathbf{0},$$

which means that $\mathbf{u} + \mathbf{v}$ is a solution of $A\mathbf{x} = \mathbf{0}$. Also, for k a scalar,

$$A(k\mathbf{u}) = k(A\mathbf{u}) = k\mathbf{0} = \mathbf{0},$$

which means that $k\mathbf{u}$ is also a solution of $A\mathbf{x} = \mathbf{0}$. Since conditions (a) and (b) of Theorem 5.2 (frame 4) both hold, W is a subspace of R^n.

Let us illustrate Theorem 5.3 for the $n = 3$ case, that is, for homogeneous systems with three variables (unknowns). Consider

$$x_1 - x_2 + x_3 = 0 \tag{1}$$

$$2x_1 + x_2 + 3x_3 = 0 \tag{2}$$

$$3x_1 - x_2 - x_3 = 0, \tag{3}$$

which is the system in the Example of frame 18 of Chapter Two, but with the right-side constants replaced by zeros. Following the method of frame 18, we find that the *unique* solution is the trivial solution $x_1 = 0$, $x_2 = 0$, $x_3 = 0$, which in vector form is $\mathbf{x} = \mathbf{0} = [0, 0, 0]^t$. (A note on the solution technique: We know from Chapter Two that our system will have the same reduced row echelon form to the left of the dividing line as for the example of frame 18. Thus, the final reduced row echelon form for the current homogeneous example is $I_3|\mathbf{0}$, which implies that $\mathbf{x} = \mathbf{0}$.) Thus, the subspace here is the zero subspace $\{\mathbf{0}\}$, which corresponds to the origin point $(0, 0, 0)$.

In Example (c) of frames 1 and 4, we had the $m = 2$, $n = 3$ system

$$x_1 + x_2 + x_3 = 0, \tag{1}$$

$$x_1 + 2x_2 - x_3 = 0 \tag{2}$$

for which the solution set consisted of the straight line through the origin given by $x_1 = -3r$, $x_2 = 2r$, $x_3 = r$ or, in column vector form, $\mathbf{x} = [-3r, 2r, r]^t = r[-3, 2, 1]^t$. Thus, the subspace here is a straight line through the origin. In vector language one would say that the subspace is the set of column vectors corresponding to the points on the straight line.

In Example (b) of frame 1, we had the $m = 1$, $n = 3$ system (i.e., only one equation in three variables) given by $x_1 - 2x_2 + 4x_3 = 0$. The subspace here *is* the plane through the origin whose scalar equation is $x_1 - 2x_2 + 4x_3 = 0$. In vector language one would say that the subspace is the set of column vectors corresponding to the points on the plane. These vectors were given by $\mathbf{x} = r[2, 1, 0]^t + s[-4, 0, 1]^t$, with r and s arbitrary.

In the previous discussion of homogeneous systems for which $n = 3$, we have the following *typical* results, described in geometrical fashion. (Remember, each equation of the form $ax_1 + bx_2 + cx_3 = 0$ is the equation of a plane in 3-space that goes through the origin.)

1. If $m = 3$ (three equations), the subspace is the origin point where the three planes intersect.
2. If $m = 2$ (two equations), the subspace is a line passing through the origin which is the line of intersection of the two planes represented by the equations.
3. If $m = 1$ (one equation), the subspace is a plane passing through the origin. The equation of the plane is the given equation.

Though the preceding results are the typical results for $n = 3$, with, by the way, comparable results holding for any n, not all situations are typical. We illustrate in the following $m = 3$, $n = 3$ example in which the subspace is a straight line through the origin.

Example (a): In the example of frame 21 of Chapter Two, we have a system for which $m = 3$ and $n = 3$:

$$2x_2 - 2x_3 = 0 \tag{1}$$

$$x_1 + x_2 + x_3 = 0 \tag{2}$$

$$x_1 + 2x_2 \qquad = 0 \tag{3}$$

The solution was given by $x_1 = -2r$, $x_2 = r$, $x_3 = r$, or in column vector form, $\mathbf{x} = [-2r, r, r]^t = r[-2, 1, 1]^t$. Thus, geometrically, the subspace is a straight line through the origin. Here the planes corresponding to the three equations intersect along a straight line, and not at a single point. Refer to Figure 2.5 in frame 18 of Chapter Two for a diagram of three planes intersecting along a straight line.

Example (b): Refer to Theorem 5.3 and its proof. The set of solutions of $A\mathbf{x} = \mathbf{b}$, where A is $m \times n$ and $\mathbf{b} \neq \mathbf{0}$, is *not* a subspace of R^n. Prove this. Note that $\mathbf{b} \neq \mathbf{0}$ means that at least one of the m components of $\mathbf{b} = [b_1, b_2, \ldots, b_m]^t$ is unequal to zero. For example, in the system of two equations $2x_1 + x_2 = 0$; $3x_1 - x_2 = 8$, the vector $\mathbf{b}$ equals $[0, 8]^t$, which means that $\mathbf{b} \neq \mathbf{0}$.

Hint: for proof. Show that one of the conditions in Theorem 5.2 (frame 4) does not hold.

Let W be the subset of solutions of $A\mathbf{x} = \mathbf{b}$ and let $\mathbf{u}$ be one such solution. Thus, $A\mathbf{u} = \mathbf{b}$. For the scalar multiple $2\mathbf{u}$ we have $A(2\mathbf{u}) = 2A\mathbf{u} = 2\mathbf{b} \neq$

b, and thus **2u** is not a solution of $A\mathbf{x} = \mathbf{b}$. Since condition (b) of Theorem 5.2 fails, W is not a subspace. (In the special case where $A\mathbf{x} = \mathbf{b}$ has no solutions, then W is the empty set. So again, W is not a subspace, since the empty set is not a subspace. Note the words *nonempty subset* in Theorem 5.2 of frame 4.)

B. LINEAR INDEPENDENCE AND DEPENDENCE

7. We now continue our discussion of vector spaces and subspaces of vector spaces. The vector space we shall focus on the most is Euclidean n-space R^n, which, as previously indicated in Chapter Four (see Definition 4.1 and Notes following it), will be considered in two ways:

1. Algebraically, as column vectors of the form $\mathbf{x} = [x_1, x_2, \ldots, x_n]^t$.
2. Geometrically, as points $(x_1, x_2, \ldots, x_n)$ in an n-dimensional space.

Of course, from a geometrical point of view, we also have the representation of $\mathbf{x} = [x_1, x_2, \ldots, x_n]^t$ as a directed line segment (arrow) from the origin to the terminal point $(x_1, x_2, \ldots, x_n)$. (See Section A of Chapter Four.) The latter idea is useful if we want to pictorially represent the addition of vectors (say, $\mathbf{x} + \mathbf{y}$), or scalar multiplication of a vector (say, $k\mathbf{x}$) in R^n. Models for our general discussions will come from R^2 (two-dimensional space) and R^3 (three-dimensional space).

Every vector space has an infinite number of vectors except for the zero vector space $\{\mathbf{0}\}$, which has only the single vector $\mathbf{0}$. (Recall that $\{\mathbf{0}\}$ is a subspace of every vector space.) To demonstrate that every other vector space has an infinite number of vectors, suppose that a nonzero vector $\mathbf{x}$ is in a vector space V. Then $k\mathbf{x}$ is in V where k is *any* real number, and so V contains an infinite number of vectors.

In this and the next section we shall show that every nonzero vector space studied in this book can be described in terms of a *finite number* of vectors. Note the wording here. A collection of such a finite number of vectors will be known as a *basis*. We will get back to the basis concept later.

Definition 5.3 (Linear Combination)

A vector $\mathbf{x}$ is defined to be a linear combination of the vectors $\mathbf{v}_1, \mathbf{v}_2, \ldots, \mathbf{v}_r$ if it can be written as

$$\mathbf{x} = c_1\mathbf{v}_1 + c_2\mathbf{v}_2 + \ldots + c_r\mathbf{v}_r,$$

where $c_1, c_2, \ldots, c_r$ are scalars.

Example (a): For R^2, suppose $\mathbf{v} = \begin{bmatrix} 2 \\ 1 \end{bmatrix}$ and $\mathbf{w} = \begin{bmatrix} 1 \\ 3 \end{bmatrix}$. Determine the linear combination $\mathbf{x} = 3\mathbf{v} + 2\mathbf{w}$ algebraically. Give geometrical representations of $\mathbf{v}$, $\mathbf{w}$, and $\mathbf{x}$ on a single diagram.

Solution:

$$\mathbf{x} = 3\mathbf{v} + 2\mathbf{w} = 3\begin{bmatrix} 2 \\ 1 \end{bmatrix} + 2\begin{bmatrix} 1 \\ 3 \end{bmatrix} = \begin{bmatrix} 3(2) + 2(1) \\ 3(1) + 2(3) \end{bmatrix} = \begin{bmatrix} 8 \\ 9 \end{bmatrix}$$

Geometrically, we indicate the addition $3\mathbf{v} + 2\mathbf{w}$ by using the parallelogram rule (frame 2 of Chapter Four). Refer to Figure 5.1 for an explanation in the current example. First, we construct geometrical representations of $\mathbf{v}$, $\mathbf{w}$, and then $3\mathbf{v}$ and $2\mathbf{w}$. Next, we form a parallelogram by drawing broken lines parallel to $2\mathbf{w}$ and $3\mathbf{v}$, respectively, from the terminal points of $3\mathbf{v}$ and $2\mathbf{w}$. The directed line segment drawn from the origin to the point of intersection of the broken lines is the geometrical representation of $\mathbf{x}$.

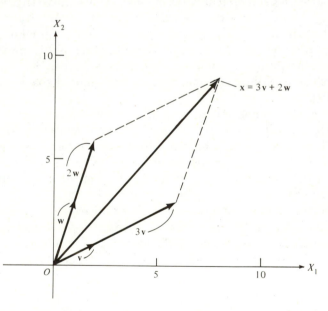

FIGURE 5.1   Illustrating the linear combination $\mathbf{x} = 3\mathbf{v} + 2\mathbf{w}$, where $\mathbf{v} = [2, 1]^t$; $\mathbf{w} = [1, 3]^t$; thus, $\mathbf{x} = [8, 9]^t$.

Given a collection of one or more vectors of a vector space V. The set of *all* linear combinations of such vectors is a subspace of V. (Later, in frame 10, we shall refer to such a set with the symbol *Span*.) The proof for the case of two vectors will be given in Example (b) below.

This provides us with a simple way of constructing subspaces of a vector space.

Example (b): Let $\mathbf{v}_1$ and $\mathbf{v}_2$ be two vectors in vector space V. Show that the set of all linear combinations of $\mathbf{v}_1$ and $\mathbf{v}_2$ is a subspace of V.

Solution: Let W be the subset of vectors of V of the form

$$c_1\mathbf{v}_1 + c_2\mathbf{v}_2,$$

where c_1 and c_2 are any real numbers (thus, every vector in W is a linear combination of $\mathbf{v}_1$ and $\mathbf{v}_2$). Let us verify conditions (a) and (b) of Theorem 5.2. Let $\mathbf{w} = a_1\mathbf{v}_1 + a_2\mathbf{v}_2$ and $\mathbf{y} = b_1\mathbf{v}_1 + b_2\mathbf{v}_2$ be vectors in W. Then,

$$\mathbf{w} + \mathbf{y} = (a_1\mathbf{v}_1 + a_2\mathbf{v}_2) + (b_1\mathbf{v}_1 + b_2\mathbf{v}_2)$$

$$= (a_1 + b_1)\mathbf{v}_1 + (a_2 + b_2)\mathbf{v}_2,$$

indicating that $\mathbf{w} + \mathbf{y}$ is in W. Also, if k is a scalar, then

$$k\mathbf{w} = k(a_1\mathbf{v}_1 + a_2\mathbf{v}_2) = (ka_1)\mathbf{v}_1 + (ka_2)\mathbf{v}_2,$$

which means that $k\mathbf{w}$ is in W. Thus, W is a subspace of V.

Example (c): Given vectors $\mathbf{u} = [-2, 7, 4]^t$ and $\mathbf{v} = [4, 8, 3]^t$ in R^3. Determine if the vector $\mathbf{w} = [2, 37, 18]^t$ is a linear combination of $\mathbf{u}$ and $\mathbf{v}$.

Solution: The vector $\mathbf{w}$ is a linear combination of $\mathbf{u}$ and $\mathbf{v}$ if there are scalars c_1 and c_2 such that

$$\mathbf{w} = c_1\mathbf{u} + c_2\mathbf{v} \tag{1}$$

Substituting for $\mathbf{u}$, $\mathbf{v}$, and $\mathbf{w}$, we obtain

$$c_1\begin{bmatrix} -2 \\ 7 \\ 4 \end{bmatrix} + c_2\begin{bmatrix} 4 \\ 8 \\ 3 \end{bmatrix} = \begin{bmatrix} 2 \\ 37 \\ 18 \end{bmatrix}. \tag{2}$$

Combining terms on the left, we get

$$\begin{bmatrix} -2c_1 + 4c_2 \\ 7c_1 + 8c_2 \\ 4c_1 + 3c_2 \end{bmatrix} = \begin{bmatrix} 2 \\ 37 \\ 18 \end{bmatrix}, \tag{3}$$

which leads to the three scalar equations:

$$-2c_1 + 4c_2 = 2 \tag{4a}$$

$$7c_1 + 8c_2 = 37 \tag{4b}$$

$$4c_1 + 3c_2 = 18 \tag{4c}$$

Now, applying the Gauss–Jordan elimination method (Method 2.1 in frame 14 of Chapter Two) to the augmented coefficient matrix $\begin{bmatrix} -2 & 4 & | & 2 \\ 7 & 8 & | & 37 \\ 4 & 3 & | & 18 \end{bmatrix}$, we obtain the following reduced row echelon form:

$$\begin{bmatrix} 1 & 0 & | & 3 \\ 0 & 1 & | & 2 \\ 0 & 0 & | & 0 \end{bmatrix}. \tag{5}$$

This indicates that the unique solution for c_1 and c_2 is $c_1 = 3$, $c_2 = 2$. This means that $\mathbf{w}$ is the following linear combination of $\mathbf{u}$ and $\mathbf{v}$:

$$\mathbf{w} = 3\mathbf{u} + 2\mathbf{v} \tag{6}$$

Example (d):   Determine if the vector $\mathbf{y} = [10, 31, 17]^t$ is a linear combination of the vectors $\mathbf{u}$ and $\mathbf{v}$ of Example (c). Proceed as in Example (c).

- - - - - - - - - -

Using Method 2.1, we obtain the following augmented matrix (not in reduced row echelon form):

$$\begin{bmatrix} 1 & 0 & | & 1 \\ 0 & 1 & | & 3 \\ 0 & 0 & | & 4 \end{bmatrix}$$

The last row indicates that there is no solution for c_1 and c_2, and thus $\mathbf{y}$ is not a linear combination of $\mathbf{u}$ and $\mathbf{v}$. Refer to frame 16 of Chapter Two for a similar situation.

8.   Notice how the material of Chapter Two is playing a role in the current development. We will see more of this and also material from Chapter Three, in the coming frames.

However, it should be noted that other methods, perhaps learned in ordinary algebra, can be used to analyze certain problems. For example, in Example (c) of the previous frame, we could do the problem as follows:

1. Solve Eqs. (4a) and (4b) for c_1 and c_2 using, say, the method of substitution. This leads to $c_1 = 3$ and $c_2 = 2$.
2. Then substitute the values just found for c_1 and c_2 into Eq. (4c). In this case we would get the consistent equation $18 = 18$, thus indicating that the answer for the linear combination is $3\mathbf{u} + 2\mathbf{v}$.

Using the same type of approach in Example (d) of frame 7 would lead to an inconsistency in step 2.

In the following we shall again employ the curled brackets symbols { } to enclose the elements of a set. Refer to frame 3. The elements in our work will almost always be vectors.

Definition 5.4

Let $S = \{\mathbf{v}_1, \mathbf{v}_2, \ldots, \mathbf{v}_r\}$ be a set of vectors in a vector space V. If every vector in V can be expressed in at least one way as a linear combination of the vectors in S, then we say that the set S *spans* V, or that V is *spanned* by S.

Note: The vector space V in Definition 5.4 may itself be a subspace of some other vector space. Remember that any subspace of a vector space is itself a vector space (Definition 5.2).

Example (a): Let V be the vector space R^2 and let $S = \{\mathbf{v}, \mathbf{w}\}$, where $\mathbf{v} = [2, 1]^t$ and $\mathbf{w} = [1, 3]^t$, as in Example (a) of frame 7. Determine whether S spans R^2.

Solution: We let $\mathbf{x} = [a, b]^t$ represent any vector in R^2 (a and b are arbitrary real numbers). We must determine if there are constants c_1 and c_2 such that

$$c_1\mathbf{v} + c_2\mathbf{w} = \mathbf{x} \tag{1}$$

This leads to

$$c_1 \begin{bmatrix} 2 \\ 1 \end{bmatrix} + c_2 \begin{bmatrix} 1 \\ 3 \end{bmatrix} = \begin{bmatrix} a \\ b \end{bmatrix}, \tag{2}$$

from which we obtain

$$\begin{bmatrix} 2c_1 + c_2 \\ c_1 + 3c_2 \end{bmatrix} = \begin{bmatrix} a \\ b \end{bmatrix}, \tag{3}$$

and thus the following system in "unknowns" c_1 and c_2:

$$2c_1 + c_2 = a \tag{4a}$$

$$c_1 + 3c_2 = b \tag{4b}$$

We find the unique solution for c_1 and c_2 (using Theorem 3.11, Cramer's rule, for example) to be

$$c_1 = \frac{3a - b}{5}, \qquad c_2 = \frac{2b - a}{5} \tag{5a, 5b}$$

Thus, $\{\mathbf{v}, \mathbf{w}\}$ spans R^2.

For example, suppose $\mathbf{x} = [5.5, 6.5]^t$, which means $a = 5.5$ and $b = 6.5$ above. We then solve for c_1 and c_2 from Eqs. (5a) and (5b) as follows:

$$c_1 = \frac{3(5.5) - 6.5}{3} = \frac{16.5 - 6.5}{5} = 2,$$

$$c_2 = \frac{2(6.5) - 5.5}{5} = \frac{7.5}{5} = 1.5$$

Thus, $\mathbf{x} = 2\mathbf{v} + 1.5\mathbf{w}$.

Let us illustrate the fact that $\{\mathbf{v}, \mathbf{w}\}$ spans R^2. Refer to Figure 5.2a, where the geometrical representations of $\mathbf{v}$ and $\mathbf{w}$ are drawn initially. If we wanted to graphically find the linear combination of $\mathbf{v}$ and $\mathbf{w}$ that equals $\mathbf{x} = \begin{bmatrix} 5.5 \\ 6.5 \end{bmatrix}$, we would next draw the geometrical representation of $\mathbf{x}$ as shown.

Now we use a variation of the parallelogram rule. We draw broken lines through the terminal point of $\mathbf{x}$, which are parallel to $\mathbf{v}$ and $\mathbf{w}$. Then we draw lines through vectors $\mathbf{v}$ and $\mathbf{w}$ (i.e., we "extend" vectors $\mathbf{v}$ and $\mathbf{w}$) until the latter lines intersect the broken lines. The vectors from the origin to the intersection points are $2\mathbf{v}$ and $1.5\mathbf{w}$, respectively. Here $2\mathbf{v}$ is twice as long as $\mathbf{v}$ and $1.5\mathbf{w}$ is 1.5 times as long as $\mathbf{w}$. The parallelogram rule thus indicates that $\mathbf{x} = 2\mathbf{v} + 1.5\mathbf{w}$.

Example (b): Suppose $\mathbf{v}$ and $\mathbf{w}$ are as in Example (a). For $\mathbf{y} = \begin{bmatrix} -2 \\ 4 \end{bmatrix}$, determine c_1 and c_2 such that $\mathbf{y} = c_1\mathbf{v} + c_2\mathbf{w}$. Then geometrically interpret as in Figure 5.2a.

Hint: Use the equations from Example (a).

— — — — — — — — — —

Since $a = -2$ and $b = 4$,

$$c_1 = \frac{3a - b}{5} = \frac{-6 - 4}{5} = -2 \qquad \text{and}$$

$$c_2 = \frac{2b - a}{5} = \frac{8 - (-2)}{5} = 2$$

Thus, $\mathbf{y} = -2\mathbf{v} + 2\mathbf{w}$.

The geometrical interpretation is given in Figure 5.2b. Note how we extend the broken lines (which are parallel to $\mathbf{w}$ and $\mathbf{v}$) until they meet the vector extensions of $\mathbf{v}$ and $\mathbf{w}$; the latter are $-2\mathbf{v}$ and $2\mathbf{w}$, respectively.

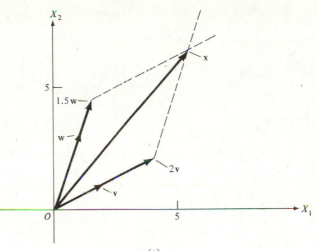

(a)

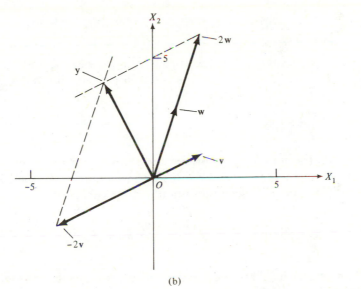

(b)

FIGURES 5.2a and 5.2b $v = [2, 1]'$ and $w = [1, 3]'$. In 5.2a, $x = [5.5, 6.5]'$, and thus $x = 2v + 1.5w$; In 5.2b, $y = [-2, 4]'$, and thus $y = -2v + 2w$.

9.   The two vectors v and w of the previous frame form a spanning set for R^2 in the sense that any vector in R^2 can be expressed *uniquely* in terms of v and w. Note the words *"at least"* in Definition 5.4. Typically (but not always), a set of three vectors would form a spanning set for R^2 in that there would be more than one solution for c_1, c_2, and c_3 when trying to express an arbitrary vector x of R^2 as

$$x = c_1 v_1 + c_2 v_2 + c_3 v_3$$

for the spanning set $S = \{v_1, v_2, v_3\}$.

Example (a): The set $\hat{S} = \{\mathbf{v}, \mathbf{w}, \mathbf{z}\}$ is a spanning set for R^2, where $\mathbf{v} = \begin{bmatrix} 2 \\ 1 \end{bmatrix}$, and $\mathbf{w} = \begin{bmatrix} 1 \\ 3 \end{bmatrix}$, as in frame 8, and $\mathbf{z} = \begin{bmatrix} 1 \\ 1 \end{bmatrix}$.

Determine c_1, c_2, and c_3 such that any vector $\mathbf{x} = \begin{bmatrix} a \\ b \end{bmatrix}$ is expressible as

$$\mathbf{x} = c_1\mathbf{v} + c_2\mathbf{w} + c_3\mathbf{z} \tag{1}$$

Solution: Proceeding as in Example (a) of frame 8, we obtain

$$\begin{bmatrix} 2c_1 + c_2 + c_3 \\ c_1 + 3c_2 + c_3 \end{bmatrix} = \begin{bmatrix} a \\ b \end{bmatrix}, \tag{2}$$

and thus

$$2c_1 + c_2 + c_3 = a \tag{3a}$$

$$c_1 + 3c_2 + c_3 = b, \tag{3b}$$

a system of two equations in the three unknowns c_1, c_2, and c_3.
   Using Method 2.1, we obtain the reduced row echelon form:

$$\begin{bmatrix} \underline{1} & 0 & 2/5 & (3a - b)/5 \\ 0 & \underline{1} & 1/5 & (2b - a)/5 \end{bmatrix}, \tag{4}$$

which yields the following solutions for c_1 and c_2 in terms of c_3 (c_1 and c_2 are leading variables and c_3 is a nonleading variable; see frame 15 of Chapter Two for definitions of leading and nonleading variables):

$$c_1 = \frac{-2c_3 + 3a - b}{5} \qquad c_2 = \frac{-c_3 + 2b - a}{5} \tag{5a) (5b}$$

Here c_3 is arbitrary. Thus, we have infinitely many solutions for c_1, c_2, and c_3, and thus infinitely many ways to represent an arbitrary vector $\mathbf{x}$ in terms of $\mathbf{v}$, $\mathbf{w}$, and $\mathbf{z}$.

Let us illustrate for $\mathbf{x} = \begin{bmatrix} 5.5 \\ 6.5 \end{bmatrix}$, the sample vector in Example (a) of frame 8. Thus, $a = 5.5$ and $b = 6.5$. If we let $c_3 = 0$, say, we obtain $c_1 = 2$, and $c_2 = 1.5$, and thus $\mathbf{x} = 2\mathbf{v} + 1.5\mathbf{w} + 0\mathbf{z} = 2\mathbf{v} + 1.5\mathbf{w}$, as in frame 8.

Example (b):   Suppose again that $x = \begin{bmatrix} 5.5 \\ 6.5 \end{bmatrix}$ and that $\mathbf{v}$, $\mathbf{w}$, and $\mathbf{z}$ are as in Example (a). Determine c_1 and c_2 if we set $c_3 = 1$.

_ _ _ _ _ _ _ _ _ _ _

Solution: From Eqs. (5a) and (5b) we have

$$c_1 = \frac{-2 + 16.5 - 6.5}{5} = \frac{8}{5} = 1.6 \quad \text{and}$$

$$c_2 = \frac{-1 + 13 - 5.5}{5} = \frac{6.5}{5} = 1.3,$$

and thus $\mathbf{x} = 1.6\mathbf{v} + 1.3\mathbf{w} + \mathbf{z}$.

10. It should be clear that for a set of vectors S to span R^2, S must contain at least two vectors. If S contained a single nonzero vector, say, $S = \{\mathbf{v}\}$, where $\mathbf{v} = [2, 1]^t$, then the totality of linear combinations of $\mathbf{v}$ could not possibly span R^2. In fact, the set of all possible linear combinations of $\mathbf{v}$, which would have the form $c\mathbf{v} = \begin{bmatrix} 2c \\ c \end{bmatrix}$, would span a straight line that passes through the origin. (A scalar equation would be $2x_2 - x_1 = 0$.) See frame 16 of Chapter Four also. Similar statements apply for R^n. Thus, a set of vectors that spans R^3 must contain at least three vectors, and a set that spans R^n must contain at least n vectors.

Remember in frame 7, right after Example (a), we said that for a collection of one or more vectors of a vector space V, the set of *all* linear combinations of such vectors will be a subspace of V. We restate this now in theorem form.

Theorem 5.4

If $\mathbf{v}_1$, $\mathbf{v}_2$, . . ., $\mathbf{v}_p$ are vectors in a vector space V, then the set W of all linear combinations of $\mathbf{v}_1$, $\mathbf{v}_2$, . . ., $\mathbf{v}_p$ is a subspace of V.

Suppose we denote the set of vectors $\mathbf{v}_1$, $\mathbf{v}_2$, . . ., $\mathbf{v}_p$ by $S = \{\mathbf{v}_1, \mathbf{v}_2, . . ., \mathbf{v}_p\}$. Let us define the set of all linear combinations of vectors $\mathbf{v}_1$, $\mathbf{v}_2$, . . ., $\mathbf{v}_p$ by the symbol

Span $(\{\mathbf{v}_1, \mathbf{v}_2, . . ., \mathbf{v}_p\})$ or   Span (S)

We interpret this as the subspace spanned by $\{\mathbf{v}_1, \mathbf{v}_2, . . ., \mathbf{v}_p\}$.

Thus, Theorem 5.4 says that Span (S), which equals W in the theorem, is a subspace of V. The symbol "Span" is meaningful here. We are saying that the set of vectors $\{\mathbf{v}_1, \mathbf{v}_2, . . ., \mathbf{v}_p\}$ spans a subspace of V.

To illustrate Theorem 5.4 in R^2, we know that for $\mathbf{v} = \begin{bmatrix} 2 \\ 1 \end{bmatrix}$, the set of all linear combinations of $\mathbf{v}$, namely $\{c\begin{bmatrix} 2 \\ 1 \end{bmatrix}\}$, where c can take on any scalar value, constitutes a straight line through the origin. The latter is a subspace of R^2.

Several points on this straight line are $(0, 0)$ for $c = 0$, $(2, 1)$ for $c = 1$, $(-2, -1)$ for $c = -1$, and $(30, 15)$ for $c = 15$.

In R^3, the set of all linear combinations of a single vector $\mathbf{w}$, say, $\mathbf{w} = [2, 3, 1]^t$, also constitutes a straight line through the origin, but this time in 3-space. See frame 16 of Chapter Four. This set of vectors of the form $c\mathbf{w} = [2c, 3c, c]^t$ is a subspace of R^3. In short, Span $(\{\mathbf{w}\})$ is a subspace of R^3.

Example (a): In R^3, show that the subspace spanned by the vectors $\mathbf{v}_1 = [2, 3, 1]^t$ and $\mathbf{v}_2 = [4, 8, 3]^t$—in short, Span $(\{\mathbf{v}_1, \mathbf{v}_2\})$—represents a plane through the origin of R^3.

Solution: The symbol Span $(\{\mathbf{v}_1, \mathbf{v}_2\})$ stands for the set of all linear combinations of the form

$$c_1\mathbf{v}_1 + c_2\mathbf{v}_2 = c_1 \begin{bmatrix} 2 \\ 3 \\ 1 \end{bmatrix} + c_2 \begin{bmatrix} 4 \\ 8 \\ 3 \end{bmatrix} = \begin{bmatrix} 2c_1 + 4c_2 \\ 3c_1 + 8c_2 \\ c_1 + 3c_2 \end{bmatrix}, \tag{1}$$

where c_1 and c_2 take on all scalar values. Theorem 5.4 indicates that Span $(\{\mathbf{v}_1, \mathbf{v}_2\})$ is a subspace of R^3. Our task is to show for any vector $\mathbf{x} = [x_1, x_2, x_3]^t$ expressible in the form $c_1\mathbf{v}_1 + c_2\mathbf{v}_2$ that the components x_1, x_2, and x_3 are related by a scalar equation of the form

$$a_1 x_1 + a_2 x_2 + a_3 x_3 = 0 \tag{2}$$

This, after all, is the equation for a plane in 3-space that passes through the origin. Refer to the discussion on hyperplanes following Definition 4.5 (frame 11 of Chapter Four). Note that the origin coordinates $(0, 0, 0)$ satisfy Eq. (2).

Thus, we set $\mathbf{x} = [x_1, x_2, x_3]^t$, the column vector for a point (x_1, x_2, x_3) on the plane, equal to the expression from (1) above.

$$\begin{bmatrix} 2c_1 + 4c_2 \\ 3c_1 + 8c_2 \\ c_1 + 3c_2 \end{bmatrix} = \begin{bmatrix} x_1 \\ x_2 \\ x_3 \end{bmatrix} \tag{3}$$

Thus, we obtain the following three scalar equations:

$$2c_1 + 4c_2 = x_1 \tag{4a}$$

$$3c_1 + 8c_2 = x_2 \tag{4b}$$

$$c_1 + 3c_2 = x_3 \tag{4c}$$

Now we "solve for" c_1 and c_2 from Eqs. (4a) and (4b), say. Cramer's rule (Theorem 3.11) is useful here. We get

$$c_1 = \frac{8x_1 - 4x_2}{4} \quad \text{and} \quad c_2 = \frac{2x_2 - 3x_1}{4} \tag{5a} \tag{5b}$$

Substituting c_1 and c_2 from Eqs. (5a) and (5b) into Eq. (4c) yields

$$x_1 - 2x_2 + 4x_3 = 0, \tag{6}$$

which is in the desired form.

We check by noting that the points corresponding to v_1 and v_2, namely, $(2, 3, 1)$ and $(4, 8, 3)$, satisfy scalar equation (6).

Example (b): In Example (a), determine the vector $x = c_1 v_1 + c_2 v_2$ for which $c_1 = -11$ and $c_2 = 5$. Show that the components of the vector satisfy Eq. (6) above.

— — — — — — — — — —

From Eq. (3) in Example (a), we have

$$\mathbf{x} = \begin{bmatrix} 2c_1 + 4c_2 \\ 3c_1 + 8c_2 \\ c_1 + 3c_2 \end{bmatrix} = \begin{bmatrix} -22 + 20 \\ -33 + 40 \\ -11 + 15 \end{bmatrix} = \begin{bmatrix} -2 \\ 7 \\ 4 \end{bmatrix}$$

The point $(-2, 7, 4)$ corresponds to vector $\mathbf{x}$. Substituting $x_1 = -2$, $x_2 = 7$, $x_3 = 4$ into Eq. (6), we obtain

$$-2 - 14 + 16 = 0,$$

which indicates that the equation is satisfied.

11. Let $S = \{v_1, v_2, \ldots, v_r\}$ be a set of vectors in a vector space V. We know that the equation

$$c_1 v_1 + c_2 v_2 + \cdots + c_r v_r = 0 \tag{I}$$

will certainly have the solution $c_1 = c_2 = \ldots = c_r = 0$. (The reason: from Theorem 5.1 in frame 4, $0\mathbf{x} = 0$; also, $0 + 0 + \cdots + 0 = 0$ from comment after Theorem 5.2 in frame 4.)

Definition 5.5 (Linear Independence and Dependence)

If the *only* solution of Eq. (I) is $c_1 = c_2 = \ldots = c_r = 0$, then S is called a *linearly independent* set. If there are other solutions for the c_i's, then S is called a *linearly dependent* set.

Notes: (i) If the set S *is* linearly independent, we also say that the vectors $v_1, v_2, \ldots, v_r$ *are* linearly independent (similarly, if S is a linearly dependent set).

(ii)   Other wordings when S is linearly independent are the following:

"If Eq. (I) holds only when $c_1 = c_2 = \ldots = c_r = 0$, then S is a linearly independent set."

"If Eq. (I) implies that $c_1 = c_2 = \ldots = c_r = 0$, then S is a linearly independent set."

(iii)   When we speak of "other" solutions when the set S is linearly dependent, we mean solutions of Eq. (I) for which at least one of the c_i's is *unequal* to zero. Another wording in the dependent case is "S is linearly dependent if there exist constants $c_1, c_2, \ldots, c_r$, not all zero, such that Eq. (I) holds."

Example (a):   Refer to Example (a) of frame 8. There we established that the set $S = \{\mathbf{v}, \mathbf{w}\}$, where $\mathbf{v} = \begin{bmatrix} 2 \\ 1 \end{bmatrix}$ and $\mathbf{w} = \begin{bmatrix} 1 \\ 3 \end{bmatrix}$, spans R^2. Determine if S is a linearly independent set.

Solution: We consider the equation $c_1\mathbf{v} + c_2\mathbf{w} = \mathbf{0}$, that is,

$$c_1 \begin{bmatrix} 2 \\ 1 \end{bmatrix} + c_2 \begin{bmatrix} 1 \\ 3 \end{bmatrix} = \begin{bmatrix} 0 \\ 0 \end{bmatrix} \tag{1}$$

We thus obtain

$$\begin{bmatrix} 2c_1 + c_2 \\ c_1 + 3c_2 \end{bmatrix} = \begin{bmatrix} 0 \\ 0 \end{bmatrix}, \tag{2}$$

and Eq. (2) leads to the following homogeneous system of equations:

$$2c_1 + c_2 = 0 \tag{3a}$$

$$c_1 + 3c_2 = 0 \tag{3b}$$

By using Method 2.1, for example, we see that the reduced row echelon form for Eqs. (3a) and (3b) is $\begin{bmatrix} 1 & 0 & 0 \\ 0 & 1 & 0 \end{bmatrix}$, which implies that the *only* solution is $c_1 = c_2 = 0$. (For a review refer to frames 20 and 21 of Chapter Two.) Thus, S is linearly independent.

Also, from Theorem 3.10 (frame 13 of Chapter Three), the only solution of Eqs. (3a) and (3b) is $c_1 = c_2 = 0$ since the determinant $|A| = 5$, which is unequal to zero. Here $A = \begin{bmatrix} 2 & 1 \\ 1 & 3 \end{bmatrix}$. The columns of A are $\mathbf{v}$ and $\mathbf{w}$, respectively, and the coefficients c_1 and c_2 here correspond to the components of vector $\mathbf{x}$ in Theorem 3.10. See Example (b) of frame 13 in Chapter Three.

Example (b): Refer to Example (a) of frame 9. Show that the set $\hat{S} = \{v, w,$ $z\}$, where $v = \begin{bmatrix} 2 \\ 1 \end{bmatrix}$, $w = \begin{bmatrix} 1 \\ 3 \end{bmatrix}$, and $z = \begin{bmatrix} 1 \\ 1 \end{bmatrix}$, is linearly dependent.

Solution: From

$$c_1 v + c_2 w + c_3 z = 0, \tag{1}$$

we obtain

$$\begin{bmatrix} 2c_1 + c_2 + c_3 \\ c_1 + 3c_2 + c_3 \end{bmatrix} = \begin{bmatrix} 0 \\ 0 \end{bmatrix},$$

which leads to the following homogeneous linear system:

$$2c_1 + c_2 + c_3 = 0 \tag{2a}$$

$$c_1 + 3c_2 + c_3 = 0 \tag{2b}$$

The reduced row echelon form for Eqs. (2a) and (2b) is $\begin{bmatrix} 1 & 0 & 2/5 & 0 \\ 0 & 1 & 1/5 & 0 \end{bmatrix}$, which indicates that

$$c_1 = \frac{-2c_3}{5}, \tag{3a}$$

$$c_2 = \frac{-c_3}{5}, \qquad \text{where } c_3 \text{ is arbitrary.} \tag{3b}$$

This indicates that there exist solutions other than $c_1 = c_2 = c_3 = 0$. In fact, we see there are infinitely many solutions, one for each value of c_3. For example, if $c_3 = 1$, then $c_1 = -\frac{2}{5}$ and $c_2 = -\frac{1}{5}$ and if $c_3 = -5$, then $c_1 = 2$ and $c_2 = 1$.

Thus $\hat{S}$ is a linearly dependent set.

Notes: (i) From the Note following the Corollary to Theorem 2.4 in frame 25 of Chapter Two, we see there are *automatically* infinitely many solutions for c_1, c_2, and c_3, since we have a homogeneous linear system with fewer equations (2) than variables (3). (ii) Similarly, if we have a set of vectors from R^n (each vector thus has n components) containing more than n vectors, then the set is linearly dependent.

Example (c): Determine whether the vectors $v_1 = [2, 1, 1]^t$, $v_2 = [3, 2, 2]^t$, $v_3 = [4, 1, 3]^t$ of R^3 are linearly independent. Repeat for the vectors $w_1 = [3, -1, 2]^t$, $w_2 = [-9, 3, -6]^t$, $w_3 = [1, 2, -1]^t$.

Solution: We form the equation

$$c_1 \mathbf{v}_1 + c_2 \mathbf{v}_2 + c_3 \mathbf{v}_3 = \mathbf{0}, \tag{1}$$

with $\mathbf{0} = [0, 0, 0]^t$ here, which leads to the homogeneous system

$$2c_1 + 3c_2 + 4c_3 = 0 \tag{2a}$$

$$c_1 + 2c_2 + c_3 = 0 \tag{2b}$$

$$c_1 + 2c_2 + 3c_3 = 0 \tag{2c}$$

By using Method 2.1, we see that the reduced row echelon form for this system is $\begin{bmatrix} \underline{1} & 0 & 0 & 0 \\ 0 & \underline{1} & 0 & 0 \\ 0 & 0 & \underline{1} & 0 \end{bmatrix}$, which means that the only solution is $c_1 = c_2 = c_3 = 0$. Thus, the vectors $\mathbf{v}_1$, $\mathbf{v}_2$, and $\mathbf{v}_3$ are linearly independent.

For an alternate approach, refer to Theorem 3.10 (frame 13 of Chapter Three). The coefficient matrix of the homogeneous system of Eqs. (2a), (2b), and (2c) is $A = \begin{bmatrix} 2 & 3 & 4 \\ 1 & 2 & 1 \\ 1 & 2 & 3 \end{bmatrix}$; the columns are the vectors $\mathbf{v}_1$, $\mathbf{v}_2$, and $\mathbf{v}_3$, respectively. Since the determinant $|A| = 2$, it follows that the only solution is $c_1 = c_2 = c_3 = 0$.

Now do the calculations for $\mathbf{w}_1$, $\mathbf{w}_2$, and $\mathbf{w}_3$.

— — — — — — — — — —

From the equation $c_1 \mathbf{w}_1 + c_2 \mathbf{w}_2 + c_3 \mathbf{w}_3 = \mathbf{0}$, we obtain the homogeneous system

$$3c_1 - 9c_2 + c_3 = 0 \tag{1a}$$

$$-c_1 + 3c_2 + 2c_3 = 0 \tag{1b}$$

$$2c_1 - 6c_2 - c_3 = 0 \tag{1c}$$

Let us use the determinant approach. The coefficient matrix is

$$\hat{A} = \begin{bmatrix} 3 & -9 & 1 \\ -1 & 3 & 2 \\ 2 & -6 & -1 \end{bmatrix}, \tag{2}$$

where the columns of $\hat{A}$ are $\mathbf{w}_1$, $\mathbf{w}_2$, and $\mathbf{w}_3$, respectively. Since $|\hat{A}| = 0$, there exist solutions other than $c_1 = c_2 = c_3 = 0$. (These other solutions are called nontrivial solutions; see frame 13 of Chapter Three.) Thus, $\mathbf{w}_1$, $\mathbf{w}_2$, and $\mathbf{w}_3$ are linearly dependent.

12.   Suppose that a candidate set consists of $\mathbf{u}_1 = [2, 1, 1]^t$, $\mathbf{u}_2 = [3, 2, 2]^t$, $\mathbf{u}_3 = [8, 2, 6]^t$, and $\mathbf{u}_4 = [0, 1, 1]^t$ in R^3. We immediately conclude that the set is linearly dependent, since it contains more than three vectors. See Note (ii) following Example (b) of frame 11.

If a set of vectors from a vector space contains the zero vector, the set is automatically linearly dependent. We illustrate for a special case.

> **Example (a):**   Given the set $S = \{\mathbf{v}_1, \mathbf{v}_2, \mathbf{v}_3\}$, where $\mathbf{v}_3 = \mathbf{0}$. The equation $c_1\mathbf{v}_1 + c_2\mathbf{v}_2 + c_3\mathbf{0} = \mathbf{0}$ holds with $c_1 = c_2 = 0$, and $c_3 = 1$, say. Since not all of the c_i's are equal to zero, S is linearly dependent.

Let us consider the meaning of linear dependence and independence of a set of two vectors. If the set $S = \{\mathbf{v}_1, \mathbf{v}_2\}$ is linearly dependent, then there exist scalars c_1 and c_2, not both zero, such that

$$c_1\mathbf{v}_1 + c_2\mathbf{v}_2 = \mathbf{0}$$

If $c_1 \neq 0$, then $\mathbf{v}_1 = -(c_2/c_1)\mathbf{v}_2 = k\mathbf{v}_2$. If $c_2 \neq 0$, then $\mathbf{v}_2 = -(c_1/c_2)\mathbf{v}_1 = \hat{k}\mathbf{v}_1$. Thus, one of the vectors is a scalar multiple of the other.

Conversely, if one vector is a scalar multiple of the other, then the vectors are linearly dependent. For example, if $\mathbf{v}_2 = m\mathbf{v}_1$, then $m\mathbf{v}_1 - \mathbf{v}_2 = \mathbf{0}$, and Eq. (I) of frame 11 holds (see right before Definition 5.5) with $c_1 = m$ and $c_2 = -1$.

Summarizing, the set $S = \{\mathbf{v}_1, \mathbf{v}_2\}$ is linearly dependent if and only if one of the vectors is a scalar multiple of the other. The appropriate sketches for R^2 are shown below in Figure 5.3.

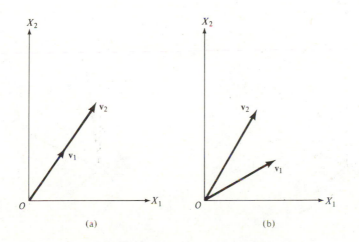

FIGURE 5.3   (a) Linearly dependent. (b) Linearly independent.

Also, in Figure 5.2a (frame 8), vectors $\mathbf{v}$ and $\mathbf{w}$ are linearly independent, whereas vectors $\mathbf{v}$ and $2\mathbf{v}$ are linearly dependent.

Let us illustrate linear dependence and independence for three vectors in R^3. Suppose $S = \{\mathbf{v}_1, \mathbf{v}_2, \mathbf{v}_3\}$ is linearly dependent. Then we can write

$$c_1\mathbf{v}_1 + c_2\mathbf{v}_2 + c_3\mathbf{v}_3 = \mathbf{0}, \tag{1}$$

where at least one of the c_i's is unequal to zero. If, say, $c_3 \neq 0$, then we can write

$$\mathbf{v}_3 = -\frac{c_1}{c_3}\mathbf{v}_1 - \frac{c_2}{c_3}\mathbf{v}_2 = k_1\mathbf{v}_1 + k_2\mathbf{v}_2, \tag{2}$$

which means that $\mathbf{v}_3$ is in the subspace of R^3 spanned by the vectors $\mathbf{v}_1$ and $\mathbf{v}_2$. This subspace is the plane through the origin determined by the vectors $\mathbf{v}_1$ and $\mathbf{v}_2$. [Refer to Examples (a) and (b) of frame 10; there the subspace of R^3 spanned by $\mathbf{v}_1 = [2, 3, 1]^t$ and $\mathbf{v}_2 = [4, 8, 3]^t$ was the plane with scalar equation $x_1 - 2x_2 + 4x_3 = 0$.] The converse is true too. Thus, three vectors in R^3 are linearly dependent if and only if their geometrical representations (directed line segments) all lie in the same plane passing through the origin. Refer to Figures 5.4a and 5.4b.

We have been concentrating on linear independence and dependence with respect to the vector space R^n. Recall, however, that in Section A we sometimes focused on the vector spaces P_n (of polynomials) and M_{mn} ($m \times n$ matrices). See, for example, frames 2 and 5.

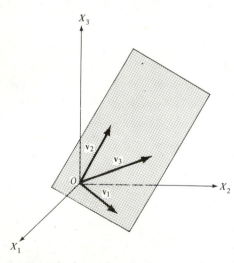

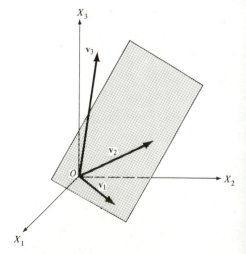

(a) Linearly dependent vectors: $\mathbf{v}_3 = k_1\mathbf{v}_1 + k_2\mathbf{v}_2$.

(b) Linearly independent vectors.

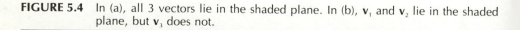

FIGURE 5.4   In (a), all 3 vectors lie in the shaded plane. In (b), $\mathbf{v}_1$ and $\mathbf{v}_2$ lie in the shaded plane, but $\mathbf{v}_3$ does not.

Example (b): The polynomials $p(t) = t^2 + t + 3$, $q(t) = -2t + 4$ and $r(t) = 2$ are vectors in the vector space P_2. Show that $p(t)$, $q(t)$, and $r(t)$ are linearly independent. Then show that they span P_2.

Solution: *Independence:*
The equation $c_1 p(t) + c_2 q(t) + c_3 r(t) = 0$ is

$$c_1(t^2 + t + 3) + c_2(-2t + 4) + c_3(2) = 0 \tag{1}$$

Collecting t^2, t, and constant terms on the left yields

$$c_1 t^2 + (c_1 - 2c_2)t + (3c_1 + 4c_2 + 2c_3) = 0 \tag{2}$$

For the polynomial of the left of Eq. (2) to equal zero for all t values, the coefficients in the t^2, t, and constant terms must each be equal to zero. Thus, we obtain

$$c_1 \qquad\qquad = 0 \tag{3a}$$

$$c_1 - 2c_2 \qquad = 0 \tag{3b}$$

$$3c_1 + 4c_2 + 2c_3 = 0 \tag{3c}$$

Thus, from Eq. (3a), $c_1 = 0$. Then, from Eq. (3b), $c_2 = 0$, and finally from Eq. (3c), $c_3 = 0$. Thus, the vectors $p(t)$, $q(t)$, and $r(t)$ are linearly independent.

Spanning:
To see if the set spans P_2, we let $v(t) = at^2 + bt + c$ be any polynomial in P_2, where a, b, and c are any scalars. We must be able to determine constants c_1, c_2, and c_3 such that

$$v(t) = c_1 p(t) + c_2 q(t) + c_3 r(t) \tag{1}$$

(Refer to Definition 5.4 of frame 8 for definition of a spanning set.) That is,

$$at^2 + bt + c = c_1(t^2 + t + 3) + c_2(-2t + 4) + c_3(2) \tag{2}$$

Collecting similar terms on the right, we get

$$at^2 + bt + c = c_1 t^2 + (c_1 - 2c_2)t + (3c_1 + 4c_2 + 2c_3) \tag{3}$$

Now Eq. (3) holds for all values of t, and if two polynomials are equal for all values of t, then the coefficients of the same respective powers of t are equal. Thus, we have

$$c_1 \qquad\qquad = a \tag{4a}$$

$$c_1 - 2c_2 \qquad\quad = b \tag{4b}$$

$$3c_1 + 4c_2 + 2c_3 = c \tag{4c}$$

Now, finish the problem by solving for the c_i's in terms of a, b, and c.

— — — — — — — — — —

From (4a), $c_1 = a$. Then we successively obtain

$$c_2 = \frac{a - b}{2} \quad \text{and} \quad c_3 = \frac{c - 5a + 2b}{2} .$$

Thus, the set of polynomials $p(t)$, $q(t)$, and $r(t)$ spans the vector space P_2 (recall that this vector space consists of all polynomials of degree 2 and lower, together with the zero polynomial).

Note that Eqs. (3a), (3b), and (3c) in the first part of this example are special cases of Eqs. (4a), (4b), and (4c) for $a = b = c = 0$.

C.   BASIS AND DIMENSION

13.   We know that we can express every vector in a vector space as a linear combination of the members of a spanning set. A natural question concerns the smallest possible number of vectors in a spanning set. The answer is provided by knowledge of what is known as a basis for a vector space.

Definition 5.6 (Basis)

A set of vectors $S = \{\mathbf{v}_1, \mathbf{v}_2, \ldots, \mathbf{v}_r\}$ in a vector space V is called a *basis* for V if

(i) S spans V, and
(ii) S is linearly independent.

Note: The plural of the word "*basis*" is "*bases*."

Example (a):   Show that the set $S = \{\mathbf{e}_1, \mathbf{e}_2, \mathbf{e}_3\}$, where $\mathbf{e}_1 = [1, 0, 0]^t$, $\mathbf{e}_2 = [0, 1, 0]^t$, and $\mathbf{e}_3 = [0, 0, 1]^t$, is a basis for R^3.

Solution: To show that S spans R^3, observe that for $\mathbf{x} = [x_1, x_2, x_3]^t$, an arbitrary vector in R^3, the equation $\mathbf{x} = c_1\mathbf{e}_1 + c_2\mathbf{e}_2 + c_3\mathbf{e}_3$ becomes

$$\begin{bmatrix} x_1 \\ x_2 \\ x_3 \end{bmatrix} = c_1 \begin{bmatrix} 1 \\ 0 \\ 0 \end{bmatrix} + c_2 \begin{bmatrix} 0 \\ 1 \\ 0 \end{bmatrix} + c_3 \begin{bmatrix} 0 \\ 0 \\ 1 \end{bmatrix} = \begin{bmatrix} c_1 \\ c_2 \\ c_3 \end{bmatrix} \tag{1}$$

That is, we have the unique solution $c_i = x_i$ for $i = 1, 2, 3$. In short,

$$\mathbf{x} = x_1\mathbf{e}_1 + x_2\mathbf{e}_2 + x_3\mathbf{e}_3 \tag{2}$$

Thus, S spans R^3.

To show that S is linearly independent, observe that $c_1\mathbf{e}_1 + c_2\mathbf{e}_2 + c_3\mathbf{e}_3 = \mathbf{0}$ leads to

$$c_1\begin{bmatrix}1\\0\\0\end{bmatrix} + c_2\begin{bmatrix}0\\1\\0\end{bmatrix} + c_3\begin{bmatrix}0\\0\\1\end{bmatrix} = \begin{bmatrix}c_1\\c_2\\c_3\end{bmatrix} = \begin{bmatrix}0\\0\\0\end{bmatrix} \tag{3}$$

The only solution in Eq. (3) is $c_1 = c_2 = c_3 = 0$, and thus S is linearly independent. Thus, S is a basis for R^3. The basis $S = \{\mathbf{e}_1, \mathbf{e}_2, \mathbf{e}_3\}$ is known as the *standard basis* of R^3; the vectors are often given as $\mathbf{i}$, $\mathbf{j}$, and $\mathbf{k}$, respectively.

Similarly, the set $S = \{\mathbf{e}_1, \mathbf{e}_2\}$, where $\mathbf{e}_1 = \begin{bmatrix}1\\0\end{bmatrix}$ and $\mathbf{e}_2 = \begin{bmatrix}0\\1\end{bmatrix}$, is the *standard basis* for R^2; often $\mathbf{e}_1$ and $\mathbf{e}_2$ are replaced by the symbols $\mathbf{i}$ and $\mathbf{j}$, respectively. Also, for $\mathbf{x} = [x_1, x_2]^t$, we have $\mathbf{x} = x_1\mathbf{e}_1 + x_2\mathbf{e}_2$.

The geometrical representations of $u = \begin{bmatrix}3\\2\end{bmatrix}$ in R^2 and $v = [3, 4, 3]^t$ in R^3

in terms of their respective standard bases are shown in Figure 5.5. The parallelogram rule is employed in both sketches; see Figures 5.1 (frame 7) and 5.2 (frame 8) for a review. Note that the $\mathbf{e}_i$'s are mutually perpendicular.

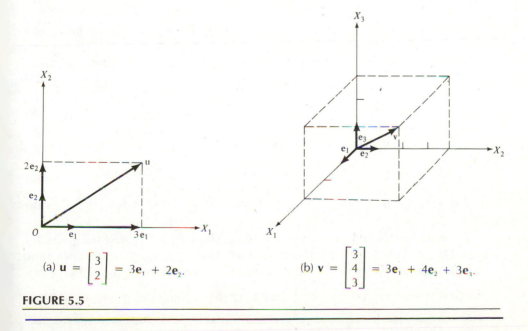

(a) $\mathbf{u} = \begin{bmatrix}3\\2\end{bmatrix} = 3\mathbf{e}_1 + 2\mathbf{e}_2.$ (b) $\mathbf{v} = \begin{bmatrix}3\\4\\3\end{bmatrix} = 3\mathbf{e}_1 + 4\mathbf{e}_2 + 3\mathbf{e}_3.$

FIGURE 5.5

In R^n, let $e_1 = [1, 0, 0, \ldots, 0]^t$, $e_2 = [0, 1, 0, \ldots, 0]^t$, $\ldots$, $e_n = [0, 0, 0, \ldots, 1]^t$. The set $S = \{e_1, e_2, \ldots, e_n\}$ spans R^n and is linearly independent; it is known as the *standard basis* for R^n. Any vector $x = [x_1, x_2, \ldots, x_n]^t$ in R^n can be written as

$$x = x_1 e_1 + x_2 e_2 + \cdots + x_n e_n.$$

Example (b): In Example (a) of frames 8 and 11 we established that the set $S = \{v, w\}$, where $v = [2, 1]^t$ and $w = [1, 3]^t$, was a spanning set for R^2 and was linearly independent. Thus, S is basis for R^2.

In Example (a) of frame 9 and Example (b) of frame 11, we established that $\acute{S} = \{v, w, z\}$, with $v = [2, 1]^t$, $w = [1, 3]^t$, and $z = [1, 1]^t$ spanned R^2 but was linearly dependent. Thus, $\acute{S}$ is not a basis for R^2.

Example (c): Let $v_1 = \begin{bmatrix} 3 \\ 2 \end{bmatrix}$ and $v_2 = \begin{bmatrix} 1 \\ 4 \end{bmatrix}$. Show that the set $S = \{v_1, v_2\}$ is a basis for R^2.

Solution: To show that S spans R^2, we must show that any vector $x = [x_1, x_2]^t$ in R^2 can be expressed as a linear combination of v_1 and v_2:

$$x = c_1 v_1 + c_2 v_2 \tag{1}$$

This leads to the system of linear equations

$$3c_1 + c_2 = x_1 \tag{2a}$$

$$2c_1 + 4c_2 = x_2 \tag{2b}$$

For S to span R^2 we must show that this system has a solution for c_1 and c_2 for any possible choice of x.

To prove that S is linearly independent, we must show that the only solution of

$$c_1 v_1 + c_2 v_2 = 0 \tag{3}$$

is $c_1 = c_2 = 0$. From Eq. (3) we obtain

$$3c_1 + c_2 = 0 \tag{4a}$$

$$2c_1 + 4c_2 = 0 \tag{4b}$$

Thus, to verify independence, we must show that the system given by Eqs. (4a) and (4b) has only the trivial solution $c_1 = c_2 = 0$. Observe that the systems given by (2a), (2b) and (4a), (4b), respectively, have the same coefficient matrix $A = \begin{bmatrix} 3 & 1 \\ 2 & 4 \end{bmatrix}$ and that the columns of A are v_1 and v_2, respectively! [Also, Eqs. (4a) and (4b) form a special case of Eqs. (2a) and (2b), with $x_1 = x_2 = 0$.] We can then represent both systems by

$$Ac = x \quad \text{and} \quad Ac = 0, \qquad\qquad (2'), (4')$$

respectively; here $c = [c_1, c_2]^t$. From Theorems 3.9 and 3.10, we can deduce the following criteria involving determinant $|A|$. (Note that the equation $Ax = b$ of Chapter Three is replaced here by $Ac = x$ and $Ax = 0$ is replaced by $Ac = 0$.)

   (I)   If $|A| \neq 0$, then $Ac = x$ has a unique solution, and $Ac = 0$ has a unique solution, namely, $c = 0$ (i.e., $c_1 = c_2 = \ldots = c_n = 0$).
   (II) If $|A| = 0$, then neither $Ac = x$ nor $Ac = 0$ has a unique solution.

Thus, it follows that S is linearly independent and spans R^2 (i.e., S is a basis), if $|A| \neq 0$. If $|A| = 0$, then $Ac = 0$ has solutions other than the trivial solution $c = 0$, and hence S is not a basis.

Since here we have $A = \begin{vmatrix} 3 & 1 \\ 2 & 4 \end{vmatrix} = 10$, it follows that S is a basis.

We can generalize the discussion of Example (c) as follows.

Theorem 5.5

For vector space R^n, given a set of n column vectors $S = \{v_1, v_2, \ldots, v_n\}$. Let A denote the $n \times n$ matrix whose columns are $v_1, v_2, \ldots,$ and v_n. Then S is a basis for R^n if and only if $|A| \neq 0$.

Example (d): Given the set $S = \{u_1, u_2, u_3\}$, where $u_1 = [1, 2, 1]^t$, $u_2 = [2, 7, 0]^t$, and $u_3 = [0, 3, -2]^t$. Is S a basis for R^3?

Solution: The matrix A is $A = \begin{bmatrix} 1 & 2 & 0 \\ 2 & 7 & 3 \\ 1 & 0 & -2 \end{bmatrix}$. Since $|A| = 0$, S is not a basis

for R^3.

14.   In the next few frames we shall state, illustrate, and occasionally prove some of the major theorems pertaining to the concepts of spanning set, linear independence, and basis.

Theorem 5.6

Given a basis $S = \{v_1, v_2, \ldots, v_n\}$ for a vector space V, then every vector in V can be written in exactly one way as a linear combination of the vectors in the basis S.

A proof is given on p. 181 of Kolman (1980). The question of how to determine the linear combination for a given vector x that holds for a particular basis is easily answered if the vector space is R^n. In frame 13 we saw for the standard basis that the correct linear combination for the column vector $x = [x_1, x_2, \ldots, x_n]^t$ was $x = x_1 e_1 + x_2 e_2 + \cdots + x_n e_n$. Matters are not quite so easy for other bases. Refer to Theorem 5.5 of frame 13. Say we know that the set $S = \{v_1, v_2, \ldots, v_n\}$ is a basis for R^n. Thus, $|A| \neq 0$, and from Theorem 3.9 (frame 12 of Chapter Three), A^{-1} exists and thus the system $Ac = x$ has the solution $c = A^{-1}x$. We summarize in the following theorem, and then we will illustrate in an example.

Theorem 5.7

Given vector space R^n, with a basis $S = \{v_1, v_2, \ldots, v_n\}$, and matrix A whose columns are the v_i's. Any column vector x in the vector space R^n can be expressed uniquely as a linear combination

$$x = c_1 v_1 + c_2 v_2 + \cdots + c_n v_n,$$

where the c_i's are the components of the column vector c, and $c = A^{-1}x$.

Example (a): Refer to Example (c) of frame 13, where a basis for R^2 is $S = \{v_1, v_2\}$, and $v_1 = \begin{bmatrix} 3 \\ 2 \end{bmatrix}$, $v_2 = \begin{bmatrix} 1 \\ 4 \end{bmatrix}$. Thus, $A = \begin{bmatrix} 3 & 1 \\ 2 & 4 \end{bmatrix}$. If $x = \begin{bmatrix} 9 \\ 16 \end{bmatrix}$, determine c_1 and c_2 in the linear combination $x = c_1 v_1 + c_2 v_2$.

Solution: We find $A^{-1} = \frac{1}{10} \begin{bmatrix} 4 & -1 \\ -2 & 3 \end{bmatrix}$, using a method from Chapter Two or Three. Thus,

$$c = A^{-1}x = \frac{1}{10} \begin{bmatrix} 4 & -1 \\ -2 & 3 \end{bmatrix} \begin{bmatrix} 9 \\ 16 \end{bmatrix}$$

$$= \frac{1}{10} \begin{bmatrix} 20 \\ 30 \end{bmatrix} = \begin{bmatrix} 2 \\ 3 \end{bmatrix},$$

and thus $c_1 = 2$, $c_2 = 3$. Thus, $x = 2v_1 + 3v_2$ for $x = [9, 16]^t$. This checks out, since $\begin{bmatrix} 9 \\ 16 \end{bmatrix}$ is equal to $2 \begin{bmatrix} 3 \\ 2 \end{bmatrix} + 3 \begin{bmatrix} 1 \\ 4 \end{bmatrix}$.

Let us consider situations pertaining to bases for other vector spaces.

Example (b): Refer to Example (a) of frame 10. There we learned that the set of all linear combinations of $v_1 = [2, 3, 1]^t$ and $v_2 = [4, 8, 3]^t$, denoted by Span ($\{v_1, v_2\}$), was a subspace of R^3. Geometrically, this subspace was given by the plane $x_1 - 2x_2 + 4x_3 = 0$. By the nature of what a spanning set is, we see that $S = \{v_1, v_2\}$ is a spanning set for the subspace. Show that S is a basis for the subspace.

Solution: From Definition 5.6 (frame 13), we have to show that S is linearly independent. It is since the only solution of $c_1 v_1 + c_2 v_2 = 0$, that is of the system

$$2c_1 + 4c_2 = 0$$

$$3c_1 + 8c_2 = 0$$

$$c_1 + 3c_2 = 0$$

is $c_1 = c_2 = 0$. (Use Method 2.1, for example.) Thus, $S = \{v_1, v_2\}$ is a basis for the subspace.

Example (c): Refer to Example (b) of frame 12. Explain why the set of polynomials $p(t) = t^2 + t + 3$, $q(t) = -2t + 4$ and $r(t) = 2$ is a basis for the polynomial vector space P_2.

-- -- -- -- -- -- -- -- --

Solution: In frame 12 we showed that $p(t)$, $q(t)$, and $r(t)$ spanned P_2 and were linearly independent.

15. Referring again to the polynomial vector space P_2, we see that another basis for P_2 is the set $S = \{1, t, t^2\}$, since every polynomial in P_2 is of the form $a_2 t^2 + a_1 t + a_0$, and $c_1 + c_2 t + c_3 t^2 = 0$ for all t implies $c_1 = c_2 = c_3 = 0$. Similarly, the set $S = \{1, t, t^2, \ldots, t^n\}$ is a basis for the vector space P_n (see frames 2 and 5). This basis is called the *standard basis* for P_n.

Example (a): Refer to Example (a) of frame 2 where we discussed M_{mn}, the vector space of all $m \times n$ matrices. Show that a basis for M_{22} is $S = \{m_1, m_2, m_3, m_4\}$, where

$$m_1 = \begin{bmatrix} 1 & 0 \\ 0 & 0 \end{bmatrix}, \quad m_2 = \begin{bmatrix} 0 & 1 \\ 0 & 0 \end{bmatrix},$$

$$m_3 = \begin{bmatrix} 0 & 0 \\ 1 & 0 \end{bmatrix}, \quad m_4 = \begin{bmatrix} 0 & 0 \\ 0 & 1 \end{bmatrix}.$$

Solution: To show that S spans M_{22} observe that if we express a typical 2×2 matrix $\begin{bmatrix} a_{11} & a_{12} \\ a_{21} & a_{22} \end{bmatrix}$ by

$$\begin{bmatrix} a_{11} & a_{12} \\ a_{21} & a_{22} \end{bmatrix} = c_1 \mathbf{m}_1 + c_2 \mathbf{m}_2 + c_3 \mathbf{m}_3 + c_4 \mathbf{m}_4, \tag{1}$$

then $c_1 = a_{11}$, $c_2 = a_{12}$, $c_3 = a_{21}$, and $c_4 = a_{22}$.

To show independence, we observe that $\mathbf{0} = \begin{bmatrix} 0 & 0 \\ 0 & 0 \end{bmatrix}$ here, and thus,

$$c_1 \mathbf{m}_1 + c_2 \mathbf{m}_2 + c_3 \mathbf{m}_3 + c_4 \mathbf{m}_4 = \mathbf{0} \tag{2}$$

is equivalent to

$$\begin{bmatrix} c_1 & c_2 \\ c_3 & c_4 \end{bmatrix} = \begin{bmatrix} 0 & 0 \\ 0 & 0 \end{bmatrix} \tag{3}$$

Equation (3) implies that the only solution for the c_i's is $c_1 = c_2 = c_3 = c_4 = 0$.

One of the major results dealing with the basis concept is given in the following theorem.

Theorem 5.8

Every basis of a vector space has the same number of vectors.

Because of this theorem it is natural to make the following definition.

Definition 5.7 (Dimension of Vector Space)

The dimension of a nonzero vector space V is the number of vectors in a basis for V. Often we write Dim (V) for the dimension of V.

Note: Since the zero vector space $\{0\}$ (frame 4) contains only the vector $\mathbf{0}$, and a set containing $\mathbf{0}$ has to be linearly dependent (frame 12), we define the dimension of the zero vector space to be zero.

For R^n, one basis is the standard basis, and it has n vectors. Thus, the dimension of R^n is n. In this and the previous frame, we saw that a basis for P_2 has three vectors. Thus, Dim $(P_2) = 3$. Likewise, Dim $(P_n) = (n + 1)$.

All the vector spaces to be studied in this book are finite dimensional. This means the dimension is a finite nonnegative integer. Note that infinite dimensional vector spaces do occur in many applications.

One important fact about finite dimensional vector spaces is that vector spaces with the same dimension have *identical algebraic properties*. The vector spaces differ only in the nature of their vectors. The technical term for these identical properties is *isomorphism*; refer to pp. 121–124 and p. 141 of Bloch and Michaels (1977). Thus, for example, the vector space of polynomials P_2 is isomorphic to R^3, since both have dimension 3.

Example (b): Refer to Example (b) of frame 14. We saw that the subspace of R^3 given by the plane $x_1 - 2x_2 + 4x_3 = 0$ had two vectors in a particular basis. Thus, the dimension of the subspace is 2. In fact, every plane through the origin of R^3, when considered as a subspace, has dimension 2. Every straight line through the origin, when regarded as a subspace, has dimension 1.

In general for R^n, if a subspace is spanned by a set containing a single nonzero vector, then its dimension is 1, and, geometrically, the subspace is a straight line through the origin (see frames 13–16 of Chapter Four). If a subspace is spanned by a set of two linearly independent vectors, the subspace has dimension 2. Similarly, if a subspace of R^n is spanned by k linearly independent vectors, the subspace has dimension k.

Example (c): Refer to Example (a) of this frame. Determine the dimension of M_{22}, the vector space of all 2×2 matrices.

_ _ _ _ _ _ _ _ _ _ _ _

Since the basis given has four vectors, then Dim $(M_{22}) = 2 \cdot 2 = 4$. Likewise, for the vector space M_{mn}, the dimension is the product $m \cdot n$.

16. The parts of the following theorem indicate some properties pertaining to the concepts linear independence, spanning set, and basis. Illustrations will be given.

Theorem 5.9

Let V be an n-dimensional vector space, and let $S = \{\mathbf{v}_1, \mathbf{v}_2, \ldots, \mathbf{v}_n\}$ be a set of n vectors.

(a) If S is linearly independent, then S is a basis for V.
(b) If S spans V, then S is a basis for V.
(c) If S spans V, then every vector $\mathbf{x}$ in V can be expressed as a linear combination $c_1\mathbf{v}_1 + c_2\mathbf{v}_2 + \cdots + c_n\mathbf{v}_n$ in exactly one way (i.e., uniquely).

Note: Part (c) follows directly from part (b) and Theorem 5.6 (frame 14).

Example (a): For the polynomial vector space P_2, the set of three polynomials $S = \{t^2 - 1, 2t, 1\}$ spans P_2, since any polynomial $v(t) = at^2 + bt + c$ in P_2 can be expressed as a linear combination $c_1(t^2 - 1) + c_2(2t) + c_3(1)$ if

$$c_1 = a, \qquad c_2 = \frac{b}{2}, \qquad \text{and} \qquad c_3 = c + a.$$

Thus, by part (b) of Theorem 5.9, S is a basis. Also, part (c) is confirmed, since the preceding solution for c_1, c_2, and c_3 is unique.

Example (b): Refer to Example (a) of frame 10 and Example (b) of frame 14. Use part (a) of Theorem 5.9 to show that $S = \{\mathbf{u}, \mathbf{w}\}$, where $\mathbf{u} = [0, 2, 1]^t$ and $\mathbf{w} = [4, 0, -1]^t$, constitutes a basis for the subspace of R^3 with equation $x_1 - 2x_2 + 4x_3 = 0$. Note that $\mathbf{u}$ results by setting $x_1 = 0$ and $x_2 = 2$ in the plane equation, and $\mathbf{w}$ results by setting $x_1 = 4$ and $x_2 = 0$ in the plane equation.

_ _ _ _ _ _ _ _ _ _

The only solution of the equation $c_1\mathbf{u} + c_2\mathbf{w} = \mathbf{0}$ is $c_1 = c_2 = 0$, and so $S = \{\mathbf{u}, \mathbf{w}\}$ is a linearly independent set. Since the dimension of the subspace is 2 (in frame 14 we found a basis containing *two* vectors), it follows that $S = \{\mathbf{u}, \mathbf{w}\}$ is also a basis.

17.   The next theorem lists some additional useful properties pertaining to the concept of a basis for a finite dimensional vector space. Remember that if a vector space has dimension n, then every basis contains n vectors.

Theorem 5.10

Given that the nonzero vector space V has finite dimension Dim (V).

(a)   For every nonzero subspace W of V, Dim $(W) \leqslant$ Dim (V), where Dim (W) is the number of vectors in any basis for W. If Dim $(W) =$ Dim (V), then W *is* V.

(b)   If Dim $(V) = n$, then any set of more than n vectors in V is linearly dependent.

(c)   If Dim $(V) = n$, then no set of fewer than n vectors in V can span V.

Example (a): For the vector space R^2, the set $S = \{\mathbf{v}, \mathbf{w}, \mathbf{z}\}$, where $\mathbf{v} = \begin{bmatrix} 2 \\ 1 \end{bmatrix}$, $\mathbf{w} = \begin{bmatrix} 1 \\ 3 \end{bmatrix}$, and $\mathbf{z} = \begin{bmatrix} 1 \\ 1 \end{bmatrix}$, is linearly dependent [Example (b) of frame 11]. Here Dim $(R^2) = 2$, so this illustrates part (b) of Theorem 5.10.

Example (b):   Explain why the set $T = \{\mathbf{e}_1, \mathbf{e}_2\}$, where $\mathbf{e}_1 = [1, 0, 0]^t$ and $\mathbf{e}_2 = [0, 1, 0]^t$, cannot possibly span R^3. Use Theorem 5.10 above.

– – – – – – – – – –

Solution: First, observe that Dim $(R^3) = 3$. There are only two vectors in T. Thus, by part (c), T cannot span R^3.

18.   Let us focus again on the set of vector solutions (or solution set) of a general homogeneous linear system $A\mathbf{x} = \mathbf{0}$, where A is $m \times n$ and $\mathbf{x}$ is an n-component column vector. We learned in Theorem 5.3 (frame 6) that this subset of R^n is a subspace of R^n. The next example illustrates the following theorem, which reintroduces the rank concept. This concept was first studied in Chapter Two (Section D, starting with frame 22).

Theorem 5.11

Given that the rank of the $m \times n$ matrix A is Rank (A). [Thus, Rank $(A|\mathbf{0})$ = Rank (A), where $A|\mathbf{0}$ is the augmented coefficient matrix of the related homogeneous system $A\mathbf{x} = \mathbf{0}$.] The dimension of the subspace of vector solutions of $A\mathbf{x} = \mathbf{0}$ is the number $[n - \text{Rank }(A)]$.

The following example will also illustrate an easy way of determining a basis for the subspace.

Example (a):   For the following homogeneous linear system, determine a basis and the dimension of the subspace of solutions:

$$\left. \begin{aligned} x_1 - 4x_2 + 2x_3 + 9x_4 &= 0 \\ 3x_1 - 12x_2 - x_3 + 13x_4 &= 0 \\ x_3 + 2x_4 &= 0 \end{aligned} \right\} \tag{1}$$

The subspace referred to is known as the *solution space* of the system of equations. Note that the overall vector space here is R^4.

Solution: Our goal is to obtain a general expression for a vector solution of the preceding system of equations. Using Method 2.1 (Chapter Two, frame 14), we find that the reduced row echelon form of $A|\mathbf{0}$, that is, of

$$\begin{bmatrix} 1 & -4 & 2 & 9 & | & 0 \\ 3 & -12 & -1 & 13 & | & 0 \\ 0 & 0 & 1 & 2 & | & 0 \end{bmatrix}, \text{ is}$$

$$\begin{bmatrix} 1 & -4 & 0 & 5 & | & 0 \\ 0 & 0 & 1 & 2 & | & 0 \\ 0 & 0 & 0 & 0 & | & 0 \end{bmatrix} \tag{2}$$

Observe that Rank (A) = Rank $(A|0)$ = 2, since the reduced row echelon form in (2) has two leading entries. From (2) we obtain the following solutions for the two leading variables x_1 and x_3 in terms of the two nonleading variables:

$$x_1 = 4x_2 - 5x_4, \qquad x_3 = -2x_4 \qquad\qquad \text{(3a), (3b)}$$

Setting $x_2 = r$ and $x_4 = s$, where both r and s are arbitrary, we obtain the following vector solution:

$$\mathbf{x} = \begin{bmatrix} x_1 \\ x_2 \\ x_3 \\ x_4 \end{bmatrix} = \begin{bmatrix} 4r - 5s \\ r \\ -2s \\ s \end{bmatrix} = r\begin{bmatrix} 4 \\ 1 \\ 0 \\ 0 \end{bmatrix} + s\begin{bmatrix} -5 \\ 0 \\ -2 \\ 1 \end{bmatrix} \qquad (4)$$

Note how we split the $\mathbf{x}$ vector into two vector terms, one containing r times a vector and the other having s times a vector. Now let the vectors that multiply r and s in Eq. (4) be $\mathbf{v}_1$ and $\mathbf{v}_2$, respectively. That is,

$$\mathbf{v}_1 = [4, 1, 0, 0]^t \qquad \text{and} \qquad \mathbf{v}_2 = [-5, 0, -2, 1]^t.$$

[Also, $\mathbf{v}_1$ is obtained by setting $r = 1$ and $s = 0$, and $\mathbf{v}_2$ by setting $r = 0$ and $s = 1$ in Eq. (4).] Thus, Eq. (4) can be written

$$\mathbf{x} = r\mathbf{v}_1 + s\mathbf{v}_2 \qquad\qquad (5)$$

From (5) we see that the subspace of vector solutions is spanned by $\mathbf{v}_1$ and $\mathbf{v}_2$. That is,

Subspace = Span $(\{\mathbf{v}_1, \mathbf{v}_2\})$

Now let us show that $\mathbf{v}_1$ and $\mathbf{v}_2$ are linearly independent. The equation $c_1\mathbf{v}_1 + c_2\mathbf{v}_2 = \mathbf{0}$ is equivalent to

$$[4c_1 - 5c_2, c_1, -2c_2, c_2]^t = [0, 0, 0, 0]^t \qquad\qquad (6)$$

By equating respective components on the left and right, we obtain $c_1 = 0$ and $c_2 = 0$ from the equations relating the second and fourth components. This implies that $\mathbf{v}_1$ and $\mathbf{v}_2$ are linearly independent, and hence that the set $\{\mathbf{v}_1, \mathbf{v}_2\}$ is a basis for the subspace of solutions of system (1). The dimension of this subspace is thus 2.

Note the correspondence between the basis vectors $\mathbf{v}_1$, $\mathbf{v}_2$, and the nonleading variables x_2 and x_4, respectively. This correspondence can be indicated by

$$x_2 \leftrightarrow [4, \underline{1}, 0, 0]^t \qquad \text{and} \qquad x_4 \leftrightarrow [-5, 0, -2, \underline{1}]^t$$

Both the second component of $\mathbf{v}_1$ and the fourth component of $\mathbf{v}_2$ are equal to 1. In Eq. (4), $\mathbf{v}_1$ is multiplied by r, which is what x_2 was set equal to. A similar relationship applies between $\mathbf{v}_2$ and x_4.

In general, the dimension of the subspace of solutions of the homogeneous system is equal to the number of nonleading variables, and that number equals $[n - \text{Rank}\,(A)]$, since Rank (A) equals the number of leading variables.

Example (b): Determine the dimension of the subspace of solutions of $A\mathbf{x} = \mathbf{0}$ if A is the following 3×3 matrix:

$$A = \begin{bmatrix} 3 & -9 & 1 \\ -1 & 3 & 2 \\ 2 & -6 & -1 \end{bmatrix} \tag{1}$$

Determine a basis for the subspace.

- - - - - - - - - -

The reduced row echelon form of $A|\mathbf{0}$ is

$$\left[\begin{array}{ccc|c} 1 & -3 & 0 & 0 \\ 0 & 0 & 1 & 0 \\ 0 & 0 & 0 & 0 \end{array}\right] \tag{2}$$

Thus, Rank $(A) = 2$, and the dimension of the subspace is $3 - 2 = 1$. From (2), we see that the vector solution of $A\mathbf{x} = \mathbf{0}$ is given by

$$\mathbf{x} = \begin{bmatrix} 3r \\ r \\ 0 \end{bmatrix} = r\begin{bmatrix} 3 \\ 1 \\ 0 \end{bmatrix} \tag{3}$$

Thus, a basis for the subspace of solutions of $A\mathbf{x} = \mathbf{0}$ is $\{\mathbf{v}_1\}$, where $\mathbf{v}_1 = [3, 1, 0]^t$. The subspace can be expressed as Span $(\{\mathbf{v}_1\})$.

19. Consider again the 3×4 matrix $A = \begin{bmatrix} 1 & -4 & 2 & 9 \\ 3 & -12 & -1 & 13 \\ 0 & 0 & 1 & 2 \end{bmatrix}$ of Example (a) of frame 18. Let us identify the row vectors that make up the rows of A:

$$\mathbf{a}_1 = [1, -4, 2, 9], \qquad \mathbf{a}_2 = [3, -12, -1, 13], \qquad \mathbf{a}_3 = [0, 0, 1, 2]$$

Each of these $\mathbf{a}_i$'s has four components. The vector space (or subspace) spanned by the set $\{\mathbf{a}_1, \mathbf{a}_2, \mathbf{a}_3\}$ [which can be denoted as Span $(\{\mathbf{a}_1, \mathbf{a}_2, \mathbf{a}_3\})$] is a subspace of R^4, and it is defined as the *row space* of matrix A. Note that the subspace is the set of all linear combinations of the vectors $\mathbf{a}_1$, $\mathbf{a}_2$, and $\mathbf{a}_3$.

Suppose $\acute{A}$ is the matrix generated by applying a sequence of elementary row operations to matrix A (see Definition 2.3 of Chapter Two, frame 8). Let the row vectors for the matrix $\hat{A}$ be denoted by $\hat{a}_1$, $\hat{a}_2$, $\hat{a}_3$. It can be shown that the row space of $\hat{A}$ [which is Span $(\{\hat{a}_1, \hat{a}_2, \hat{a}_3\})$] is identical to the row space of A. [See p. 153 of Anton (1977), p. 190 of Kolman (1980), and p. 193 of Shields (1980)]. Thus, if we take the given matrix A and transform it to reduced row echelon form A^*, the row spaces of A and A^* are identical.

Example (a): For the matrix A cited above, find the reduced row echelon form A^*. Then from the latter, determine a basis for the row space of A and its dimension.

Solution: From (2) in Example (a) of frame 18, the reduced row echelon form A^* is as follows (delete the column of zeros to the right of the dividing line):

$$A^* = \begin{bmatrix} 1 & -4 & 0 & 5 \\ 0 & 0 & 1 & 2 \\ 0 & 0 & 0 & 0 \end{bmatrix} \tag{1}$$

The vector space spanned by the three row vectors of A^* is the same as the vector space spanned by the first two row vectors of A^*, since the four-component zero row vector $\mathbf{0} = [0, 0, 0, 0]$ does not contribute anything (the reason: $c\mathbf{0} = \mathbf{0}$ for any c). Thus, the row space of A is equal to the subspace spanned by the set

$$\{\mathbf{a}_1^*, \mathbf{a}_2^*\}, \tag{2}$$

where

$$\mathbf{a}_1^* = [1, -4, 0, 5] \qquad \text{and} \qquad \mathbf{a}_2^* = [0, 0, 1, 2] \tag{3a, 3b}$$

Moreover, $\mathbf{a}_1^*$ and $\mathbf{a}_2^*$ are linearly independent, since $c_1\mathbf{a}_1^* + c_2\mathbf{a}_2^* = \mathbf{0}$, with $\mathbf{0} = [0, 0, 0,0]$, results in $c_1 = c_2 = 0$ (equate the first components on left and right, and likewise the third components). Thus, the set $\{\mathbf{a}_1^*, \mathbf{a}_2^*\}$ is a *basis* for the row space of A. Hence, the row space of A has dimension 2. (Recall that the row space of A is a subspace of R^4.)

Note: In Example (a), observe that the number of nonzero row vectors in A^* is the dimension of the row space of A. Also, this number equals the rank of matrix A (see Definition 2.6 of frame 22 of Chapter Two). This is true in general; that is, the dimension of the row space of A equals Rank (A), the rank of matrix A.

Now let us generalize the work done thus far in this frame.

Definition 5.8

For the $m \times n$ matrix

$$A = \begin{bmatrix} a_{11} & a_{12} & \cdots & a_{1n} \\ a_{21} & a_{22} & \cdots & a_{2n} \\ \cdots & \cdots & \cdots & \cdots \\ a_{m1} & a_{m2} & \cdots & a_{mn} \end{bmatrix},$$

the rows of A, considered as vectors in R^n, span a subspace of R^n, called the *row space* of A. The dimension of the row space of A is called the *row rank* of A.

Theorem 5.12

Let A^* denote the reduced row echelon form of A. The row space of A is identical to the row space of A^*. The set of nonzero rows of A^* constitutes a basis for the row space of A. The row rank of A, which equals the number of nonzero row vectors in A^*, is equal to Rank (A), the rank of matrix A.

Example (b): Refer to Example (b) of frame 18. Determine a basis for the row space of A, and find the row rank of A.

— — — — — — — — — —

Solution: The reduced row echelon form of

$$A = \begin{bmatrix} 3 & -9 & 1 \\ -1 & 3 & 2 \\ 2 & -6 & -1 \end{bmatrix} \quad \text{is} \quad A^* = \begin{bmatrix} 1 & -3 & 0 \\ 0 & 0 & 1 \\ 0 & 0 & 0 \end{bmatrix}$$

A basis for the row space of A is $\{\mathbf{a}_1^*, \mathbf{a}_2^*\}$, where $\mathbf{a}_1^* = [1, -3, 0]$ and $\mathbf{a}_2^* = [0, 0, 1]$. The row rank of A equals 2, which is also equal to Rank (A).

20. Refer to Definition 5.8 of the previous frame, and the general $m \times n$ matrix A cited there. We now focus on the columns of A, each of which has m components.

Definition 5.9

The columns of the $m \times n$ matrix A, considered as vectors in R^m, span a subspace of R^m, called the *column space* of A. The dimension of the column space is called the *column rank* of A.

Results for the column space are similar to those that hold for the row space. [See Anton (1977) and Kolman (1980).] We illustrate in the next example.

Example (a): For the matrix A of Example (a) in frames 18 and 19, find a basis for the column space of A, and find the column rank of A.

Solution: Here the column space is a subspace of R^3, since matrix A is 3×4. If we write the column vectors as row vectors, we obtain the 4×3 transpose of A:

$$A^t = \begin{bmatrix} 1 & 3 & 0 \\ -4 & -12 & 0 \\ 2 & -1 & 1 \\ 9 & 13 & 2 \end{bmatrix}$$

If we transform A^t to reduced row echelon form, we obtain

$$(A^t)^* = \begin{bmatrix} 1 & 0 & \frac{3}{7} \\ 0 & 1 & -\frac{1}{7} \\ 0 & 0 & 0 \\ 0 & 0 & 0 \end{bmatrix}$$

Thus, the vectors $[1, 0, \frac{3}{7}]$ and $[0, 1, -\frac{1}{7}]$ form a basis for the row space of A^t. This means that the vectors $\begin{bmatrix} 1 \\ 0 \\ \frac{3}{7} \end{bmatrix}$ and $\begin{bmatrix} 0 \\ 1 \\ -\frac{1}{7} \end{bmatrix}$ form a basis for the column space of A. We conclude that the column rank of A is 2.

From Example (a) of this and the prior frame we see that the row and column ranks are equal for the particular matrix A considered in those examples. Actually, this is always true, as indicated in the following theorem, which is proved in Anton (1977) and Kolman (1980).

Theorem 5.13

For any $m \times n$ matrix A, the row rank and column rank are equal; moreover, the common value is equal to Rank (A), the rank of matrix A.

Example (b): Refer to Example (b) of frames 18 and 19. Determine the column rank of matrix A.

— — — — — — — — — —

Solution:

Column rank = row rank = Rank $(A) = 2$

21. Because of the results contained in Theorem 5.13, we can now refer to *the rank* of a matrix A. This means that when we speak of Rank (A) (as defined in Definition 2.6), we will also be speaking of the row rank and column rank of matrix A.

An important characteristic of the rank is that for any matrix A, Rank (A) is equal to the number of linearly independent rows (or columns) in matrix A. Thus, in the 3×4 matrix A considered in Example (a) of the two previous frames the number of linearly independent rows (columns) is two.

We are now in a position to update and extend master Theorem 3.9 (Chapter Three, frame 12), which contains a list of equivalent statements for an $n \times n$ matrix.

Theorem 5.14

If A is an $n \times n$ matrix, then the following statements are equivalent.

(a)   Rank $(A) = n$.
(b)   The reduced row echelon form of A is I_n.
(c) A is invertible (i.e., A^{-1} exists).
(d)   The system $A\mathbf{x} = \mathbf{b}$ has a unique solution (given by $\mathbf{x} = A^{-1}\mathbf{b}$).
(e) $|A| \neq 0$.
(f)   The n row vectors of A are linearly independent.
(g)   The n column vectors of A are linearly independent.

Remember how we interpret this type of theorem. If, for a particular $n \times n$ matrix A, one of the statements in the list is true, then all the statements are true for that matrix. Similarly, if one statement is false for a particular matrix A, then all the statements are false for the particular matrix A.

Example (a): Determine if the rows (columns) of $A = \begin{bmatrix} 2 & 3 & 4 \\ 1 & 2 & 1 \\ 1 & 2 & 3 \end{bmatrix}$ are linearly independent.

Solution: Since the determinant $|A| = 2$, statement (e) is true, and thus, all the other statements listed in Theorem 5.14 are true. Thus, in particular, statements (f) and (g) are true. Thus, the three rows (columns) of A are linearly independent.

In engineering and natural science books, still another type of rank appears. See p. 491 of Wylie (1975), for example.

Definition 5.10

Given an $m \times n$ matrix A. The *determinant rank* of A is the largest value of r for which there exists an $r \times r$ submatrix whose determinant is unequal to zero.

Example (b): Find the determinant rank of the matrix A of Example (a) of frames 19 and 20.

Solution: For $A = \begin{bmatrix} 1 & -4 & 2 & 9 \\ 3 & -12 & -1 & 13 \\ 0 & 0 & 1 & 2 \end{bmatrix}$, the determinants of the 3×3 sub-

matrices are

$$\begin{vmatrix} 1 & -4 & 2 \\ 3 & -12 & -1 \\ 0 & 0 & 1 \end{vmatrix} = 0, \qquad \begin{vmatrix} 1 & -4 & 9 \\ 3 & -12 & 13 \\ 0 & 0 & 2 \end{vmatrix} = 0,$$

$$\begin{vmatrix} 1 & 2 & 9 \\ 3 & -1 & 13 \\ 0 & 1 & 2 \end{vmatrix} = 0, \qquad \begin{vmatrix} -4 & 2 & 9 \\ -12 & -1 & 13 \\ 0 & 1 & 2 \end{vmatrix} = 0$$

For the 2×2 submatrix $\begin{bmatrix} 1 & 2 \\ 3 & -1 \end{bmatrix}$ formed from rows 1 and 2 and columns

1 and 3, $\begin{vmatrix} 1 & 2 \\ 3 & -1 \end{vmatrix} \neq 0$, and thus the determinant rank equals 2. This is also

equal to Rank (A).

The result from Example (b) always holds.

Theorem 5.15

For any $m \times n$ matrix A, the determinant rank equals Rank (A).

Example (c): Refer to Example (b) of frames 18, 19, and 20. Determine the determinant rank of matrix A.

— — — — — — — — — —

Solution: From Theorem 5.15, determinant rank = Rank (A) = 2.

22.   The methods of frames 19 and 20 can be used to determine a basis for a subspace of R^n spanned by a given set of vectors.

Example: Find a basis for the subspace W of R^3 spanned by the three column vectors

$$\mathbf{v}_1 = \begin{bmatrix} 1 \\ 1 \\ 1 \end{bmatrix}, \qquad \mathbf{v}_2 = \begin{bmatrix} 1 \\ 2 \\ 0 \end{bmatrix}, \qquad \mathbf{v}_3 = \begin{bmatrix} 0 \\ 2 \\ -2 \end{bmatrix}$$

Solution: We use a method similar to that used in Example (a) of frame 20. First, we form a matrix A whose columns are $\mathbf{v}_1$, $\mathbf{v}_2$, and $\mathbf{v}_3$. Then, for the transpose

$$A^t = \begin{bmatrix} 1 & 1 & 1 \\ 1 & 2 & 0 \\ 0 & 2 & -2 \end{bmatrix}, \tag{1}$$

we find the corresponding reduced row echelon form. Now finish the problem.

_ _ _ _ _ _ _ _ _ _ _

$$(A^t)^* = \begin{bmatrix} 1 & 0 & 2 \\ 0 & 1 & -1 \\ 0 & 0 & 0 \end{bmatrix} \tag{2}$$

Thus, a basis for the subspace of R^3 spanned by $\mathbf{v}_1$, $\mathbf{v}_2$, and $\mathbf{v}_3$ consists of the two column vectors $\mathbf{w}_1 = [1, 0, 2]^t$ and $\mathbf{w}_2 = [0, 1, -1]^t$. The dimension of the subspace is 2.

Note: If we were asked to find a basis for the subspace of R^3 spanned by the three row vectors $\hat{\mathbf{v}}_1 = [1, 1, 1]$, $\hat{\mathbf{v}}_2 = [1, 2, 0]$, and $\hat{\mathbf{v}}_3 = [0, 2, -2]$, we would proceed in essentially the same fashion. Thus, for the matrix $\hat{A} = \begin{bmatrix} 1 & 1 & 1 \\ 1 & 2 & 0 \\ 0 & 2 & -2 \end{bmatrix}$, we would find the reduced row echelon form. This is given to the right of the equals sign in Eq. (2). From the latter, we would obtain the basis (row) vectors $\hat{\mathbf{w}}_1 = [1, 0, 2]$ and $\hat{\mathbf{w}}_2 = [0, 1, -1]$.

D. ORTHOGONAL BASES AND PROJECTIONS IN R^n

23. In dealing with a vector space it is often important to choose a basis that leads to a simplification in the calculations. The easiest type of basis to work with is a so-called orthogonal basis, in which every pair of vectors is mutually perpendicular. Recall from Theorem 4.4 (Chapter Four, frame 7) that two vectors $\mathbf{a}$ and $\mathbf{b}$ in R^n are perpendicular if and only if $\mathbf{a} \cdot \mathbf{b} = 0$ (dot product equals zero).

Definition 5.11

A set of vectors $S = \{\mathbf{v}_1, \mathbf{v}_2, \ldots, \mathbf{v}_k\}$ in R^n is called *orthogonal* if any two distinct vectors in S are perpendicular, that is, if $\mathbf{v}_i \cdot \mathbf{v}_j = 0$ for $i \neq j$. An orthogonal set in which each vector is a unit vector is called *orthonormal*.

Notes: (i) A unit vector is one that has unit length. See frames 5 and 10 of Chapter Four. In particular, a unit vector in the direction of vector $\mathbf{v}$ equals $\mathbf{v}/\|\mathbf{v}\|$. (ii) In this book we use the words *perpendicular* and *orthogonal* interchangeably.

Example (a): The set $S = \{\mathbf{v}_1, \mathbf{v}_2, \mathbf{v}_3\}$, where $\mathbf{v}_1 = [1, 1, -3]^t$, $\mathbf{v}_2 = [-7, 4, -1]^t$, and $\mathbf{v}_3 = [1, 2, 1]^t$, is an orthogonal set in R^3, since $\mathbf{v}_1 \cdot \mathbf{v}_2 = \mathbf{v}_1 \cdot \mathbf{v}_3 = \mathbf{v}_2 \cdot \mathbf{v}_3 = 0$. The following vectors are unit vectors in the directions of $\mathbf{v}_1$, $\mathbf{v}_2$, and $\mathbf{v}_3$, respectively:

$$\mathbf{w}_1 = [1/\sqrt{11},\ 1/\sqrt{11},\ -3/\sqrt{11}]^t,$$

$$\mathbf{w}_2 = [-7/\sqrt{66},\ 4/\sqrt{66},\ -1/\sqrt{66}]^t, \quad \text{and}$$

$$\mathbf{w}_3 = [1/\sqrt{6},\ 2/\sqrt{6},\ 1/\sqrt{6}]^t$$

Here $\mathbf{w}_1 = \mathbf{v}_1/\|\mathbf{v}_1\|$, where $\|\mathbf{v}_1\| = \sqrt{1 + 1 + 9} = \sqrt{11}$. Similarly, for $\mathbf{w}_2$ and $\mathbf{w}_3$. Thus, $\{\mathbf{w}_1, \mathbf{w}_2, \mathbf{w}_3\}$ is an orthonormal set.

The process of dividing a vector by its length to obtain a vector of length 1 (i.e., unit length) is called *normalizing* the vector.

Example (b): The standard basis for R^3, $\{\mathbf{e}_1, \mathbf{e}_2, \mathbf{e}_3\}$, where $\mathbf{e}_1 = [1, 0, 0]^t$, $\mathbf{e}_2 = [0, 1, 0]^t$, and so on, is an orthonormal set in R^3. Also, the standard basis for R^n, $\{\mathbf{e}_1, \mathbf{e}_2, \ldots, \mathbf{e}_n\}$, is an orthonormal set. See frame 13 for a review of the standard basis concept.

Example (c): Show that the set $S = \{\mathbf{v}_1, \mathbf{v}_2\}$, where $\mathbf{v}_1 = [1, 1, 1]^t$ and $\mathbf{v}_2 = [1, -3, 2]^t$, is an orthogonal set. Derive an orthonormal set corresponding to S.

— — — — — — — — — —

Solution: $\mathbf{v}_1 \cdot \mathbf{v}_2 = 1 - 3 + 2 = 0$, and so S is orthogonal. The set $T = \{\mathbf{w}_1, \mathbf{w}_2\}$, where $\mathbf{w}_1 = [1/\sqrt{3}, 1/\sqrt{3}, 1/\sqrt{3}]^t$ and $\mathbf{w}_2 = [1/\sqrt{14}, -3/\sqrt{14}, 2/\sqrt{14}]^t$, is orthonormal. Here $\mathbf{w}_1 = \mathbf{v}_1/\|\mathbf{v}_1\|$ and $\mathbf{w}_2 = \mathbf{v}_2/\|\mathbf{v}_2\|$.

24. The following theorem indicates an important property of an orthogonal set of vectors.

Theorem 5.16

Given the orthogonal set $S = \{\mathbf{v}_1, \mathbf{v}_2, \ldots, \mathbf{v}_k\}$ of k nonzero vectors in R^n. Suppose vector $\mathbf{x}$ is a linear combination of the $\mathbf{v}_i$'s, that is,

$$\mathbf{x} = c_1\mathbf{v}_1 + c_2\mathbf{v}_2 + \cdots + c_k\mathbf{v}_k \tag{1}$$

[In other words, $\mathbf{x}$ is a vector in the subspace of R^n spanned by $\{\mathbf{v}_1, \mathbf{v}_2, \ldots, \mathbf{v}_k\}$.] Then the c_i's are given by the dot product formula:

$$c_i = \frac{\mathbf{x} \cdot \mathbf{v}_i}{\mathbf{v}_i \cdot \mathbf{v}_i} \qquad \text{for } i = 1, 2, \ldots, k. \tag{2}$$

Note: The denominator can be written as $\|\mathbf{v}_i\|^2$.

The expression for $\mathbf{x}$ given by Eq. (1), where the c_i's are given in (2), is known as an *expansion* of vector $\mathbf{x}$ in terms of an orthogonal set.

(Partial) Proof of Theorem 5.16

Illustrating the calculation of c_1, we take the dot product of both sides of Eq. (1) with $\mathbf{v}_1$:

$$\mathbf{x} \cdot \mathbf{v}_1 = c_1(\mathbf{v}_1 \cdot \mathbf{v}_1) + c_2(\mathbf{v}_2 \cdot \mathbf{v}_1) + \cdots + c_k(\mathbf{v}_k \cdot \mathbf{v}_1)$$

All terms on the right side, except the first, equal zero, since $\mathbf{v}_i \cdot \mathbf{v}_1 = 0$ if $i \neq 1$. Dividing the resulting two terms by $(\mathbf{v}_1 \cdot \mathbf{v}_1)$ results in

$$c_1 = \frac{\mathbf{x} \cdot \mathbf{v}_1}{\mathbf{v}_1 \cdot \mathbf{v}_1}$$

The calculation for any other c_i is similar.

The following corollary to Theorem 5.16 is very useful.

Corollary (A) of Theorem 5.16

Given the orthogonal set $S = \{\mathbf{v}_1, \mathbf{v}_2, \ldots, \mathbf{v}_k\}$ of k nonzero vectors in R^n. Then S is linearly independent.

The proof here is instructive and easy.

Proof of Corollary (A)

We wish to show that $c_1\mathbf{v}_1 + c_2\mathbf{v}_2 + \cdots + c_k\mathbf{v}_k = \mathbf{0}$ implies that $c_1 = c_2 = \cdots = c_k = 0$. Our equation here is merely Eq. (1) in Theorem 5.16, but with $\mathbf{x}$ equal to $\mathbf{0}$. Thus, the c_i's are given by

$$c_i = \frac{\mathbf{0} \cdot \mathbf{v}_i}{\mathbf{v}_i \cdot \mathbf{v}_i} = \frac{0}{\mathbf{v}_i \cdot \mathbf{v}_i} = 0 \qquad \text{for } i = 1, 2, \ldots, k.$$

Corollary (B) of Theorem 5.16

Given the orthogonal set $S = \{\mathbf{v}_1, \mathbf{v}_2, \ldots, \mathbf{v}_n\}$ of n nonzero vectors in R^n. The set S is a basis for R^n. The expansion formula given in Theorem 5.16, which now applies for an arbitrary vector $\mathbf{x}$ in R^n, is then known as the expansion of a vector $\mathbf{x}$ in terms of an orthogonal basis.

The major part of this corollary (which states that the set S is a basis) is valid since the set S is linearly independent [Corollary (A)], and it contains n vectors [see Theorem 5.9, part (a) of frame 16].

Example: Refer to Example (a) of frame 23. From Corollary (B), the set $S = \{\mathbf{v}_1, \mathbf{v}_2, \mathbf{v}_3\}$ is a basis for R^3. For $\mathbf{x} = [12, 4, -2]^t$, use Theorem 5.16 to calculate the c_i's in the linear combination $\mathbf{x} = c_1\mathbf{v}_1 + c_2\mathbf{v}_2 + c_3\mathbf{v}_3$.

Solution: Since $\mathbf{x} \cdot \mathbf{v}_1 = 12(1) + 4(1) + (-2)(-3) = 22$ and $\mathbf{v}_1 \cdot \mathbf{v}_1 = 1^2 + 1^2 + (-3)^2 = 11$, it follows that

$$c_1 = \frac{\mathbf{x} \cdot \mathbf{v}_1}{\mathbf{v}_1 \cdot \mathbf{v}_1} = \frac{22}{11} = 2.$$

Now finish the calculations.

- - - - - - - - - -

$\mathbf{x} \cdot \mathbf{v}_2 = -66,$ $\qquad \mathbf{v}_2 \cdot \mathbf{v}_2 = 66,$ $\qquad$ and $\quad c_2 = -1.$

$\mathbf{x} \cdot \mathbf{v}_3 = 18,$ $\qquad \mathbf{v}_3 \cdot \mathbf{v}_3 = 6,$ $\qquad$ and $\quad c_3 = 3.$

Thus, $\mathbf{x} = 2\mathbf{v}_1 - \mathbf{v}_2 + 3\mathbf{v}_3.$

25. It is useful to compare the work involved in computing the c_i's in the example of frame 24 with the work required to find the c_i's if we were given a nonorthogonal basis. See frame 14 and, in particular, Theorem 5.7. If we were to use Theorem 5.7 for a nonorthogonal basis, we would have to invert a matrix before we could calculate the c_i's. (Of course, we could also solve for the c_i's by another method—for example, Method 2.1—if the basis were nonorthogonal.) See Example (a) of frame 14. Thus, we see that an orthogonal basis is easier to work with, at least with respect to calculating c_i's.

Next we turn to the important subject of the so-called *orthogonal projection* of a vector $\mathbf{x}$ onto a subspace.

Theorem 5.17

Let $S = \{\mathbf{v}_1, \mathbf{v}_2, \ldots, \mathbf{v}_k\}$ be an orthogonal set of k nonzero vectors in R^n, where $k \leq n$. Let W denote the subspace spanned by $\{\mathbf{v}_1, \mathbf{v}_2, \ldots, \mathbf{v}_k\}$. Then every vector $\mathbf{x}$ in R^n can be expressed in the form

$$\mathbf{x} = \mathbf{w}_1 + \mathbf{w}_2, \tag{1}$$

where $\mathbf{w}_1$ is in W and $\mathbf{w}_2$ is perpendicular to W (this means that $\mathbf{w}_2$ is perpendicular to every vector in W). Also,

$$\mathbf{w}_1 = c_1\mathbf{v}_1 + c_2\mathbf{v}_2 + \cdots + c_k\mathbf{v}_k, \tag{2}$$

where

$$c_i = \frac{\mathbf{x} \cdot \mathbf{v}_i}{\mathbf{v}_i \cdot \mathbf{v}_i} \qquad \text{for } i = 1, 2, \ldots, k. \tag{3}$$

Then to calculate $\mathbf{w}_2$, we use $\mathbf{w}_2 = \mathbf{x} - \mathbf{w}_1$.

Refer to Figure 5.6 for a diagram in which R^n is R^3 and W is a two-dimensional subspace. [Thus, $S = \{\mathbf{v}_1, \mathbf{v}_2\}$.] Also, refer to Chapter Four for a discussion on the triangle rules for vector addition and subtraction (for example, see Figures 4.4 and 4.5 in frame 2).

Notes: (i) Note the similarity between the expressions for the c_i's in Theorem 5.17 and 5.16. (ii) Using Figure 5.6 for motivation, we call $\mathbf{w}_1$ the *orthogonal projection of $\mathbf{x}$ on W* and denote it by $\text{Proj}_W\,\mathbf{x}$. The vector $\mathbf{w}_2 = \mathbf{x} - \text{Proj}_W\,\mathbf{x}$ is sometimes called the *component of $\mathbf{x}$ that is perpendicular to W* and denoted by $\text{Comp}_W\,\mathbf{x}$. Also, $\mathbf{w}_2$ is sometimes called the *complementary projection with respect to W.* (iii) If $\mathbf{x}$ itself happened to be in subspace W, then $\mathbf{w}_2$ would equal $\mathbf{0}$ and $\mathbf{x} = \mathbf{w}_1$. Then Eq. (2) of Theorem 5.17 would reduce to Eq. (1) of Theorem 5.16. (iv) For a given vector $\mathbf{x}$, the vectors $\mathbf{w}_1$ and $\mathbf{w}_2$ in Theorem 5.17 are unique. The proofs of this fact and of Theorem 5.17 are discussed in Anton (1977).

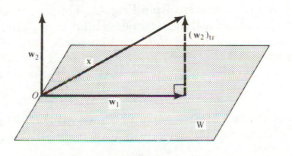

FIGURE 5.6

Example: The set $S = \{v_1, v_2\}$, where $v_1 = [1, 0, 0]^t$ and $v_2 = [0, 3, 4]^t$, is an orthogonal set of vectors in R^3. Let W be the subspace spanned by $\{v_1, v_2\}$. [In short, $W = $ Span $(\{v_1, v_2\})$.] For $x = [3, 2, 11]^t$, determine the vectors $w_1 = \text{Proj}_W x$ (the orthogonal projection of x on W), and $w_2 = x - w_1$, the component of x that is perpendicular to W (also known as $\text{Comp}_W x$).

Solution:

$$w_1 = \text{Proj}_W x = c_1 v_1 + c_2 v_2,$$

where $c_1 = \dfrac{x \cdot v_1}{v_1 \cdot v_1} = \dfrac{3}{1} = 3$ and $c_2 = \dfrac{x \cdot v_2}{v_2 \cdot v_2} = \dfrac{50}{25} = 2$. Thus,

$$w_1 = 3v_1 + 2v_2 = [3, 6, 8]^t.$$

Now finish the calculations.

———————————

Now, w_2 or $\text{Comp}_W x$ is given by

$$w_2 = x - w_1 = [3-3, \ 2-6, \ 11-8]^t = [0, -4, 3]^t$$

Note that w_2 is perpendicular to both v_1 and v_2. Thus, w_2 is perpendicular to each vector in the subspace spanned by $\{v_1, v_2\}$. One such vector is w_1. That is, we should have $w_2 \cdot w_1 = 0$. This is a check on the calculations.

26.   Refer again to Theorem 5.17. Remember that in frame 5 of Chapter Four we said that the distance between two vectors a and b was given by $\|b - a\|$, the length of the vector $b - a$. It can be shown that of all the vectors w in the subspace W, that the vector w^* that has the property that the distance between it and x is the minimum possible distance is none other than the orthogonal projection of x on W; that is, $w^* = \text{Proj}_W x$. In symbols,

$$\|x - \text{Proj}_W x\| < \|x - w\| \tag{I}$$

for every vector w in W different from $\text{Proj}_W x$. The length on the left is the length of the component of x that is perpendicular to W. That is, the left side can also be written as $\|\text{Comp}_W x\|$. Refer to Figures 5.7(a) and (b).

The result given in (I) is of great practical value. It is used by some as the starting point in least squares approximation theory. See Shields (1980) and Rorres and Anton (1979).

Example: Refer to the example of frame 25. Of all the distances between x and vectors in W, determine the minimum possible distance.

———————————

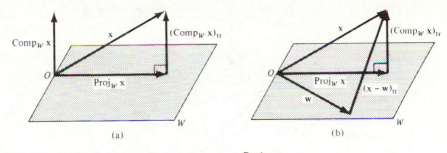

FIGURE 5.7 In the diagrams, $\text{Comp}_W \mathbf{x} = \mathbf{x} - \text{Proj}_W \mathbf{x}$.

Solution: From the Example of frame 25, $\mathbf{x} - \text{Proj}_W \mathbf{x} = [0, -4, 3]^t$. Thus, the minimum possible distance is

$$\|\mathbf{x} - \text{Proj}_W \mathbf{x}\| = \sqrt{0^2 + (-4)^2 + 3^2} = \sqrt{25} = 5.$$

27. We can make use of Theorem 5.17 to prove the following theorem.

Theorem 5.18 (Gram–Schmidt process)

Let W be a nonzero subspace of R^n with basis $S = \{\mathbf{u}_1, \mathbf{u}_2, \ldots, \mathbf{u}_k\}$. Then there exists an orthogonal basis $T = \{\mathbf{v}_1, \mathbf{v}_2, \ldots, \mathbf{v}_k\}$ for W.

Notes: (i)   J. P. Gram (1850–1916) was a Danish actuary, and Erhard Schmidt (1876–1959) was a German mathematician who made contributions to the subjects partial differential equations and integral equations. (ii)   The subspace W in Theorem 5.18 could be the vector space R^n, in which case k would be n.

We shall give an outline of the proof. The constructive technique used in the proof is very similar to the method to be used in computational problems.

Outline of Proof of Theorem 5.18

Let

$$\mathbf{v}_1 = \mathbf{u}_1 \tag{1}$$

Now let W_1 be the subspace spanned by $\mathbf{v}_1$ [in short, $W_1 = \text{Span}(\{\mathbf{v}_1\})$]. Let

$$\mathbf{v}_2 = \mathbf{u}_2 - \text{Proj}_{W_1} \mathbf{u}_2. \tag{2}$$

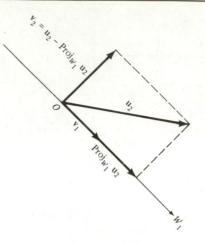

FIGURE 5.8

From Theorem 5.17,

$$\text{Proj}_{W_1} \mathbf{u}_2 = c_1\mathbf{v}_1, \qquad \text{where } c_1 = \frac{\mathbf{u}_2 \cdot \mathbf{v}_1}{\mathbf{v}_1 \cdot \mathbf{v}_1}.$$

In words, $\mathbf{v}_2$ is the component of $\mathbf{u}_2$ that is perpendicular to the subspace W_1. See Figure 5.8.

Now let W_2 be the subspace spanned by $\mathbf{v}_1$ and $\mathbf{v}_2$ [in short, $W_2 = \text{Span}(\{\mathbf{v}_1, \mathbf{v}_2\})$]. Then let

$$\mathbf{v}_3 = \mathbf{u}_3 - \text{Proj}_{W_2} \mathbf{u}_3. \tag{3}$$

From Theorem 5.17,

$$\text{Proj}_{W_2} \mathbf{u}_3 = \hat{c}_1\mathbf{v}_1 + \hat{c}_2\mathbf{v}_2, \qquad \text{where } \hat{c}_i = \frac{\mathbf{u}_3 \cdot \mathbf{v}_i}{\mathbf{v}_i \cdot \mathbf{v}_i} \qquad \text{for } i = 1, 2.$$

In words, $\mathbf{v}_3$ is the component of $\mathbf{u}_3$ that is perpendicular to the subspace W_2. See Figure 5.9.

Next we seek a vector $\mathbf{v}_4$ that is perpendicular to the subspace W_3 spanned by $\mathbf{v}_1$, $\mathbf{v}_2$, and $\mathbf{v}_3$. Here

$$\mathbf{v}_4 = \mathbf{u}_4 - \text{Proj}_{W_3} \mathbf{u}_4, \tag{4}$$

where

$$\text{Proj}_{W_3} \mathbf{u}_4 = \frac{\mathbf{u}_4 \cdot \mathbf{v}_1}{\mathbf{v}_1 \cdot \mathbf{v}_1}\mathbf{v}_1 + \frac{\mathbf{u}_4 \cdot \mathbf{v}_2}{\mathbf{v}_2 \cdot \mathbf{v}_2}\mathbf{v}_2 + \frac{\mathbf{u}_4 \cdot \mathbf{v}_3}{\mathbf{v}_3 \cdot \mathbf{v}_3}\mathbf{v}_3$$

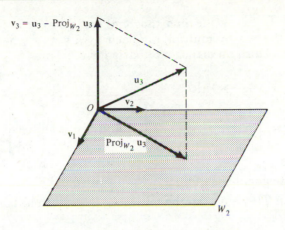

$v_3 = u_3 - \text{Proj}_{W_2} u_3$

u_3

v_2

O

v_1

$\text{Proj}_{W_2} u_3$

W_2

FIGURE 5.9

from Theorem 5.17. We continue in this way until we obtain an orthogonal set $T = \{\mathbf{v}_1, \mathbf{v}_2, \ldots, \mathbf{v}_k\}$ of k vectors. Thus, T is an orthogonal basis for the subspace W.

Example: Given the nonorthogonal basis $S = \{\mathbf{u}_1, \mathbf{u}_2, \mathbf{u}_3\}$ for R^3, where $\mathbf{u}_1 = [1, 1, 0]^t$, $\mathbf{u}_2 = [1, 0, 1]^t$, and $\mathbf{u}_3 = [0, 1, 1]^t$. Transform S into an orthogonal basis T. Then obtain an orthonormal basis.

Solution: First, let

$$\mathbf{v}_1 = \mathbf{u}_1 = [1, 1, 0]^t.$$

Then,

$$\mathbf{v}_2 = \mathbf{u}_2 - \text{Proj}_{W_1} \mathbf{u}_2 = \mathbf{u}_2 - \frac{\mathbf{u}_2 \cdot \mathbf{v}_1}{\mathbf{v}_1 \cdot \mathbf{v}_1} \mathbf{v}_1$$

$$= [1, 0, 1]^t - \left(\tfrac{1}{2}\right)[1, 1, 0]^t = \left[\tfrac{1}{2}, -\tfrac{1}{2}, 1\right]^t.$$

Now finish the calculations.

- - - - - - - - - - -

$$\mathbf{v}_3 = \mathbf{u}_3 - \text{Proj}_{W_2} \mathbf{u}_3$$

$$= \mathbf{u}_3 - \frac{\mathbf{u}_3 \cdot \mathbf{v}_1}{\mathbf{v}_1 \cdot \mathbf{v}_1} \mathbf{v}_1 - \frac{\mathbf{u}_3 \cdot \mathbf{v}_2}{\mathbf{v}_2 \cdot \mathbf{v}_2} \mathbf{v}_2$$

$$= [0, 1, 1]^t - \left(\tfrac{1}{2}\right)[1, 1, 0]^t - \left(\tfrac{1}{3}\right)\left[\tfrac{1}{2}, -\tfrac{1}{2}, 1\right]^t$$

$$= \left[-\tfrac{2}{3}, \tfrac{2}{3}, \tfrac{2}{3}\right]^t$$

Thus, $T = \{\mathbf{v}_1, \mathbf{v}_2, \mathbf{v}_3\}$ is an orthogonal basis for R^3. Let us now clear the fractions in the $\mathbf{v}_i$'s by multiplying each by a scalar when necessary. The resulting set is also an orthogonal basis for R^3. Thus,

$$T' = \{\mathbf{v}_1', \mathbf{v}_2', \mathbf{v}_3'\},$$

$$\text{where } \mathbf{v}_1' = \begin{bmatrix} 1 \\ 1 \\ 0 \end{bmatrix}, \quad \mathbf{v}_2' = \begin{bmatrix} 1 \\ -1 \\ 2 \end{bmatrix}, \quad \text{and} \quad \mathbf{v}_3' = \begin{bmatrix} -1 \\ 1 \\ 1 \end{bmatrix},$$

is also an orthogonal basis for R^3 (we multiplied $\mathbf{v}_3$ by $\frac{3}{2}$ here). If we normalize the vectors in T', we obtain the following orthonormal basis:

$$T'' = \{\mathbf{y}_1, \mathbf{y}_2, \mathbf{y}_3\},$$

$$\text{where } \mathbf{y}_1 = \begin{bmatrix} 1/\sqrt{2} \\ 1/\sqrt{2} \\ 0 \end{bmatrix}, \quad \mathbf{y}_2 = \begin{bmatrix} 1/\sqrt{6} \\ -1/\sqrt{6} \\ 2/\sqrt{6} \end{bmatrix}, \quad \text{and} \quad \mathbf{y}_3 = \begin{bmatrix} -1/\sqrt{3} \\ 1/\sqrt{3} \\ 1/\sqrt{3} \end{bmatrix}.$$

SELF-TEST

This Self-Test will help you determine whether or not you have mastered the chapter objectives and are ready to go on to another chapter. Correct answers are given at the end of the test.

1.   Which of the following sets, together with the given operations, is a vector space? If not, why not? Here you are expected to check out the axioms of Definition 5.1.

   (a)   The set of all three-component column vectors $[x_1, x_2, x_3]^t$ with the operations $[x_1, x_2, x_3]^t + [y_1, y_2, y_3]^t = [y_1, y_2, x_3 + y_3]^t$ and $k[x_1, x_2, x_3]^t = [kx_1, kx_2, kx_3]^t$.

   (b)   The set of all three-component row vectors of the form $[x_1, x_2, 2x_1 + 3x_2]$ where the rules for the addition and scalar multiplication operations are the usual rules for matrix addition and scalar multiplication of a matrix. (Here we can say the row vectors are of the form $[x_1, x_2, x_3]$, where $x_3 = 2x_1 + 3x_2$.)

2.   Which of the following subsets of R^3 are subspaces of R^3? Explain.

   (a)   The vector solutions of the system $x_1 - x_3 = 0$; $x_1 + 2x_2 - 3x_3 = 0$.

   (b)   The set of all column vectors of the form $[x_1, x_2, x_3]^t$ where $x_2 = x_1 - 2x_3$.

   (c)   The set of all column vectors of the form $[x_1, x_2, x_3]^t$ where $x_3 = x_1 + 3$.

3. Consider the set of vectors from R^2, $S = \{\mathbf{v}, \mathbf{w}\}$, where $\mathbf{v} = \begin{bmatrix} 3 \\ 4 \end{bmatrix}$ and $\mathbf{w} = \begin{bmatrix} 2 \\ 5 \end{bmatrix}$. In parts (a) and (b), use methods from Section B.

    (a) Show that S spans R^2.

    (b) Show that S is linearly independent.

4. Consider the set of vectors from R^3, $S = \{\mathbf{v}_1, \mathbf{v}_2\}$, where $\mathbf{v}_1 = [3, -2, 1]^t$ and $\mathbf{v}_2 = [5, 2, 3]^t$.

    (a) Show that the subspace spanned by S represents a plane through the origin of R^3. State an equation for the plane. *Hint:* See Example (a) of frame 10.

    (b) Show that S is linearly independent. *Hint:* See Example (b) of frame 14.

5. (a) In question 3 indicate why the set $S = \{\mathbf{v}, \mathbf{w}\}$ is a basis for R^2.

    (b) In question 4 indicate why the set $S = \{\mathbf{v}_1, \mathbf{v}_2\}$ is a basis for the subspace spanned by S. What is the dimension of the subspace?

6. Refer to the set $S = \{\mathbf{v}, \mathbf{w}\}$ given in question 3.

    (a) Use Theorem 5.5 to show that S is a basis for R^2.

    (b) Use Theorem 5.7 to determine c_1 and c_2 in the linear combination

$$\mathbf{x} = c_1 \mathbf{v} + c_2 \mathbf{w} \text{ if } \mathbf{x} \text{ is any vector } \begin{bmatrix} a \\ b \end{bmatrix} \text{ in } R^2.$$

    (c) Do part (b) for the specific vector $\mathbf{x} = \begin{bmatrix} 7 \\ -14 \end{bmatrix}$.

7. Use the techniques of frame 18 to determine a general expression for a vector solution $\mathbf{x}$ and a basis (if it exists) for the subspace of solutions for the following homogeneous linear systems. What is the dimension of the subspace?

    (a)
$$\begin{aligned} x_1 + x_2 + 5x_3 &= 0 \\ 4x_1 - 3x_2 - x_3 &= 0 \\ x_1 - x_2 - x_3 &= 0 \end{aligned}$$

    (b)
$$\begin{aligned} 2x_1 + x_2 - x_3 + x_4 &= 0 \\ 4x_1 - 3x_2 - 3x_3 + x_4 &= 0 \end{aligned}$$

    (c)
$$\begin{aligned} x_1 - x_2 + x_3 &= 0 \\ 2x_1 + x_2 + 3x_3 &= 0 \\ 3x_1 - x_2 - x_3 &= 0 \end{aligned}$$

    (d) The system consisting of the single equation $x_1 + 3x_2 - 4x_3 = 0$. The situation here is the reverse of the situation in question 4.

8. Given the matrix $A = \begin{bmatrix} 2 & 1 & -1 & 1 \\ 3 & -1 & -2 & 1 \\ 1 & -2 & -1 & 0 \end{bmatrix}$.

    (a) Find a basis for the row space of A, and the row rank of A.

(b) Find a basis for the column space of A, and the column rank of A.

(c) Find the determinant rank of A.

9. Given the set $S = \{\mathbf{v}_1, \mathbf{v}_2, \mathbf{v}_3\}$ of column vectors from R^3, where $\mathbf{v}_1 = [3, 0, 0]^t$, $\mathbf{v}_2 = [0, 3, -4]^t$, $\mathbf{v}_3 = [0, 4, 3]^t$.

(a) Show that S is orthogonal.

(b) Derive an orthonormal set T from S.

(c) Refer to part (a). Find c_1, c_2, and c_3 in $\mathbf{x} = c_1\mathbf{v}_1 + c_2\mathbf{v}_2 + c_3\mathbf{v}_3$ if $\mathbf{x} = [6, 6, 17]^t$.

10. Let W be the subspace of R^3 spanned by the orthogonal set $\{\mathbf{v}_1, \mathbf{v}_2\}$, where $\mathbf{v}_1 = [0, 1, 0]^t$ and $\mathbf{v}_2 = [3, 0, -4]^t$. Also, $\mathbf{x} = [15, 3, 5]^t$.

(a) Determine $\text{Proj}_W \mathbf{x}$.

(b) Determine the component of $\mathbf{x}$ that is perpendicular to W.

(c) Of all the distances between $\mathbf{x}$ and vectors in W, determine the minimum distance. For what vector $\mathbf{w}^*$ in W is the distance a minimum?

11. (a) Given the nonorthogonal basis for R^3, $S = \{\mathbf{u}_1, \mathbf{u}_2, \mathbf{u}_3\}$, where $\mathbf{u}_1 = [1, 1, 0]^t$, $\mathbf{u}_2 = [1, 1, 1]^t$, $\mathbf{u}_3 = [0, 1, 0]^t$. Transform S into an orthogonal basis T. Then obtain an orthonormal basis T'.

(b) Let W be the subspace of R^3 with basis $S = \{\mathbf{u}_1, \mathbf{u}_2\}$, where $\mathbf{u}_1 = [-4, 2, 2]^t$ and $\mathbf{u}_2 = [6, -1, 1]^t$. Transform S into an orthogonal basis T. Then obtain an orthonormal basis T'.

ANSWERS TO SELF-TEST

If your answers to the test questions do not agree with the ones that follow, review the frames indicated in parentheses after each answer before you go on to another chapter.

1. (a) Not a vector space. For example, (v) and (viii) of Definition 5.1 do not hold. Also, there is no zero vector. That is, there is no unique column vector $[z_1, z_2, z_3]^t$ such that $[x_1, x_2, x_3]^t + [z_1, z_2, z_3]^t = [x_1, x_2, x_3]^t$, regardless of what x_1, x_2, and x_3 are.

(b) Is a vector space. Scalar equation is $x_3 - 2x_1 - 3x_2 = 0$, a plane in R^3 through the origin.   (frames 1–3)

2. (a) Yes. Vector solutions are of form $\mathbf{x} = [r, r, r]^t = r[1, 1, 1]^t$, indicating a straight line through the origin.

(b) Yes. Set represents plane through origin.

(c) No. $\mathbf{x} + \mathbf{y} = [x_1 + y_1, x_2 + y_2, x_1 + y_1 + 6]^t$, and the third component is not $x_1 + y_1 + 3$.   (frames 4–6)

3. (a) For any vector $\mathbf{x} = \begin{bmatrix} a \\ b \end{bmatrix}$ in R^2, we have $c_1 = (5a - 2b)/7$ and $c_2 = (-4a + 3b)/7$ for $\mathbf{x} = c_1\mathbf{v} + c_2\mathbf{w}$.

(b) $c_1\mathbf{v} + c_2\mathbf{w} = \begin{bmatrix} 0 \\ 0 \end{bmatrix}$ implies $c_1 = c_2 = 0$.   (frames 7–11)

4. (a)   An equation of the plane is $2x_1 + x_2 - 4x_3 = 0$. (b) The vector equation $c_1\mathbf{v}_1 + c_2\mathbf{v}_2 = \mathbf{0}$ leads to the system $3c_1 + 5c_2 = 0$, $-2c_1 + 2c_2 = 0$, $c_1 + 3c_2 = 0$. The reduced row echelon form is

$$\begin{bmatrix} 1 & 0 & | & 0 \\ 0 & 1 & | & 0 \\ 0 & 0 & | & 0 \end{bmatrix},$$ which means that $c_1 = c_2 = 0$ is the only solution.
                    (frames 7–11, 14)

5. (a) $S = \{\mathbf{v}, \mathbf{w}\}$ spans R^2 and is linearly independent.

   (b) $S = \{\mathbf{v}_1, \mathbf{v}_2\}$ spans the subspace (by definition of what the subspace is) and is linearly independent. Dimension is 2.   (frames 13–15)

6. (a)   For $A = \begin{bmatrix} 3 & 2 \\ 4 & 5 \end{bmatrix}$, $|A| = 7 \neq 0$.

   (b)   Here, $A^{-1} = \dfrac{1}{7}\begin{bmatrix} 5 & -2 \\ -4 & 3 \end{bmatrix}$, and $\mathbf{c} = A^{-1}\mathbf{x}$ with $\mathbf{x} = \begin{bmatrix} a \\ b \end{bmatrix}$. Thus, $c_1 = (5a - 2b)/7$ and $c_2 = (-4a + 3b)/7$.

   (c) $c_1 = 9$, $c_2 = -10$.   (frames 13–15)

7. (a) $\mathbf{x} = r\mathbf{v}_1$ where $\mathbf{v}_1 = [-2, -3, 1]^t$. $S = \{\mathbf{v}_1\}$ is basis. Dimension equals 1.

   (b) $\mathbf{x} = r\mathbf{v}_1 + s\mathbf{v}_2$, where $\mathbf{v}_1 = [\frac{3}{5}, -\frac{1}{5}, 1, 0]^t$ and $\mathbf{v}_2 = [-\frac{2}{5}, -\frac{1}{5}, 0, 1]^t$. $S = \{\mathbf{v}_1, \mathbf{v}_2\}$ is basis. Dimension is 2.

   (c) $\mathbf{x} = \mathbf{0}$. Subspace is $\{\mathbf{0}\}$, so no basis. Dimension equals 0.

   (d) $\mathbf{x} = r\mathbf{v}_1 + s\mathbf{v}_2$ where $\mathbf{v}_1 = [-3, 1, 0]^t$ and $\mathbf{v}_2 = [4, 0, 1]^t$. $S = \{\mathbf{v}_1, \mathbf{v}_2\}$ is basis. Dimension equals 2.   (frames 17–18)

8. (a)   Basis for row space is $\{\mathbf{a}_1^*, \mathbf{a}_2^*\}$ where $\mathbf{a}_1^* = [1, 0, -\frac{3}{5}, \frac{2}{5}]$ and $\mathbf{a}_2^* = [0, 1, \frac{1}{5}, \frac{1}{5}]$. Here $\mathbf{a}_1^*$ and $\mathbf{a}_2^*$ are first two rows of reduced row echelon form of A. Row rank $= 2$

   (b)   Basis for column space is $\{\mathbf{w}_1, \mathbf{w}_2\}$ where $\mathbf{w}_1 = [1, 0, -1]^t$ and $\mathbf{w}_2 = [0, 1, 1]^t$. Column rank $= 2$. Here $[1, 0, -1]$ and $[0, 1, 1]$ are first two rows of reduced row echelon form of A^t.

   (c)   The determinants of all four 3×3 submatrices equal zero. For sub-matrix $\begin{bmatrix} 2 & 1 \\ 3 & -1 \end{bmatrix}$ from rows 1, 2 and columns 1, 2, we have $\begin{vmatrix} 2 & 1 \\ 3 & -1 \end{vmatrix} = -5 \neq 0$. Determinant rank equals 2.

                    (frames 19–21)

9. (a) $\mathbf{v}_1 \cdot \mathbf{v}_2 = \mathbf{v}_1 \cdot \mathbf{v}_3 = \mathbf{v}_2 \cdot \mathbf{v}_3 = 0.$

    (b) $T = \{\mathbf{w}_1, \mathbf{w}_2, \mathbf{w}_3\}$ where $\mathbf{w}_1 = [1, 0, 0]^t$; $\mathbf{w}_2 = \mathbf{v}_2/\|\mathbf{v}_2\| = \left[0, \frac{3}{5}, -\frac{4}{5}\right]^t$

    and $\mathbf{w}_3 = \left[0, \frac{4}{5}, \frac{3}{5}\right]^t$.

    (c) $\mathbf{x} = 2\mathbf{v}_1 - 2\mathbf{v}_2 + 3\mathbf{v}_3.$ (frames 23–24)

10. (a) $\text{Proj}_W \mathbf{x} = 3\mathbf{v}_1 + \mathbf{v}_2 = [3, 3, -4]^t.$

    (b) The vector is $\text{Comp}_W \mathbf{x} = \mathbf{x} - \text{Proj}_W \mathbf{x} = [12, 0, 9]^t.$

    (c) Min. distance $= \|\mathbf{x} - \text{Proj}_W \mathbf{x}\| = 15$; $\mathbf{w}^* = \text{Proj}_W \mathbf{x} = [3, 3, -4]^t.$
                                                          (frames 25–26)

11. (a) $T = \{\mathbf{v}_1, \mathbf{v}_2, \mathbf{v}_3\}$ where $\mathbf{v}_1 = [1, 1, 0]^t$, $\mathbf{v}_2 = [0, 0, 1]^t$, and $\mathbf{v}_3 = \left[-\frac{1}{2}, \frac{1}{2}, 0\right]^t$. $T' = \{\mathbf{w}_1, \mathbf{w}_2, \mathbf{w}_3\}$, where $\mathbf{w}_1 = [1/\sqrt{2}, 1/\sqrt{2}, 0]^t$, $\mathbf{w}_2 = [0, 0, 1]^t$, and $\mathbf{w}_3 = [-1/\sqrt{2}, 1/\sqrt{2}, 0]^t$.

    (b) $T = \{\mathbf{v}_2, \mathbf{v}_2\}$, where $\mathbf{v}_1 = \mathbf{u}_1 = [-4, 2, 2]^t$ and $\mathbf{v}_2 = \mathbf{u}_2 - \left(\dfrac{\mathbf{u}_2 \cdot \mathbf{v}_1}{\mathbf{v}_1 \cdot \mathbf{v}_1}\right)\mathbf{v}_1 = \mathbf{u}_2 + \mathbf{v}_1 = [2, 1, 3]^t$. $T' = \{\mathbf{w}_1, \mathbf{w}_2\}$, where $\mathbf{w}_1 = [-2/\sqrt{6}, 1/\sqrt{6}, 1/\sqrt{6}]^t$ and $\mathbf{w}_2 = [2/\sqrt{14}, 1/\sqrt{14}, 3/\sqrt{14}]^t$.
                                                          (frame 27)

CHAPTER SIX
Linear Transformations

In this chapter we shall study functions that are known as linear transformations. A linear transformation that is useful in analytic geometry, engineering, and physics is the rotational linear transformation; this rotates vectors in R^2 (two-dimensional space). Another linear transformation that has practical value is the transformation that generates the orthogonal (perpendicular) projection of a vector onto a subspace. (We have already been exposed to orthogonal projections; see frames 25–27 of Chapter Five.) Both the rotational and orthogonal projection linear transformations are among those studied in the current chapter.

In the most typical case, for our purposes, the linear transformation will act on column vectors in R^n (i.e., column vectors with n real components) and "transform" them into column vectors in R^m. Often this will be accomplished by an equation of the form $\mathbf{y} = A\mathbf{x}$, where $\mathbf{x}$ is an n-component column vector, $\mathbf{y}$ is an m-component column vector, and A is an $m \times n$ matrix.

A practical example of a linear transformation occurs in Section G of Chapter Two. There we had the equation $\mathbf{c} = (I - A)\mathbf{x}$, or

$$\mathbf{c} = B\mathbf{x},$$

if we let $B = (I - A)$. (See answer in frame 34 of Chapter Two.) In Section G both $\mathbf{c}$ and $\mathbf{x}$ were three-component column vectors, and B was a 3×3 matrix. The preceding equation expresses a linear transformation of the total output vector $\mathbf{x}$ into the consumer demand vector $\mathbf{c}$. Also, we can convert the preceding equation into

$$\mathbf{x} = B^{-1}\mathbf{c}$$

which expresses a linear transformation of the consumer demand vector $\mathbf{c}$ into the total output vector $\mathbf{x}$. [See frames 34–36 of Chapter Two; there, $B^{-1} = (I - A)^{-1}$.] The latter transformation equation is very useful in economic analyses.

OBJECTIVES

When you complete this chapter, you should be able to

- Determine if a transformation from one vector space into another vector space is linear.
- Calculate and use the standard matrix A for a linear transformation L from R^n into R^m.
- Determine the kernel and range of a linear transformation.
- Establish if a linear transformation is, respectively, one-to-one or onto (or both).
- Interpret and use the equation that states that for a linear transformation $L: V \rightarrow W$, the sum of the dimensions of the kernel and range of L is equal to the dimension of V. (Our main emphasis will be on linear transformations for which V is R^n and W is R^m.)
- Calculate the *coordinate column vector* of a vector $\mathbf{x}$ of V with respect to a basis S of V.
- Calculate and interpret the matrix A of linear transformation $L: V \rightarrow W$ with respect to bases S and T of V and W, respectively.

A. DEFINITION OF A LINEAR TRANSFORMATION

1. In this section we shall begin the study of those functions known as linear transformations. A linear transformation is a special type of function, of the form $\mathbf{w} = F(\mathbf{v})$, where the independent variable $\mathbf{v}$ and the dependent variable $\mathbf{w}$ are both vectors in vector spaces. Linear transformations are very important in many areas of mathematics and in applications to engineering and the natural and social sciences.

Definition 6.1 (Linear Transformation)

Let V and W be vector spaces. A *linear transformation L* from V into W is a function that assigns a unique vector $L(\mathbf{x})$ in W to each vector $\mathbf{x}$ in V, where

$$L(c_1\mathbf{x} + c_2\mathbf{y}) = c_1 L(\mathbf{x}) + c_2 L(\mathbf{y})$$

for every pair $\mathbf{x}$ and $\mathbf{y}$ in V and for every pair of scalars (real numbers) c_1 and c_2. We shall indicate that L from V into W is a linear transformation by writing $L: V \rightarrow W$.

Notes: (i) The expression $c_1\mathbf{x} + c_2\mathbf{y}$ is known as a *linear combination* of $\mathbf{x}$ and $\mathbf{y}$. (ii) The preceding definition indicates that a linear transformation L has the *linearity property* or follows the *linearity rule*. [A general function f has the linearity property if $f(c_1x + c_2y)$ equals $c_1 f(x)$ plus $c_2 f(y)$; here c_1

and c_2 denote constants.] (iii)   If $V = W$, the linear transformation $L: V \to V$ is known as a *linear operator* on V. (iv)   In $L: V \to W$, the vector space V is known as the *domain* or *input space* and W is known as the *codomain*. The vectors in V are known as input vectors. If a vector $\mathbf{y}$ in W is such that $\mathbf{y} = L(\mathbf{x})$ for some vector $\mathbf{x}$ in V, then $\mathbf{y}$ is called the *image* of $\mathbf{x}$ under L, or an *output vector* corresponding to $\mathbf{x}$.

The set of all vectors in W that are images of vectors in V is called the *range* of the linear transformation L. Thus, the range is a subset of the codomain. If the range turns out to be the codomain, then L is called an *onto* linear transformation. We shall discuss the concepts *range* and *onto* more in Section C of this chapter.

We shall concentrate mostly on vector spaces of the form R^n. Remember that we consider a typical element of R^n to be the column vector $\mathbf{x}$, where

$$\mathbf{x} = \begin{bmatrix} x_1 \\ x_2 \\ \vdots \\ x_n \end{bmatrix} = [x_1, x_2, \ldots, x_n]^t \qquad (1)$$

Here $x_1, x_2, \ldots, x_n$ are all real numbers (see Chapters Four and Five). As usual, the transpose notation will be used as a space-saving device.

In a typical case, V shall be R^n and W shall be R^m.

Example (a):   Suppose L from R^2 into R^2 is defined by

$$L\left(\begin{bmatrix} x_1 \\ x_2 \end{bmatrix}\right) = \begin{bmatrix} 2x_1 + x_2 \\ x_1 + 3x_2 \end{bmatrix} \qquad (I)$$

Note that the left side expresses $L(\mathbf{x})$, where $\mathbf{x} = [x_1, x_2]^t$ is a typical column vector in R^2.

Determine if L is a linear transformation.

Solution: Let $\mathbf{x} = [x_1, x_2]^t$ and $\mathbf{y} = [y_1, y_2]^t$ be typical column vectors in R^2 and let c_1 and c_2 be typical scalars. First, we express $c_1\mathbf{x} + c_2\mathbf{y}$ as a single column vector in R^2.

$$c_1\mathbf{x} + c_2\mathbf{y} = c_1\begin{bmatrix} x_1 \\ x_2 \end{bmatrix} + c_2\begin{bmatrix} y_1 \\ y_2 \end{bmatrix} = \begin{bmatrix} c_1x_1 + c_2y_1 \\ c_1x_2 + c_2y_2 \end{bmatrix} \qquad (1)$$

Here, and in the following pages we make use of rules for matrix addition and scalar multiplication (Definitions 1.1 and 1.2). A column vector is the same as a matrix with one column (see frames 5 and 14 of Chapter One).

Thus, from defining rule (I), we have

$$L(c_1\mathbf{x} + c_2\mathbf{y}) = \begin{bmatrix} 2(c_1x_1 + c_2y_1) + (c_1x_2 + c_2y_2) \\ (c_1x_1 + c_2y_1) + 3(c_1x_2 + c_2y_2) \end{bmatrix}$$

$$= \begin{bmatrix} c_1(2x_1 + x_2) + c_2(2y_1 + y_2) \\ c_1(x_1 + 3x_2) + c_2(y_1 + 3y_2) \end{bmatrix} \tag{2}$$

Also, from (I), we have

$$c_1L(\mathbf{x}) = c_1\begin{bmatrix} 2x_1 + x_2 \\ x_1 + 3x_2 \end{bmatrix} = \begin{bmatrix} 2c_1x_1 + c_1x_2 \\ c_1x_1 + 3c_1x_2 \end{bmatrix}, \tag{3}$$

and

$$c_2L(\mathbf{y}) = c_2\begin{bmatrix} 2y_1 + y_2 \\ y_1 + 3y_2 \end{bmatrix} = \begin{bmatrix} 2c_2y_1 + c_2y_2 \\ c_2y_1 + 3c_2y_2 \end{bmatrix}. \tag{4}$$

Adding (3) and (4) yields the same result as the right side of (2), thus establishing that L is a linear transformation.

The key idea of Definition 6.1 can be extended to a linear combination of more than two vectors from V.

Theorem 6.1

If L is a linear transformation from V into W, then

$$L(c_1\mathbf{x}_1 + c_2\mathbf{x}_2 + \cdots + c_k\mathbf{x}_k) = c_1L(\mathbf{x}_1) + c_2L(\mathbf{x}_2) + \cdots + c_kL(\mathbf{x}_k)$$

for any vectors $\mathbf{x}_1, \mathbf{x}_2, \ldots, \mathbf{x}_k$ in V and any scalars $c_1, c_2, \ldots, c_k$.

The property given by Theorem 6.1 is described as a *general linearity property* or *rule*.

Example (b): Let T from R^3 into R^2 be defined by

$$T(\mathbf{x}) = T([x_1, x_2, x_3]^t) = \begin{bmatrix} 2x_1 + 1 \\ x_2 + x_3 \end{bmatrix} \tag{II}$$

Is T a linear transformation?

Solution: For typical column vectors of R^3, $\mathbf{x} = [x_1, x_2, x_3]^t$ and $\mathbf{y} = [y_1, y_2, y_3]^t$, we have

$$c_1\mathbf{x} + c_2\mathbf{y} = \begin{bmatrix} c_1x_1 + c_2y_1 \\ c_1x_2 + c_2y_2 \\ c_1x_3 + c_2y_3 \end{bmatrix} \tag{1}$$

Thus, from defining rule (II), we have

$$T(c_1\mathbf{x} + c_2\mathbf{y}) = \begin{bmatrix} 2(c_1x_1 + c_2y_1) + 1 \\ (c_1x_2 + c_2y_2) + (c_1x_3 + c_2y_3) \end{bmatrix} \qquad (2)$$

Now complete the problem.

— — — — — — — — — —

$$c_1T(\mathbf{x}) = c_1 \begin{bmatrix} 2x_1 + 1 \\ x_2 + x_3 \end{bmatrix} = \begin{bmatrix} 2c_1x_1 + c_1 \\ c_1x_2 + c_1x_3 \end{bmatrix} \qquad (3)$$

$$c_2T(\mathbf{y}) = c_2 \begin{bmatrix} 2y_1 + 1 \\ y_2 + y_3 \end{bmatrix} = \begin{bmatrix} 2c_2y_1 + c_2 \\ c_2y_2 + c_2y_3 \end{bmatrix} \qquad (4)$$

Adding (3) and (4) leads to

$$c_1T(\mathbf{x}) + c_2T(\mathbf{y}) = \begin{bmatrix} 2(c_1x_1 + c_2y_1) + (c_1 + c_2) \\ c_1x_2 + c_2y_2 + c_1x_3 + c_2y_3 \end{bmatrix} \qquad (5)$$

Comparing the right sides of (2) and (5), we see that T is not linear, since $(c_1 + c_2)$ is not equal to 1, in general (c_1 and c_2 can, individually, be any scalars whatsoever).

2.   Some authors use the following two-part definition of a linear transformation, which is equivalent to Definition 6.1.

Definition 6.1' (Linear Transformation)

Let V and W be vector spaces. A *linear transformation L* of V into W is such that

(a) $L(\mathbf{x} + \mathbf{y}) = L(\mathbf{x}) + L(\mathbf{y})$, for every $\mathbf{x}$ and $\mathbf{y}$ in V, and
(b) $L(c\mathbf{x}) = cL(\mathbf{x})$ for every $\mathbf{x}$ in V and every scalar c.

An important type of linear transformation from R^n into R^m is defined by the product $A\mathbf{x}$ of the $m \times n$ matrix A times the column vector $\mathbf{x}$ from R^n. Thus,

$$L(\mathbf{x}) = A\mathbf{x} \qquad (I)$$

Observe that $A\mathbf{x}$ is $m \times 1$, or, equivalently, a column vector from R^m.

Let us show that we have a linear transformation. Let $\mathbf{x}$ and $\mathbf{y}$ be column vectors from R^n, and let c_1 and c_2 be any scalars.

$$L(c_1\mathbf{x} + c_2\mathbf{y}) = A(c_1\mathbf{x} + c_2\mathbf{y}), \qquad \text{by definition of } L, \tag{1}$$

$$= Ac_1\mathbf{x} + Ac_2\mathbf{y}, \qquad \begin{array}{l}\text{by distributive law} \\ \text{[Theorem 1.3, part (b)]},\end{array} \tag{2}$$

$$= c_1 A\mathbf{x} + c_2 A\mathbf{y}, \qquad \text{by Theorem 1.3, part (d)}, \tag{3}$$

$$= c_1 L(\mathbf{x}) + c_2 L(\mathbf{y}), \qquad \text{by definition of } L. \tag{4}$$

Thus, Definition 6.1 has been verified, and we have a linear transformation.

Example: Given the matrix $A = \begin{bmatrix} 1 & 3 \\ 2 & 4 \end{bmatrix}$. Let us define $L(\mathbf{x})$ as $A\mathbf{x}$, where $\mathbf{x} = [x_1, x_2]^t$. Calculate $L(\mathbf{x})$ if (a) $\mathbf{x} = [5, -2]^t$, (b) $\mathbf{x} = [1, 0]^t$, and (c) $\mathbf{x} = [0, 1]^t$.

Solution:

(a) $L(\mathbf{x}) = A\mathbf{x} = \begin{bmatrix} 1 & 3 \\ 2 & 4 \end{bmatrix}\begin{bmatrix} 5 \\ -2 \end{bmatrix} = \begin{bmatrix} 5-6 \\ 10-8 \end{bmatrix} = \begin{bmatrix} -1 \\ 2 \end{bmatrix}.$

(b) $L(\mathbf{x}) = A\mathbf{x} = \begin{bmatrix} 1 & 3 \\ 2 & 4 \end{bmatrix}\begin{bmatrix} 1 \\ 0 \end{bmatrix} = \begin{bmatrix} 1 \\ 2 \end{bmatrix}.$

Now finish the problem.

— — — — — — — — — —

(c) $L(\mathbf{x}) = A\mathbf{x} = \begin{bmatrix} 1 & 3 \\ 2 & 4 \end{bmatrix}\begin{bmatrix} 0 \\ 1 \end{bmatrix} = \begin{bmatrix} 3 \\ 4 \end{bmatrix}.$

Notes: (i)   The vectors in parts (b) and (c) are the vectors $\mathbf{e}_1 = [1, 0]^t$ and $\mathbf{e}_2 = [0, 1]^t$ of the *standard* basis for R^2. (ii)   From the answers to parts (b) and (c), we see that $L(\mathbf{e}_1)$ and $L(\mathbf{e}_2)$ are equal to the first and second columns of A, respectively. This is a special case of a general property that will be explained further in Section B.

3.   If $L: R^n \rightarrow R^m$ is defined by $L(\mathbf{x}) = A\mathbf{x}$, where A is an $m \times n$ matrix, then we call this type of linear transformation a *matrix transformation*. Let us consider some more examples.

Example (a) (Rotation in R^2): Suppose we rotate a vector $\mathbf{x} = [x_1, x_2]^t$ through an angle ϕ as shown in Figure 6.1. By convention, ϕ is positive if it is counterclockwise. It can be shown, using trigonometry or analytic geometry arguments [see, for example, Anton (1977)], that the new components of the resultant vector $\mathbf{x}' = [x_1', x_2']^t$ are related to the components of $\mathbf{x}$ by the following equations:

$$x_1' = x_1 \cos \phi - x_2 \sin \phi \tag{1a}$$

$$x_2' = x_1 \sin \phi + x_2 \cos \phi \tag{1b}$$

We can write this in matrix–vector form as follows:

$$\begin{bmatrix} x_1' \\ x_2' \end{bmatrix} = \begin{bmatrix} \cos \phi & -\sin \phi \\ \sin \phi & \cos \phi \end{bmatrix} \begin{bmatrix} x_1 \\ x_2 \end{bmatrix} \tag{2}$$

Thus, $\mathbf{x}' = A\mathbf{x}$, where

$$A = \begin{bmatrix} \cos \phi & -\sin \phi \\ \sin \phi & \cos \phi \end{bmatrix}. \tag{3}$$

Letting $L(\mathbf{x}) = A\mathbf{x}$, the linear transformation $L(\mathbf{x})$ is called a *rotation in* R^2 through the angle ϕ. Note that $L(\mathbf{x}) = \mathbf{x}' = [x_1', x_2']^t$.

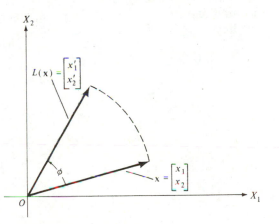

FIGURE 6.1   Rotation linear transformation.

For a sample calculation, suppose $\mathbf{x} = \begin{bmatrix} 2 \\ 1 \end{bmatrix}$ and $\phi = 30° = \pi/6$ radians. Then

$$x_1' = 2 \cos (\pi/6) - \sin (\pi/6) = 1.232$$

and $x_2' = 2 \sin (\pi/6) + \cos (\pi/6) = 1.866$

Put differently, $L(\mathbf{x}) = \mathbf{x}' = \begin{bmatrix} 1.232 \\ 1.866 \end{bmatrix}.$

Example (b): If $L: R^n \rightarrow R^n$ is defined by $L(\mathbf{x}) = k\mathbf{x}$, then it is easy to show that L is a linear transformation. We shall call this the scalar multiple linear transformation. If $0 < k < 1$, L is called a *contraction* and if $k > 1$, L is called a *dilation*. If $k < 0$, then L is called a *direction reversal*.

See Figure 6.2 for diagrams pertaining to R^2.

Example (c) (Projection): It can be shown for any vector $\mathbf{x}$ in R^2 that the orthogonal projection of $\mathbf{x}$ onto a straight line through the origin is a linear transformation. The straight line represents a one-dimensional subspace of R^2. (See frame 25 of Chapter Five for a review of the orthogonal projection concept.) To be specific, consider the straight line with equation $x_1 - 2x_2 = 0$. To obtain a basis vector for the line, we let $x_2 = 1$ (any other nonzero number will do), and find that $x_1 = 2$. Thus, $\mathbf{v} = [2, 1]^t$ is a basis vector for the line, or put differently, $S = \{\mathbf{v}\}$ is a basis set for the line. Thus, every vector for the line can be represented as $c\mathbf{v}$. Using Theorem 5.17 (frame 25), we see that $L(\mathbf{x}) = \text{Proj}_\ell \mathbf{x}$ is given by

$$\text{Proj}_\ell \mathbf{x} = \left(\frac{\mathbf{x} \cdot \mathbf{v}}{\mathbf{v} \cdot \mathbf{v}}\right) \mathbf{v} \tag{1}$$

Here the subspace is the line with equation $x_1 - 2x_2 = 0$; it is denoted by "ℓ" in Eq. (1), and its dimension equals 1. Carrying out the computations for $\mathbf{x} = [x_1, x_2]^t$, we have

$$\mathbf{x} \cdot \mathbf{v} = 2x_1 + x_2, \qquad \mathbf{v} \cdot \mathbf{v} = 2^2 + 1^2 = 5$$

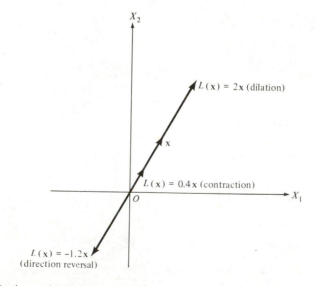

FIGURE 6.2 Scalar multiple linear transformation.

and thus

$$\text{Proj}_\ell \ \mathbf{x} = \frac{(2x_1 + x_2)}{5} \begin{bmatrix} 2 \\ 1 \end{bmatrix} = \begin{bmatrix} (4x_1 + 2x_2)/5 \\ (2x_1 + x_2)/5 \end{bmatrix} = \frac{1}{5} \begin{bmatrix} 4x_1 + 2x_2 \\ 2x_1 + x_2 \end{bmatrix} \tag{2}$$

Refer to Figure 6.3 for a diagram showing a typical $\mathbf{x}$ and $\text{Proj}_\ell \ \mathbf{x}$.

Example (d): Refer to Example (c). Suppose $\mathbf{x} = [5, 5]^t$. Determine $\text{Proj}_\ell \ \mathbf{x}$.

- - - - - - - - - - -

Solution: From Eq. (2) in Example (c),

$$\text{Proj}_\ell \ \mathbf{x} = \frac{1}{5} \begin{bmatrix} 4(5) + 2(5) \\ 2(5) + 5 \end{bmatrix} = \frac{1}{5} \begin{bmatrix} 30 \\ 15 \end{bmatrix} = \begin{bmatrix} 6 \\ 3 \end{bmatrix}$$

The $\mathbf{x}$ and $\text{Proj}_\ell \ \mathbf{x}$ on Figure 6.3 are consistent with this calculation.

4. The results of Examples (c) and (d) of the previous frame can be extended. Refer to frame 25 of Chapter Five and, in particular, to Figure 5.6. Thus, if $S = \{\mathbf{v}_1, \mathbf{v}_2, \ldots, \mathbf{v}_k\}$ is an orthogonal basis for a subspace of W of R^n, then $\text{Proj}_W \ \mathbf{x}$ is a linear transformation. Here $\mathbf{x}$ represents any vector in R^n. As in Theorem 5.17 of frame 25, $\text{Proj}_W \ \mathbf{x}$ is given by

$$\text{Proj}_W \ \mathbf{x} = c_1\mathbf{v}_1 + c_2\mathbf{v}_2 + \cdots + c_k\mathbf{v}_k,$$

where $\quad c_i = \dfrac{\mathbf{x} \cdot \mathbf{v}_i}{\mathbf{v}_i \cdot \mathbf{v}_i} \quad$ for $i = 1, 2, \ldots, k$.

Here the symbol "$\text{Proj}_W \ \mathbf{x}$" refers to an *orthogonal* projection of $\mathbf{x}$ onto a subspace W.

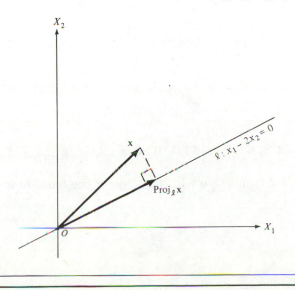

FIGURE 6.3

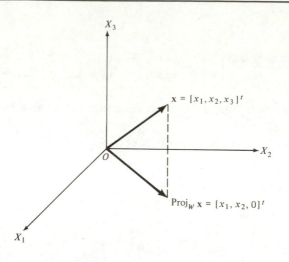

FIGURE 6.4

Example (a): Suppose R^n is R^3 and that $S = \{e_1, e_2\}$, where $e_1 = [1, 0, 0]^t$ and $e_2 = [0, 1, 0]^t$. Thus, S is an orthonormal basis for the X_1X_2 plane. (The latter is, of course, a subspace of R^3.) Determine the expression for $\text{Proj}_W x$, where $x = [x_1, x_2, x_3]^t$ and W stands for the X_1X_2 plane. From the equations above, we have

$$c_1 = \frac{x \cdot e_1}{e_1 \cdot e_1} = x \cdot e_1 = x_1 \qquad \text{and} \qquad c_2 = x \cdot e_2 = x_2$$

and hence

$$\text{Proj}_W x = x_1[1, 0, 0]^t + x_2[0, 1, 0]^t = [x_1, x_2, 0]^t.$$

Refer to Figure 6.4. The effect of projecting x onto the X_1X_2 plane is to transform its third component to zero while keeping the first two components unchanged.

Example (b): In Example (a), suppose $x = [3, 4, 5]^t$. Determine $\text{Proj}_W x$.

— — — — — — — — — —

$\text{Proj}_W x = [3, 4, 0]^t$. Here $[3, 4, 0]^t$ is a vector in the X_1X_2 plane.

5. In the next few frames we will list some of the basic properties of linear transformations.

Theorem 6.2

Suppose L: $V \to W$ is a linear transformation from V into W. Then

(a) $L(\mathbf{0}_V) = \mathbf{0}_W$, where $\mathbf{0}_V$ and $\mathbf{0}_W$ are the zero vectors in V and W, respectively.

(b) $L(-\mathbf{x}) = -L(\mathbf{x})$ for any $\mathbf{x}$ in V.

(c) $L(\mathbf{x} - \mathbf{y}) = L(\mathbf{x}) - L(\mathbf{y})$ for any $\mathbf{x}$ and $\mathbf{y}$ in V.

Once we know what a linear transformation does to the vectors in a basis of a vector space, then we can easily determine the images of all other vectors in the vector space. [Remember, $L(\mathbf{x})$, which is in W, is the "image" of $\mathbf{x}$ under L.]

Theorem 6.3

Suppose L: $V \to W$ is a linear transformation and V is n-dimensional. Let $S = \{\mathbf{v}_1, \mathbf{v}_2, \ldots, \mathbf{v}_n\}$ be a basis for V. For any vector $\mathbf{x}$ in V, $L(\mathbf{x})$ is completely determined by the set $\{L(\mathbf{v}_1), L(\mathbf{v}_2), \ldots, L(\mathbf{v}_n)\}$.

Proof of Theorem 6.3

Since $\mathbf{x}$ is in V, we can write $\mathbf{x} = c_1\mathbf{v}_1 + c_2\mathbf{v}_2 + \cdots + c_n\mathbf{v}_n$, where the c_i's are uniquely determined. Then

$$L(\mathbf{x}) = L(c_1\mathbf{v}_1 + c_2\mathbf{v}_2 + \cdots + c_n\mathbf{v}_n) \tag{1}$$

From Theorem 6.1 this becomes

$$L(\mathbf{x}) = c_1 L(\mathbf{v}_1) + c_2 L(\mathbf{v}_2) + \cdots + c_n L(\mathbf{v}_n). \tag{2}$$

Equation (2) indicates that $L(\mathbf{x})$ is completely determined by the set $\{L(\mathbf{v}_1), L(\mathbf{v}_2), \ldots, L(\mathbf{v}_n)\}$.

Example: Given the basis $S = \{\mathbf{v}_1, \mathbf{v}_2, \mathbf{v}_3\}$ for R^3, where $\mathbf{v}_1 = [1, 1, 0]^t$, $\mathbf{v}_2 = [1, 0, 1]^t$, and $\mathbf{v}_3 = [0, 1, 1]^t$. Suppose L: $R^3 \to R^2$ is a linear transformation where

$$L(\mathbf{v}_1) = \begin{bmatrix} 5 \\ -1 \end{bmatrix}, \qquad L(\mathbf{v}_2) = \begin{bmatrix} 1 \\ 5 \end{bmatrix}, \qquad \text{and} \qquad L(\mathbf{v}_3) = \begin{bmatrix} 2 \\ 2 \end{bmatrix} \tag{1}$$

Find $L(\mathbf{x})$ if $\mathbf{x} = [-1, 6, 1]^t$.

Solution: First, we express $\mathbf{x}$ as a linear combination of the $\mathbf{v}_i$'s:

$$\begin{bmatrix} -1 \\ 6 \\ 1 \end{bmatrix} = c_1 \begin{bmatrix} 1 \\ 1 \\ 0 \end{bmatrix} + c_2 \begin{bmatrix} 1 \\ 0 \\ 1 \end{bmatrix} + c_3 \begin{bmatrix} 0 \\ 1 \\ 1 \end{bmatrix} \tag{2}$$

This leads to three scalar equations in terms of the c_i's. Solving, we obtain $c_1 = 2$, $c_2 = -3$, and $c_3 = 4$, and thus $\mathbf{x} = 2\mathbf{v}_1 - 3\mathbf{v}_2 + 4\mathbf{v}_3$. Now finish the problem.

- - - - - - - - - - -

From the proof of Theorem 6.3, we have

$$L(\mathbf{x}) = 2L(\mathbf{v}_1) - 3L(\mathbf{v}_2) + 4L(\mathbf{v}_3) \tag{3}$$

Using the given information from (1), we have

$$L(\mathbf{x}) = 2 \begin{bmatrix} 5 \\ -1 \end{bmatrix} - 3 \begin{bmatrix} 1 \\ 5 \end{bmatrix} + 4 \begin{bmatrix} 2 \\ 2 \end{bmatrix} = \begin{bmatrix} 15 \\ -9 \end{bmatrix} \tag{4}$$

B.   THE MATRIX OF A LINEAR TRANSFORMATION (PRELIMINARY)

6.   In frame 2 we learned that if A is an $m \times n$ matrix, then we can define a linear transformation L from R^n into R^m by $L(\mathbf{x}) = A\mathbf{x}$ for $\mathbf{x}$ in R^n.

Suppose that L is a linear transformation from vector space V into vector space W. We shall learn how to associate a unique matrix with L that will enable us to find $L(\mathbf{x})$ for $\mathbf{x}$ in V by performing matrix multiplication. The main emphasis in our work will be on linear transformations from vector space R^n into vector space R^m.

Example: Consider a specific linear transformation L from R^3 to R^2 (or, $L\colon R^3 \to R^2$, for short) given by

$$L\left(\begin{bmatrix} x_1 \\ x_2 \\ x_3 \end{bmatrix}\right) = \begin{bmatrix} 2x_1 + x_2 - x_3 \\ x_1 + 3x_2 + 2x_3 \end{bmatrix} \tag{1}$$

Let $\{\mathbf{e}_1, \mathbf{e}_2, \mathbf{e}_3\}$ be the standard basis for R^3 ($\mathbf{e}_1 = [1, 0, 0]^t$, etc.), and let A be the 2×3 matrix having $L(\mathbf{e}_1)$, $L(\mathbf{e}_2)$, and $L(\mathbf{e}_3)$ as its columns. Thus,

$$L(\mathbf{e}_1) = L\left(\begin{bmatrix} 1 \\ 0 \\ 0 \end{bmatrix}\right) = \begin{bmatrix} 2 + 0 + 0 \\ 1 + 0 + 0 \end{bmatrix} = \begin{bmatrix} 2 \\ 1 \end{bmatrix} \tag{2}$$

$$L(\mathbf{e}_2) = L\left(\begin{bmatrix} 0 \\ 1 \\ 0 \end{bmatrix}\right) = \begin{bmatrix} 0 + 1 - 0 \\ 0 + 3 + 0 \end{bmatrix} = \begin{bmatrix} 1 \\ 3 \end{bmatrix} \qquad (3)$$

$$L(\mathbf{e}_3) = L\left(\begin{bmatrix} 0 \\ 0 \\ 1 \end{bmatrix}\right) = \begin{bmatrix} -1 \\ 2 \end{bmatrix} \qquad (4)$$

Thus,

$$A = \begin{bmatrix} 2 & 1 & -1 \\ 1 & 3 & 2 \end{bmatrix} \qquad (5)$$

$$\uparrow \qquad \uparrow \qquad \uparrow$$
$$L(\mathbf{e}_1) \quad L(\mathbf{e}_2) \quad L(\mathbf{e}_3)$$

For the example just done, show that $L(\mathbf{x}) = A\mathbf{x}$.

– – – – – – – – –

$$A\mathbf{x} = \begin{bmatrix} 2 & 1 & -1 \\ 1 & 3 & 2 \end{bmatrix} \begin{bmatrix} x_1 \\ x_2 \\ x_3 \end{bmatrix} = \begin{bmatrix} 2x_1 + x_2 - x_3 \\ x_1 + 3x_2 + 2x_3 \end{bmatrix}$$

7.   We can generalize the results of frame 6. The key ideas are summarized in the following theorem.

Theorem 6.4

Given that L is any linear transformation from R^n into R^m and $\mathbf{x} = [x_1, x_2, \ldots, x_n]^t$ is any column vector in R^n. Then, $L(\mathbf{x})$ can be expressed by the matrix–vector product $L(\mathbf{x}) = A\mathbf{x}$, where A is the $m \times n$ matrix whose columns are $L(\mathbf{e}_1), L(\mathbf{e}_2), \ldots, L(\mathbf{e}_n)$. Here $\{\mathbf{e}_1, \mathbf{e}_2, \ldots, \mathbf{e}_n\}$ is the standard basis for R^n. The matrix A will be called the *standard matrix* for L.

We shall demonstrate the theorem for the special case of $L\colon R^3 \rightarrow R^2$. Here $\mathbf{e}_1 = [1, 0, 0]^t$, $\mathbf{e}_2 = [0, 1, 0]^t$, and $\mathbf{e}_3 = [0, 0, 1]^t$ comprise the standard basis of R^3. Our approach will be related to our later general approach, which will deal with arbitrary bases for general vector spaces V and W, respectively (Section D).

Suppose that

$$L(\mathbf{e}_1) = \begin{bmatrix} a_{11} \\ a_{21} \end{bmatrix}, \qquad L(\mathbf{e}_2) = \begin{bmatrix} a_{12} \\ a_{22} \end{bmatrix}, \qquad L(\mathbf{e}_3) = \begin{bmatrix} a_{13} \\ a_{23} \end{bmatrix} \qquad (1a), \quad (1b), \quad (1c)$$

All we are saying here is that $L(\mathbf{e}_1)$, $L(\mathbf{e}_2)$, and $L(\mathbf{e}_3)$ are column vectors in R^2. Then let

$$
A = \begin{bmatrix} a_{11} & a_{12} & a_{13} \\ a_{21} & a_{22} & a_{23} \end{bmatrix} \tag{2}
$$
$$
\quad\ \uparrow \qquad \uparrow \qquad \uparrow
$$
$$
\quad\ L(\mathbf{e}_1) \quad L(\mathbf{e}_2) \quad L(\mathbf{e}_3)
$$

On occasion, we will indicate this by writing $A = [L(\mathbf{e}_1) \,\vdots\, L(\mathbf{e}_2) \,\vdots\, L(\mathbf{e}_3)]$. Now we shall show that $L: R^3 \to R^2$ is equivalent to multiplication of A from Eq. (2) times $\mathbf{x} = [x_1, x_2, x_3]^t$, a typical column vector of R^3.

First, note that

$$
\mathbf{x} = \begin{bmatrix} x_1 \\ x_2 \\ x_3 \end{bmatrix} = \begin{bmatrix} x_1 \\ 0 \\ 0 \end{bmatrix} + \begin{bmatrix} 0 \\ x_2 \\ 0 \end{bmatrix} + \begin{bmatrix} 0 \\ 0 \\ x_3 \end{bmatrix} = x_1 \begin{bmatrix} 1 \\ 0 \\ 0 \end{bmatrix} + x_2 \begin{bmatrix} 0 \\ 1 \\ 0 \end{bmatrix} + x_3 \begin{bmatrix} 0 \\ 0 \\ 1 \end{bmatrix}, \tag{3a}
$$

after making use of rules for matrix addition and scalar multiplication (Definitions 1.1 and 1.2). Thus,

$$
\mathbf{x} = x_1 \mathbf{e}_1 + x_2 \mathbf{e}_2 + x_3 \mathbf{e}_3 \tag{3b}
$$

Now by the linearity property of L (Theorem 6.1 in frame 1),

$$
L(\mathbf{x}) = x_1 L(\mathbf{e}_1) + x_2 L(\mathbf{e}_2) + x_3 L(\mathbf{e}_3) \tag{4}
$$

But from (2) we have

$$
A\mathbf{x} = \begin{bmatrix} a_{11} & a_{12} & a_{13} \\ a_{21} & a_{22} & a_{23} \end{bmatrix} \begin{bmatrix} x_1 \\ x_2 \\ x_3 \end{bmatrix} = \begin{bmatrix} a_{11}x_1 + a_{12}x_2 + a_{13}x_3 \\ a_{21}x_1 + a_{22}x_2 + a_{23}x_3 \end{bmatrix} \tag{5}
$$

From the rules of matrix addition and scalar multiplication,

$$
\begin{bmatrix} a_{11}x_1 + a_{12}x_2 + a_{13}x_3 \\ a_{21}x_1 + a_{22}x_2 + a_{23}x_3 \end{bmatrix} = x_1 \begin{bmatrix} a_{11} \\ a_{21} \end{bmatrix} + x_2 \begin{bmatrix} a_{12} \\ a_{22} \end{bmatrix} + x_3 \begin{bmatrix} a_{13} \\ a_{23} \end{bmatrix} \tag{6}
$$

From (5) and (6), we have

$$
A\mathbf{x} = x_1 \begin{bmatrix} a_{11} \\ a_{21} \end{bmatrix} + x_2 \begin{bmatrix} a_{12} \\ a_{22} \end{bmatrix} + x_3 \begin{bmatrix} a_{13} \\ a_{23} \end{bmatrix} \tag{7}
$$

Substituting (1a), (1b), and (1c) into (7) yields

$$
A\mathbf{x} = x_1 L(\mathbf{e}_1) + x_2 L(\mathbf{e}_2) + x_3 L(\mathbf{e}_3) \tag{8}
$$

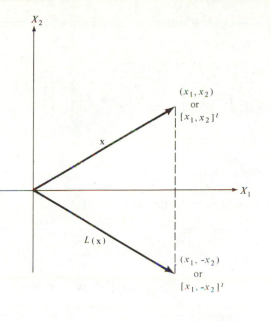

FIGURE 6.5

Since the right sides of (4) and (8) are identical, the respective left sides of (4) and (8) must be equal. That is, we have shown that

$$L(\mathbf{x}) = A\mathbf{x} \tag{9}$$

Note: Observe that writing $\mathbf{x} = [x_1, x_2, x_3]^t$ is equivalent to writing $\mathbf{x}$ in terms of the standard basis for R^3 as $\mathbf{x} = x_1\mathbf{e}_1 + x_2\mathbf{e}_2 + x_3\mathbf{e}_3$ [see Eq. (3b)]. By the same token, we can, from Eq. (5), write $L(\mathbf{x})$ in terms of the standard basis for R^2. For R^2, the standard basis consists of $\mathbf{e}_1 = \begin{bmatrix} 1 \\ 0 \end{bmatrix}$ and $\mathbf{e}_2 = \begin{bmatrix} 0 \\ 1 \end{bmatrix}$. Let $(A\mathbf{x})_i = a_{i1}x_1 + a_{i2}x_2 + a_{i3}x_2$ denote component i of the two-component vector $A\mathbf{x}$ in Eq. (5). Then, from Eq. (5), we can write $L(\mathbf{x}) = (A\mathbf{x})_1\mathbf{e}_1 + (A\mathbf{x})_2\mathbf{e}_2$ as the desired form.

Example: Let $L: R^2 \to R^2$ be the linear transformation that maps each vector $\mathbf{x}$ into its symmetric image about the X_1 axis. See Figure 6.5. Find the standard matrix for L by using Theorem 6.4. Here L is also called a reflection with respect to the X_1 axis.

Solution: The effect of the reflection is to keep $\mathbf{e}_1$ the same and to transform $\mathbf{e}_2$ into $\begin{bmatrix} 0 \\ -1 \end{bmatrix}$. That is,

$$L(\mathbf{e}_1) = L\left(\begin{bmatrix} 1 \\ 0 \end{bmatrix}\right) = \begin{bmatrix} 1 \\ 0 \end{bmatrix}, \qquad L(\mathbf{e}_2) = L\left(\begin{bmatrix} 0 \\ 1 \end{bmatrix}\right) = \begin{bmatrix} 0 \\ -1 \end{bmatrix}$$

Now finish the problem.

- - - - - - - - - - -

Since $L(\mathbf{e}_1)$ and $L(\mathbf{e}_2)$ are the columns of A,

$$A = \begin{bmatrix} 1 & 0 \\ 0 & -1 \end{bmatrix}$$

Checking for any vector $\mathbf{x} = [x_1, x_2]^t$ in R^2, we have

$$A\mathbf{x} = \begin{bmatrix} 1 & 0 \\ 0 & -1 \end{bmatrix}\begin{bmatrix} x_1 \\ x_2 \end{bmatrix} = \begin{bmatrix} x_1 \\ -x_2 \end{bmatrix},$$

which shows that $L(\mathbf{x}) = A\mathbf{x}$ is indeed the symmetric image (or reflection) of $\mathbf{x}$ with respect to the X_1 axis.

8. It should be noted for $L: R^n \rightarrow R^m$ defined by $L(\mathbf{x}) = A\mathbf{x}$, where A is an $m \times n$ matrix (frame 3, matrix transformation), that the matrix A is automatically the standard matrix for L.

In Eqs. (5) and (6) of frame 7 we encountered a useful way of expressing $A\mathbf{x}$ if A is 2×3. Let us label the columns of the matrix A as $\mathbf{a}_1$, $\mathbf{a}_2$, and $\mathbf{a}_3$; that is,

$$\mathbf{a}_1 = \begin{bmatrix} a_{11} \\ a_{21} \end{bmatrix}, \qquad \mathbf{a}_2 = \begin{bmatrix} a_{12} \\ a_{22} \end{bmatrix}, \qquad \mathbf{a}_3 = \begin{bmatrix} a_{13} \\ a_{23} \end{bmatrix}$$

Then we can express $A\mathbf{x}$ as

$$A\mathbf{x} = x_1\mathbf{a}_1 + x_2\mathbf{a}_2 + x_3\mathbf{a}_3$$

The same type of result holds in general.

Theorem 6.5

Given that A is an $m \times n$ matrix and $\mathbf{x} = [x_1, x_2, \ldots, x_n]^t$. Suppose the n columns of A are successively labeled $\mathbf{a}_1, \mathbf{a}_2, \ldots, \mathbf{a}_n$, where each $\mathbf{a}_j$ is an m component column vector. (We may write $A = [\mathbf{a}_1 \mid \mathbf{a}_2 \mid \ldots \mid \mathbf{a}_n]$.) Then the product $A\mathbf{x}$ can be expressed as the following linear combination of $\mathbf{a}_1$, $\mathbf{a}_2$, etc.:

$$A\mathbf{x} = x_1\mathbf{a}_1 + x_2\mathbf{a}_2 + \cdots + x_n\mathbf{a}_n$$

This result will prove to be valuable.

Let us continue our discussion of the standard matrix for a linear transformation L from R^n into R^m.

Example: Given the linear transformation from R^2 into R^2 cited in Example (a) of frame 1. That is, $L(\mathbf{x})$ is given by

$$L\left(\begin{bmatrix} x_1 \\ x_2 \end{bmatrix}\right) = \begin{bmatrix} 2x_1 + x_2 \\ x_1 + 3x_2 \end{bmatrix}$$

Determine the matrix A for which $L(\mathbf{x}) = A\mathbf{x}$. ($A$ is the standard matrix for L.)

- - - - - - - - - - -

Solution:

$$L(\mathbf{e}_1) = L\left(\begin{bmatrix} 1 \\ 0 \end{bmatrix}\right) = \begin{bmatrix} 2 \\ 1 \end{bmatrix}, \qquad L(\mathbf{e}_2) = L\left(\begin{bmatrix} 0 \\ 1 \end{bmatrix}\right) = \begin{bmatrix} 1 \\ 3 \end{bmatrix}$$

Thus, $A = \left[L(\mathbf{e}_1) \mid L(\mathbf{e}_2) \right] = \begin{bmatrix} 2 & 1 \\ 1 & 3 \end{bmatrix}$.

9. Let us do some more examples on finding the standard matrix for a linear transformation.

Example (a): Refer to Example (c) of frame 3. There, the linear transformation (orthogonal projection onto the line $x_1 - 2x_2 = 0$) was given by

$$\text{Proj}_\ell \, \mathbf{x} = \begin{bmatrix} (4x_1 + 2x_2)/5 \\ (2x_1 + x_2)/5 \end{bmatrix}$$

Determine the standard matrix A such that $\text{Proj}_\ell \, \mathbf{x} = A\mathbf{x}$.

Solution:

$$L(e_1) = L\left(\begin{bmatrix} 1 \\ 0 \end{bmatrix}\right) = \begin{bmatrix} \frac{4}{5} \\ \frac{2}{5} \end{bmatrix}, \qquad L(e)_2 = L\left(\begin{bmatrix} 0 \\ 1 \end{bmatrix}\right) = \begin{bmatrix} \frac{2}{5} \\ \frac{1}{5} \end{bmatrix}$$

Thus, $A = \dfrac{1}{5}\begin{bmatrix} 4 & 2 \\ 2 & 1 \end{bmatrix}$.

Consider again the last three examples where L was from R^2 into R^2. Note that whenever $L: R^2 \rightarrow R^2$ is specified in the form

$$L\left(\begin{bmatrix} x_1 \\ x_2 \end{bmatrix}\right) = \begin{bmatrix} a_1 x_1 + a_2 x_2 \\ b_1 x_1 + b_2 x_2 \end{bmatrix},$$

then it automatically follows that $A = \begin{bmatrix} a_1 & a_2 \\ b_1 & b_2 \end{bmatrix}$. Similarly, for the general case where L goes from R^n into R^m.

Example (b):   Refer to Example (a) of frame 4. There the linear transformation (an orthogonal projection) from R^3 into R^3 was given by $\text{Proj}_W \mathbf{x} = [x_1, x_2, 0]^t$ for $\mathbf{x} = [x_1, x_2, x_3]^t$. The subspace W is the $X_1 X_2$ plane. Determine the standard matrix A such that $\text{Proj}_W \mathbf{x} = A\mathbf{x}$.

— — — — — — — — — — —

Solution:

$$L(e_1) = L\left(\begin{bmatrix} 1 \\ 0 \\ 0 \end{bmatrix}\right) = \begin{bmatrix} 1 \\ 0 \\ 0 \end{bmatrix}, \qquad L(e_2) = L\left(\begin{bmatrix} 0 \\ 1 \\ 0 \end{bmatrix}\right) = \begin{bmatrix} 0 \\ 1 \\ 0 \end{bmatrix},$$

$$L(e)_3 = L\left(\begin{bmatrix} 0 \\ 0 \\ 1 \end{bmatrix}\right) = \begin{bmatrix} 0 \\ 0 \\ 0 \end{bmatrix}$$

Thus, $A = \begin{bmatrix} 1 & 0 & 0 \\ 0 & 1 & 0 \\ 0 & 0 & 0 \end{bmatrix}$. Checking, we see that matrix multiplication of A times $\mathbf{x}$ yields $A\mathbf{x} = [x_1, x_2, 0]^t$.

C. THE KERNEL AND RANGE OF A LINEAR TRANSFORMATION

10. In this section we shall study further properties of linear transformations, in particular, the concepts of "one-to-one" and "onto" linear transformations. Also, for $L: V \to W$, we shall learn about the kernel of L, which is an important subspace of V, and the range of L, which is a subspace of W.

Definition 6.2

A linear transformation $L: V \to W$ is said to be one-to-one if for vectors $\mathbf{x}$ and $\mathbf{x}'$ in V, $\mathbf{x} \neq \mathbf{x}'$ implies that $L(\mathbf{x}) \neq L(\mathbf{x}')$. Equivalently, L is one-to-one if $L(\mathbf{x}) = L(\mathbf{x}')$ implies that $\mathbf{x} = \mathbf{x}'$.

Note: The first sentence says that L is one-to-one if $L(\mathbf{x})$ and $L(\mathbf{x}')$ are distinct when $\mathbf{x}$ and $\mathbf{x}'$ are distinct. The second sentence in Definition 6.2 is the contrapositive form of the first sentence.

Refer to Figure 6.6 for illustrations of what is and what is not a one-to-one linear transformation.

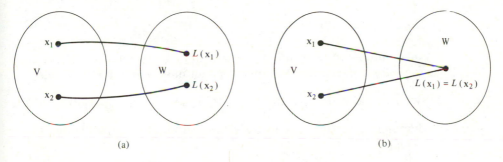

FIGURE 6.6   (a) L is one-to-one. (b) L is not one-to-one.

Example (a): In Example (a) of frame 1, $L: R^2 \to R^2$ was given by

$$L([x_1, x_2]^t) = \begin{bmatrix} 2x_1 + x_2 \\ x_1 + 3x_2 \end{bmatrix}$$

Determine whether L is one-to-one.

Solution: Let $\mathbf{x} = [x_1, x_2]^t$ and $\mathbf{y} = [y_1, y_2]^t$. Then if $L(\mathbf{x}) = L(\mathbf{y})$, we have

$$2x_1 + x_2 = 2y_1 + y_2 \tag{1}$$

$$x_1 + 3x_2 = y_1 + 3y_2 \tag{2}$$

If we multiply Eq. (1) by 3 and then subtract Eq. (2) from the resulting equation, we obtain $5x_1 = 5y_1$ and thus $x_1 = y_1$. From Eq. (1) we then have that $x_2 = y_2$. Thus, $\mathbf{x} = \mathbf{y}$, and L is one-to-one.

Example (b): In the Example of frame 6, $L: R^3 \rightarrow R^2$ was given by

$$L([x_1, x_2, x_3]^t) = \begin{bmatrix} 2x_1 + x_2 - x_3 \\ x_1 + 3x_2 + 2x_3 \end{bmatrix}$$

Determine whether L is one-to-one.

Solution: If we set $L(\mathbf{x}) = \mathbf{b}$ for a particular vector $\mathbf{b} = [b_1, b_2]^t$ in R^2 and solve for the x_i's in terms of b_1 and b_2, we obtain (by using Method 2.1, say) the following pair of equations in which x_1 and x_2 are leading variables:

$$x_1 = x_3 + \tfrac{1}{5}(3b_1 - b_2) \tag{1}$$

$$x_2 = -x_3 + \tfrac{1}{5}(2b_2 - b_1) \tag{2}$$

Thus, since nonleading variable x_3 is arbitrary, there are infinitely many solutions for $\mathbf{x}$ for a particular $\mathbf{b}$. Thus, L is not one-to-one. For example, if $\mathbf{b} = \begin{bmatrix} 5 \\ 5 \end{bmatrix}$, we have $\mathbf{x}' = [4, -1, 2]^t$ if $x_3 = 2$, and $\mathbf{x}'' = [6, -3, 4]^t$ if $x_3 = 4$. We have demonstrated that $L(\mathbf{x}') = L(\mathbf{x}'')$ even though $\mathbf{x}' \neq \mathbf{x}''$. Here $L(\mathbf{x}') = L(\mathbf{x}'') = \begin{bmatrix} 5 \\ 5 \end{bmatrix}$.

Example (c): In Example (a) of frame 4, an orthogonal projection linear transformation from R^3 into R^3 was given by $L([x_1, x_2, x_3]^t) = [x_1, x_2, 0]^t$.

Show that L is not one-to-one. Use the following approach: Determine two distinct vectors $\mathbf{x}$ and $\mathbf{y}$ such that $L(\mathbf{x}) = L(\mathbf{y})$.

– – – – – – – – – – –

Solution: We observe that for both $\mathbf{x} = [3, 2, 4]^t$ and $\mathbf{y} = [3, 2, 7]^t$, we have $L(\mathbf{x}) = L(\mathbf{y}) = [3, 2, 0]^t$. Since $\mathbf{x} \neq \mathbf{y}$, L is not one-to-one.

11.   We can develop a more efficient method for determining whether a linear transformation is one-to-one.

Definition 6.3 (Kernel)

For the linear transformation $L: V \rightarrow W$, the *kernel of L*, denoted by Ker (L), is the subset of V consisting of all vectors $\mathbf{x}$ such that $L(\mathbf{x}) = \mathbf{0}_W$.

Notes: (i)   In short, we say that a vector $\mathbf{x}$ of V is in the kernel if $\mathbf{0}_W$ is the image of $\mathbf{x}$; that is, if $L(\mathbf{x}) = \mathbf{0}_W$. (Here $\mathbf{0}_W$ is the zero vector of W.) (ii)   Another name for kernel is *nullspace*.

Example (a): For $L: R^2 \rightarrow R^2$ in Example (a) of frame 10, find Ker (L), the kernel of L.

Solution: Here $L(\mathbf{x}) = \begin{bmatrix} 2x_1 + x_2 \\ x_1 + 3x_2 \end{bmatrix}$, so setting $L(\mathbf{x})$ equal to $\mathbf{0} = \begin{bmatrix} 0 \\ 0 \end{bmatrix}$ leads to the following homogeneous linear system.

$$2x_1 + x_2 = 0$$

$$x_1 + 3x_2 = 0$$

The only solution is $x_1 = x_2 = 0$; that is, $\mathbf{x} = \mathbf{0}$. Thus, Ker $(L) = \{\mathbf{0}\}$. Note that $|A| = 5 \neq 0$, which implies that $\mathbf{x} = \mathbf{0}$ is the only solution.

Example (b): For $L: R^3 \rightarrow R^2$ in Example (b) of frame 10, find Ker (L).

Solution: First, we set $L(\mathbf{x}) = \mathbf{0}$, where $\mathbf{0} = \begin{bmatrix} 0 \\ 0 \end{bmatrix}$, since L is from R^3 into R^2.

Thus, we obtain

$$2x_1 + x_2 - x_3 = 0 \tag{1}$$

$$x_1 + 3x_2 + 2x_3 = 0 \tag{2}$$

[By the way, the standard matrix is given by $A = \begin{bmatrix} 2 & 1 & -1 \\ 1 & 3 & 2 \end{bmatrix}$, and this is the matrix of coefficients in the system given by Eqs. (1) and (2).] Solving by using Method 2.1, we first obtain the reduced row echelon form $\begin{bmatrix} 1 & 0 & -1 & | & 0 \\ 0 & 1 & 1 & | & 0 \end{bmatrix}$. This means that Ker (L) consists of all vectors of the form $\mathbf{x} = [r, -r, r]^t = r[1, -1, 1]^t$, where r is arbitrary.

Example (c): Suppose $L: R^4 \rightarrow R^3$ is defined by $L(\mathbf{x}) = A\mathbf{x}$, where $A = \begin{bmatrix} 1 & -4 & 2 & 9 \\ 3 & -12 & -1 & 13 \\ 0 & 0 & 1 & 2 \end{bmatrix}$. Determine Ker (L). The matrix here is the coefficient matrix in Example (a) of frame 18 of Chapter Five. *Hint:* Find the vector solutions of $A\mathbf{x} = \mathbf{0}$ where $\mathbf{0} = [0, 0, 0]^t$.

– – – – – – – – – – –

As in frame 18 of Chapter 5, we see that the reduced row echelon form

of $A\mathbf{x} = \mathbf{0}$ is $\begin{bmatrix} 1 & -4 & 0 & 5 & | & 0 \\ 0 & 0 & 1 & 2 & | & 0 \\ 0 & 0 & 0 & 0 & | & 0 \end{bmatrix}$, and Ker (L) consists of all vectors of the

form $\mathbf{x} = r[4, 1, 0, 0]^t + s[-5, 0, -2, 1]^t$, where r and s are arbitrary.

12.   From Theorem 6.2, part (a)—see frame 5—we see, for any L, that Ker (L) will contain $\mathbf{0}_V$ since $L(\mathbf{0}_V) = \mathbf{0}_W$. [Here, $\mathbf{0}_V$ is the zero vector of V.]

Theorem 6.6

If $L: V \rightarrow W$ is a linear transformation, then Ker (L) is a subspace of V.

Note: We denote the dimension of Ker (L) by Dim [Ker (L)]. The dimension here is the number of vectors in a basis for Ker (L); it is also known as the *nullity of L.*

Proof of Theorem 6.6

We have to verify conditions (a) and (b) of Theorem 5.2 (frame 4). Let $\mathbf{x}_1$ and $\mathbf{x}_2$ be two vectors in Ker (L). Then, since L is a linear transformation,

$$L(\mathbf{x}_1 + \mathbf{x}_2) = L(\mathbf{x}_1) + L(\mathbf{x}_2) = \mathbf{0}_W + \mathbf{0}_W = \mathbf{0}_W .$$

Thus, $\mathbf{x}_1 + \mathbf{x}_2$ is in Ker (L). If k is a scalar, then since L is a linear transformation,

$$L(k\mathbf{x}_1) = kL(\mathbf{x}_1) = k\mathbf{0}_W = \mathbf{0}_W .$$

Thus, $k\mathbf{x}_1$ is in Ker (L). Hence, Ker (L) is a subspace of V.

Example (a):   In Example (a) of frame 11, Ker (L) is the subspace $\{\mathbf{0}\}$, whose dimension is zero.

In Example (b) of frame 11, a basis for Ker (L) is $\{\mathbf{v}\}$, where $\mathbf{v} = [1, -1, 1]^t$, and thus Dim [Ker (L)] $= 1$. The kernel here is a subspace of R^3 (which is V in this example).

In Example (c) of frame 11, a basis for Ker (L) is $\{\mathbf{v}_1, \mathbf{v}_2\}$, where $\mathbf{v}_1 = [4, 1, 0, 0]^t$ and $\mathbf{v}_2 = [-5, 0, -2, 1]^t$. Thus, Dim [Ker (L)] $= 2$. Here Ker (L) is a subspace of R^4.

For the special case where V is R^n and W is R^m, suppose the standard matrix for linear transformation L is the $m \times n$ matrix A. Then, as in the Examples of frame 11, the kernel of L consists of all solutions of $A\mathbf{x} = \mathbf{0}$. Thus, the kernel is the same as the solution set of the system of equations given by $A\mathbf{x} = \mathbf{0}$. The solution set is a subspace of R^n, as is indicated in Theorem 5.3 (frame 6). Refer also to frame 18 of Chapter 5, and in particular to Example (a) of frame 18. (Henceforth, we call the solution set the solution space.)

Because of the following theorem, we can use the kernel concept to help us determine if a linear transformation $L: V \rightarrow W$ is one-to-one.

Theorem 6.7

A linear transformation $L: V \rightarrow W$ is one-to-one if and only if Ker (L) = $\{\mathbf{0}\}$.

Example (b): Use Theorem 6.7 to determine which linear transformations in the examples of frame 11 are one-to-one.

— — — — — — — — — —

Solution: The linear transformation in Example (a) is one-to-one. The linear transformations in Examples (b) and (c) are not.

13.   For the linear transformation $L: V \rightarrow W$, we shall now focus on an important subset of W (which, as we shall see, is also a *subspace* of W).

Definition 6.4

Given the linear transformation $L: V \rightarrow W$. The set of all vectors in W that are images under L of at least one vector in V is called the *range of L*. (The image vectors are also called output vectors.) We denote the range of L by Range (L).

If Range (L) is identical to W, then we say that the linear transformation L is *onto*.

Observe that a vector $\mathbf{y}$ is in Range (L) if we can find some vector $\mathbf{x}$ in V such that $L(\mathbf{x}) = \mathbf{y}$.

Theorem 6.8

If $L: V \rightarrow W$ is a linear transformation, then Range (L) is a subspace of W.

The proof of Theorem 6.8 is instructive.

Proof of Theorem 6.8

We have to verify conditions (a) and (b) of Theorem 5.2. Let $\mathbf{y}_1$ and $\mathbf{y}_2$ be in Range (L). Thus, $\mathbf{y}_1 = L(\mathbf{x}_1)$ and $\mathbf{y}_2 = L(\mathbf{x}_2)$ for $\mathbf{x}_1$ and $\mathbf{x}_2$ in V. Now,

$$\mathbf{y}_1 + \mathbf{y}_2 = L(\mathbf{x}_1) + L(\mathbf{x}_2) = L(\mathbf{x}_1 + \mathbf{x}_2).$$

This says that $\mathbf{y}_1 + \mathbf{y}_2$ is the image of $\mathbf{x}_1 + \mathbf{x}_2$, and thus $\mathbf{y}_1 + \mathbf{y}_2$ is in Range (L). Also, if k is a scalar, then $k\mathbf{y}_1 = kL(\mathbf{x}_1) = L(k\mathbf{x}_1)$. Thus, $k\mathbf{y}_1$ is the image of $k\mathbf{x}_1$, and hence is in Range (L). Thus, Range (L) is a subspace of W.

An important special case occurs when V is R^n and W is R^m. Suppose the standard matrix for the linear transformation $L: R^n \to R^m$ is the $m \times n$ matrix A. An important fact is that the range of L is identical to the column space of matrix A. We spoke briefly about the column space of a matrix in frame 20 of Chapter 5, and we shall discuss it more after the statement of Theorem 6.9.

Theorem 6.9

Given $L: R^n \to R^m$ with standard matrix A. Then the range of L is identical to the column space of A.

To reinforce the idea of a column space, consider the special case of a 2×3 standard matrix A. First, it is useful to review the symbols $\mathbf{a}_1$, $\mathbf{a}_2$, and $\mathbf{a}_3$ for the columns of A—see frame 8. Thus,

$$\mathbf{a}_1 = \begin{bmatrix} a_{11} \\ a_{21} \end{bmatrix}, \qquad \mathbf{a}_2 = \begin{bmatrix} a_{12} \\ a_{22} \end{bmatrix}, \qquad \mathbf{a}_3 = \begin{bmatrix} a_{13} \\ a_{23} \end{bmatrix}$$

The column space is then the set of *all* linear combinations of the form

$$s_1\mathbf{a}_1 + s_2\mathbf{a}_2 + s_3\mathbf{a}_3,$$

where the s_j's denote scalars. This set is then a subspace of R^2—see Theorem 5.4 in frame 10 (note that each $\mathbf{a}_j$ has two components). In the general case, the column space for the $m \times n$ matrix A is the set of all linear combinations of the form:

$$s_1\mathbf{a}_1 + s_2\mathbf{a}_2 + \cdots + s_n\mathbf{a}_n$$

and this vector space is a subspace of R^m. Here, as in Theorem 6.5 of frame 8, the $\mathbf{a}_j$'s denote the columns of A.

Proof of Theorem 6.9

Since $L: R^n \to R^m$ has standard matrix A, it follows that $L(\mathbf{x}) = A\mathbf{x}$ for any column vector $\mathbf{x} = [x_1, x_2, \ldots, x_n]^t$ in R^n.

Now, if $\mathbf{y}$ is in the range of L, this means that $\mathbf{y} = A\mathbf{x}$ for some $\mathbf{x}$ in R^n. But from Theorem 6.5 (frame 8), we can write $A\mathbf{x}$ as

$$x_1\mathbf{a}_1 + x_2\mathbf{a}_2 + \cdots + x_n\mathbf{a}_n \tag{1}$$

This means that $A\mathbf{x}$ is a vector in the column space of A.

Working in reverse, any vector in the column space of A has the form

$$s_1\mathbf{a}_1 + s_2\mathbf{a}_2 + \cdots + s_n\mathbf{a}_n \tag{2}$$

But, from Theorem 6.5 again, this can be written as $A\mathbf{s}$, with $\mathbf{s} = [s_1, s_2, \ldots, s_n]^t$. But $A\mathbf{s}$ is in the range of L since $A\mathbf{s}$ is the image of $\mathbf{s}$.

We already know that the dimension of the column space of matrix A is equal to the rank of matrix A (Definition 5.9 and Theorem 5.13 of frame 20). Thus it follows that the dimension of the range of L is also equal to the rank of matrix A. That is,

$$\text{Dim}[\text{Range }(L)] = \text{Rank }(A) \tag{I}$$

Example: Suppose $L: R^4 \to R^3$ is given by $L(\mathbf{x}) = A\mathbf{x}$ as in Example (c) of frame 11. Thus, $A = \begin{bmatrix} 1 & -4 & 2 & 9 \\ 3 & -12 & -1 & 13 \\ 0 & 0 & 1 & 2 \end{bmatrix}$. Determine the dimension of the range of L.

— — — — — — — — — —

Solution: The reduced row echelon form for A is $\begin{bmatrix} 1 & -4 & 0 & 5 \\ 0 & 0 & 1 & 2 \\ 0 & 0 & 0 & 0 \end{bmatrix}$. See the solution in frame 11 (delete the last column there). Since Rank $(A) = 2$ as there are two leading entries here, it follows from (I) above that

$$\text{Dim }[\text{Range }(L)] = 2$$

14. It is important to remember that for $L: V \to W$, the kernel of L is a subspace of V and the range of L is a subspace of W. The diagrams in Figure 6.7 are helpful. Thus, Ker (L) consists of all vectors $\mathbf{x}$ in V such that $L(\mathbf{x}) = \mathbf{0}_W$. In

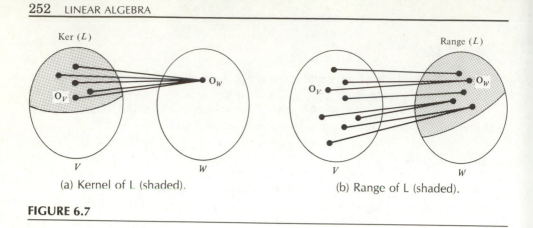

(a) Kernel of L (shaded).   (b) Range of L (shaded).

FIGURE 6.7

the special case where Ker (L) consists only of $\mathbf{0}_V$, the linear transformation L is one-to-one (Theorem 6.7 of frame 12). The converse is valid also.

Range (L) consists of all those vectors in W that are images of vectors in V. That is, $\mathbf{y}$ is in Range (L) if $\mathbf{y} = L(\mathbf{x})$ for at least one vector $\mathbf{x}$ in V. In the special case where Range (L) is equal to W, L is an "onto" linear transformation. In this case, *each* vector in W is the image of at least one vector in V.

We can make use of part (a) of Theorem 5.10 (frame 17 of Chapter 5), and some of the results of frame 13 of this chapter to determine if $L: R^n \to R^m$ is onto. [Part (a) of Theorem 5.10 indicates that for every nonzero subspace U of W, Dim $(U) \leq$ Dim (W), and that if Dim $(U) =$ Dim (W), then U is W.] First, we know that Range (L) is a subspace of R^m. Also, Dim [Range (L)] = Rank (A), where A is the standard matrix for L. The next criterion follows directly from part (a) of Theorem 5.10.

Theorem 6.10

Suppose the linear transformation $L: R^n \to R^m$ is given by $L(\mathbf{x}) = A\mathbf{x}$, where A (the standard matrix for L) is $m \times n$.

(a) If Rank $(A) = m$, then Range (L) is identical to R^m. That is, L is onto.

(b) If Rank $(A) < m$, then Range (L) is a proper subset of R^m, and L is not onto.

Notes: (i) Since it is not possible for Rank (A) to exceed m (see Section D of Chapter Two), the converses of parts (a) and (b) of Theorem 6.10 are also valid. (ii) A set A is a proper subset of set B if A is a subset of B and if there are elements of B that do not belong to A. For example, for sets that contain integers, the set $C = \{1, 2, 5\}$ is a proper subset of set $D = \{1, 2, 5, 6, 8\}$.

Example (a): In the Example of frame 13, L is not onto since Rank $(A) = 2$, and $m = 3$.

Example (b):   For $L: R^4 \to R^3$, the standard matrix is given by

$$A = \begin{bmatrix} 3 & -1 & 2 & 1 \\ 6 & -2 & 4 & -1 \\ 9 & -3 & -1 & 5 \end{bmatrix}$$

That is, $L(\mathbf{x}) = A\mathbf{x}$ for $\mathbf{x}$ in R^4. Use Theorem 6.10 to determine if L is onto.

$- - - - - - - - - -$

Solution: The reduced row echelon form for A is $\begin{bmatrix} 1 & -\frac{1}{3} & 0 & 0 \\ 0 & 0 & 1 & 0 \\ 0 & 0 & 0 & 1 \end{bmatrix}$. Since

Rank $(A) = m = 3$, L is onto.

15. A very important equation relating the dimensions of the kernel and range of L is given in the following theorem [proved on p. 227 of Kolman (1980), and on p. 218 of Anton (1977)].

Theorem 6.11

For the linear transformation $L: V \to W$,

Dim [Ker (L)] + Dim [Range (L)] = Dim (V)

This is easily demonstrated for $L: R^n \to R^m$ as we shall now show. Let $L(\mathbf{x})$ be given by $A\mathbf{x}$ where A is the $m \times n$ standard matrix for L. We have established that

$$\text{Dim [Range } (L)] = \text{Rank } (A) \tag{1}$$

in frame 13. Also, in frame 18 of Chapter Five, we established that the dimension of the solution space of $A\mathbf{x} = \mathbf{0}$ was equal to $n - \text{Rank } (A)$. Of course, we know that the solution space of $A\mathbf{x} = \mathbf{0}$ is *identical* to the kernel of L. (For a review, see the comments of frame 12, and the examples of frame 11.) Thus, we have Dim [Ker (L)] = $n - \text{Rank } (A)$, or

$$\text{Dim [Ker } (L)] + \text{Rank } (A) = n \tag{2}$$

This is agreement with Theorem 6.11 because of Eq. (1) above, and because Dim $(V) = $ Dim $(R^n) = n$ here.

Example (a): In Example (c) of frame 11, and in the Example of frame 13, we had $L: R^4 \to R^3$ where $L(\mathbf{x}) = A\mathbf{x}$, and $A = \begin{bmatrix} 1 & -4 & 2 & 9 \\ 3 & -12 & -1 & 13 \\ 0 & 0 & 1 & 2 \end{bmatrix}$.

For this matrix, we found that Dim [Ker (L)] $= 2$, and Rank $(A) = 2$, which checks out since $n = 4$.

Example (b): For Example (b) of frame 14, determine the dimension of the kernel of L by employing Eq. (2) above.

— — — — — — — — — —

Solution: A is 3×4 and thus $n = 4$. From the solution in frame 14, Rank $(A) = 3$. Thus, from Eq. (2), Dim [Ker (L)] $= 4 - 3 = 1$.

In the language of Chapter Five we say that the solution space of $A\mathbf{x} = \mathbf{0}$, for the above A, is a one-dimensional subspace of R^4.

16. In Theorems 6.7 (frame 12) and 6.10 (frame 14), we established criteria for determining whether a linear transformation is one-to-one or onto, respectively. For the class of linear transformations, any of the following four possibilities can occur.

    (i) L is neither one-to-one nor onto.
   (ii) L is one-to-one but not onto.
  (iii) L is onto but not one-to-one.
  (iv) L is both onto and one-to-one.

Possibility (iv) can occur only when Dim $(V) =$ Dim (W). Stated differently, if L is both onto and one-to-one, then Dim $(V) =$ Dim (W). Also, if Dim $(V) =$ Dim (W), then the only possible cases are (i) and (iv). [We assume here, and elsewhere in this book, that Dim (V) and Dim (W) are both finite.] This fact is indicated in the following theorem.

Theorem 6.12

Given the linear transformation $L: V \to W$, where Dim $(V) =$ Dim (W).

  (a)   If L is one-to-one, then L is onto.
  (b)   If L is onto, then L is one-to-one.

Also, if Dim $(V) >$ Dim (W), the only possible cases are (i) and (iii) in the preceding list of four possibilities. If Dim $(V) <$ Dim (W), the only possible cases are (i) and (ii).

Example: Refer to Example (a) of frame 11. Use Theorem 6.12 to show that L is onto.

— — — — — — — — — — —

Solution: Since L was from R^2 into R^2, Dim (V) = Dim (W) = 2, and Theorem 6.12 applies. From the fact that Ker (L) = $\{0\}$, we showed already (using Theorem 6.7 of frame 12) that L is one-to-one. Thus, from Theorem 6.12, part (a), it follows that L is also onto.

17. Consider the special case $L\colon R^n \to R^n$ where $L(\mathbf{x})$ = $A\mathbf{x}$. (The linear transformation L in this case is known as a linear operator on R^n.) The standard matrix A is $n \times n$ here.

From the comments of the previous frame, we know that the linear transformation is either both onto and one-to-one [case (iv)] or that the linear transformation has neither of these properties [case (i)]. In fact, it is easy to characterize whether or not A is both one-to-one and onto by making use of Rank (A) and the determinant $|A|$. From master Theorem 5.14 (frame 21 of Chapter 5), we know that Rank (A) = n if and only if $|A| \neq 0$. Now, Eq. (2) of frame 15 is

$$\text{Dim}[\text{Ker }(L)] + \text{Rank }(A) = n \tag{2}$$

and this indicates that Dim [Ker (L)] = 0 if and only if Rank (A) = n. But Dim [Ker (L)] = 0 is equivalent to Ker (L) = $\{0\}$, and the latter, in turn, is equivalent to $L\colon R^n \to R^n$ being one-to-one. (Theorem 6.7 in frame 12). Also, we know from Theorem 6.10 that Rank (A) being equal to n is equivalent to L being onto. We can summarize the preceding discussion in the following theorem, which is related to master Theorem 5.14.

Theorem 6.13

If A is an $n \times n$ matrix, then the following statements are equivalent:

(a) Rank (A) = n.
(e) $|A| \neq 0$.
(h) For the linear transformation given by $L(\mathbf{x})$ = $A\mathbf{x}$, L is both one-to-one and onto.

Here, we included only two of the equivalent statements from Theorem 5.14. [They were labeled (a) and (e) there.] We could have listed any of the others. Recall that if, for a particular $n \times n$ matrix A, one of the statements in the list is true, then all the statements are true. Similarly, if one statement is false.

By relating statements (e) and (h), we have an easy criterion involving the determinant for determining whether or not $L: R^n \to R^n$ is one-to-one and onto.

Example (a): For $L: R^3 \to R^3$ given by $L(\mathbf{x}) = A\mathbf{x}$, where $A = \begin{bmatrix} 2 & 1 & 3 \\ 1 & 1 & 2 \\ 3 & 2 & 5 \end{bmatrix}$, determine if L is one-to-one and onto by employing the determinant $|A|$.

Solution: Here $|A| = 0$, and hence L is neither one-to-one nor onto.

Example (b): Repeat Example (a) if $A = \begin{bmatrix} 2 & 3 & 4 \\ 1 & 2 & 1 \\ 1 & 2 & 3 \end{bmatrix}$.

— — — — — — — — —

Solution: Here $|A| = 2$, and hence L is both one-to-one and onto.

D. THE MATRIX OF A LINEAR TRANSFORMATION (CONTINUED)

18. In our preliminary discussion in Section B for $L: R^n \to R^m$, we obtained the standard matrix A such that $L(\mathbf{x}) = A\mathbf{x}$. It must be remembered that specifying a vector $\mathbf{x} = [x_1, x_2, \ldots, x_n]^t$ in R^n is equivalent to expressing $\mathbf{x}$ in terms of the standard basis for R^n. That is, we can write

$$\mathbf{x} = x_1 \mathbf{e}_1 + x_2 \mathbf{e}_2 + \cdots + x_n \mathbf{e}_n \tag{1}$$

The standard matrix A can be indicated by

$$A = [L(\mathbf{e}_1) \mid L(\mathbf{e}_2) \mid \ldots \mid L(\mathbf{e}_n)] \tag{2}$$

where each $L(\mathbf{e}_j)$ is an m-component column vector. (See frame 7, Theorem 6.4.)

Also, specifying the m-component column vector $A\mathbf{x}$ is equivalent to expressing $A\mathbf{x}$ in terms of the standard basis for R^m. Let us show this. Let $(A\mathbf{x})_i$ denote component i of $A\mathbf{x}$. [$(A\mathbf{x})_i = a_{i1}x_1 + a_{i2}x_2 + \cdots + a_{in}x_n$.] Then we have

$$A\mathbf{x} = \begin{bmatrix} (A\mathbf{x})_1 \\ (A\mathbf{x})_2 \\ \vdots \\ (A\mathbf{x})_m \end{bmatrix} = (A\mathbf{x})_1 \mathbf{e}_1 + (A\mathbf{x})_2 \mathbf{e}_2 + \cdots + (A\mathbf{x})_m \mathbf{e}_m. \tag{3}$$

Example (a): In a typical example in this chapter so far, we had something like $L: R^3 \rightarrow R^2$ given by the expression:

$$L\left(\begin{bmatrix} x_1 \\ x_2 \\ x_3 \end{bmatrix}\right) = \begin{bmatrix} a_1 x_1 + a_2 x_2 + a_3 x_3 \\ b_1 x_1 + b_2 x_2 + b_3 x_3 \end{bmatrix}$$

It is easy to show that standard matrix A (which is 2×3) can be obtained by successive listing of the coefficients of the x_j's in the two-component vector on the right. Thus,

$$A = \begin{bmatrix} a_1 & a_2 & a_3 \\ b_1 & b_2 & b_3 \end{bmatrix}$$

A similar result holds if L goes from R^n into R^m.

We now extend our ideas to allow for other bases for R^n and R^m, respectively, and also to handle various bases in arbitrary vector spaces V and W. First, let us consider the concept of a coordinate column vector.

Definition 6.5 (Coordinate Column Vector)

Let V be any n-dimensional vector space with basis $S = \{\mathbf{v}_1, \mathbf{v}_2, \ldots, \mathbf{v}_n\}$. If $\mathbf{x}$ is any vector in V where

$$\mathbf{x} = c_1 \mathbf{v}_1 + c_2 \mathbf{v}_2 + \cdots + c_n \mathbf{v}_n, \tag{1}$$

then the n-component column vector

$$[\mathbf{x}]_S = \begin{bmatrix} c_1 \\ c_2 \\ \vdots \\ c_n \end{bmatrix}$$

is called the *coordinate column vector* of $\mathbf{x}$ with respect to the basis S. The components of $[\mathbf{x}]_S$ are called the *coordinates of $\mathbf{x}$ with respect to S*.

Example (b): Let $S = \{\mathbf{v}_1, \mathbf{v}_2\}$ be a basis for R^2, where $\mathbf{v}_1 = \begin{bmatrix} 2 \\ 1 \end{bmatrix}$ and $\mathbf{v}_2 = \begin{bmatrix} -1 \\ 3 \end{bmatrix}$. If $\mathbf{x} = \begin{bmatrix} 8 \\ 11 \end{bmatrix}$, determine $[\mathbf{x}]_S$.

Solution: To find $[\mathbf{x}]_S$, we must determine c_1 and c_2 such that

$$\begin{bmatrix} 8 \\ 11 \end{bmatrix} = c_1\mathbf{v}_1 + c_2\mathbf{v}_2 = c_1\begin{bmatrix} 2 \\ 1 \end{bmatrix} + c_2\begin{bmatrix} -1 \\ 3 \end{bmatrix}. \tag{1}$$

This leads to

$$\begin{bmatrix} 2c_1 - c_2 \\ c_1 + 3c_2 \end{bmatrix} = \begin{bmatrix} 8 \\ 11 \end{bmatrix}, \tag{2}$$

which results in the following linear system of two equations in the two unknowns c_1 and c_2:

$$2c_1 - c_2 = 8 \tag{3a}$$

$$c_1 + 3c_2 = 11 \tag{3b}$$

Now finish the problem.

— — — — — — — — —

Solving for c_1 and c_2 (by using Theorem 3.11, Cramer's rule, for example) yields $c_1 = 5$ and $c_2 = 2$, and so $[\mathbf{x}]_S = \begin{bmatrix} 5 \\ 2 \end{bmatrix}$.

19. If E denotes the standard basis for R^n and $\mathbf{x}$ is a column vector from R^n, then $[\mathbf{x}]_E = \mathbf{x}$. For example, in Example (b) of frame 18, $\mathbf{x} = \begin{bmatrix} 8 \\ 11 \end{bmatrix} = 8\mathbf{e}_1 + 11\mathbf{e}_2$, and thus $[\mathbf{x}]_E = \begin{bmatrix} 8 \\ 11 \end{bmatrix}$ also.

Example (a): Let $S = \{\mathbf{v}_1, \mathbf{v}_2, \mathbf{v}_3\}$ be any basis for a three-dimensional vector space V. Then since $\mathbf{v}_1 = 1\mathbf{v}_1 + 0\mathbf{v}_2 + 0\mathbf{v}_3$, we have $[\mathbf{v}_1]_S = \begin{bmatrix} 1 \\ 0 \\ 0 \end{bmatrix}$. That is, $[\mathbf{v}_1]_S = \mathbf{e}_1$.

Similarly, $[\mathbf{v}_2]_S = \begin{bmatrix} 0 \\ 1 \\ 0 \end{bmatrix} = \mathbf{e}_2$ and $[\mathbf{v}_3]_S = \begin{bmatrix} 0 \\ 0 \\ 1 \end{bmatrix} = \mathbf{e}_3$.

Generalizing, if $S = \{\mathbf{v}_1, \mathbf{v}_2, \ldots, \mathbf{v}_n\}$ is a basis for an n-dimensional vector space V, then $[\mathbf{v}_j]_S = \mathbf{e}_j$, where $\{\mathbf{e}_1, \mathbf{e}_2, \ldots, \mathbf{e}_n\}$ is the standard basis for R^n.

Example (b): Consider the vector space P_1 of all polynomials of degree 1, together with the zero polynomial. (See frames 2, 5, and 15 of Chapter Five for a review.) Consider the basis $S = \{1, t\}$ for P_1; this is the so-called standard basis for P_1. For $p(t) = 13t - 10$, we have $[p(t)]_S = \begin{bmatrix} -10 \\ 13 \end{bmatrix}$. For the basis $S' = \{t - 1, 2 - 3t\}$, we can solve

$$13t - 10 = c_1(t - 1) + c_2(2 - 3t)$$

to obtain $c_1 = 4$ and $c_2 = -3$. (For example, respectively equate coefficients of t and constant terms on left and right.) Now determine $[p(t)]_{S'}$.

_ _ _ _ _ _ _ _ _ _

Solution: Since $13t - 10 = 4(t - 1) - 3(2t - 3t)$, we have $[p(t)]_{S'} = \begin{bmatrix} 4 \\ -3 \end{bmatrix}$.

20. Next, we have a very important theorem that allows us to use a product of a matrix times a column vector to determine $L(\mathbf{x})$ for $L: V \to W$, where V and W can be *any* finite dimensional vector spaces (here, $\mathbf{x}$ is a vector in V).

Theorem 6.14

Suppose $L: V \to W$ is a linear transformation of an n-dimensional vector space V into an m-dimensional vector space W ($n \geqslant 1$ and $m \geqslant 1$) and that $S = \{\mathbf{v}_1, \mathbf{v}_2, \ldots, \mathbf{v}_n\}$ and $T = \{\mathbf{w}_1, \mathbf{w}_2, \ldots, \mathbf{w}_m\}$ are bases for V and W, respectively.

The $m \times n$ matrix A has for its jth column the coordinate column vector $[L(\mathbf{v}_j)]_T$ of $L(\mathbf{v}_j)$. That is, A is *defined to be*

$$A = \begin{bmatrix} [L(\mathbf{v}_1)]_T \vdots [L(\mathbf{v}_2)]_T \vdots \ldots \vdots [L(\mathbf{v}_n)]_T \end{bmatrix}. \tag{I}$$

Now let $\mathbf{y} = L(\mathbf{x})$ for some $\mathbf{x}$ in V ($\mathbf{y}$ is thus in W). The quantity $[\mathbf{y}]_T$ is given by

$$[\mathbf{y}]_T = A[\mathbf{x}]_S. \tag{II}$$

Here $[\mathbf{x}]_S$ *and* $[\mathbf{y}]_T$ are coordinate column vectors of $\mathbf{x}$ and $\mathbf{y}$ with respect to the respective bases S and T. It should be noted that the matrix A is unique. The left side of (II) can also be written as $[L(\mathbf{x})]_T$, since $\mathbf{y} = L(\mathbf{x})$.

Notes: (i) In Section B (see Theorem 6.4 in frame 7) we have a special case of Theorem 6.14, where V is R^n and W is R^m. Also, in Theorem 6.4, the bases for R^n and R^m are the respective standard bases for these vector spaces.
(ii) The matrix A above is called the matrix of L with respect to the bases S and T.

It is useful to illustrate Theorem 6.14 by means of examples. A proof of the theorem for a special case will be given in frame 21.

Example (a): Let $L: R^3 \to R^2$ be defined by

$$L\left(\begin{bmatrix} x_1 \\ x_2 \\ x_3 \end{bmatrix}\right) = \begin{bmatrix} 2x_1 + x_2 - x_3 \\ x_1 + 3x_2 + 2x_3 \end{bmatrix} \tag{1}$$

The bases S and T for R^3 and R^2, respectively, are as follows:

$$S = \{\mathbf{v}_1, \mathbf{v}_2, \mathbf{v}_3\} \qquad \text{and} \qquad T = \{\mathbf{w}_1, \mathbf{w}_2\},$$

where $\mathbf{v}_1 = [1, 1, 0]^t$, $\mathbf{v}_2 = [1, 0, 1]^t$, $\mathbf{v}_3 = [0, 1, 1]^t$, $\mathbf{w}_1 = [1, 1]^t$ and $\mathbf{w}_2 = [0, 1]^t$. (We use transpose notation here to save space.) Determine the matrix A of L with respect to bases S and T.

Solution: From the definition of L.

$$L(\mathbf{v}_1) = \begin{bmatrix} 2 + 1 - 0 \\ 1 + 3 + 0 \end{bmatrix} = \begin{bmatrix} 3 \\ 4 \end{bmatrix} = 3\mathbf{w}_1 + \mathbf{w}_2;$$

thus, $[L(\mathbf{v}_1)]_T = \begin{bmatrix} 3 \\ 1 \end{bmatrix}$.

[The calculation of the coefficients of $\mathbf{w}_1$ and $\mathbf{w}_2$ is similar to that done in Example (b) of frame 18.] Similarly,

$$L(\mathbf{v}_2) = \begin{bmatrix} 1 \\ 3 \end{bmatrix} = \mathbf{w}_1 + 2\mathbf{w}_2, \qquad \text{and} \qquad [L(\mathbf{v}_2)]_T = \begin{bmatrix} 1 \\ 2 \end{bmatrix},$$

and

$$L(\mathbf{v}_3) = \begin{bmatrix} 0 \\ 5 \end{bmatrix} = 0\mathbf{w}_1 + 5\mathbf{w}_2, \qquad \text{and} \qquad [L(\mathbf{v}_3)]_T = \begin{bmatrix} 0 \\ 5 \end{bmatrix}.$$

Thus, the matrix for L with respect to S and T is

$$A = \begin{bmatrix} 3 & 1 & 0 \\ 1 & 2 & 5 \end{bmatrix}, \tag{2}$$

and Eq. (II) in Theorem 6.14 becomes

$$[L(\mathbf{x})]_T = \begin{bmatrix} 3 & 1 & 0 \\ 1 & 2 & 5 \end{bmatrix} [\mathbf{x}]_S. \tag{3}$$

Let us illustrate how Eq. (3) can be used. For example, for $\mathbf{x} = [3, 1, -2]^t$, the defining equation of L [Eq. (1)] yields

$$L(\mathbf{x}) = \begin{bmatrix} 2(3) + 1 - (-2) \\ 3 + 3(1) + 2(-2) \end{bmatrix} = \begin{bmatrix} 9 \\ 2 \end{bmatrix}. \tag{4}$$

Now we can verify that $\mathbf{x} = \begin{bmatrix} 3 \\ 1 \\ -2 \end{bmatrix} = 3\mathbf{v}_1 + 0\mathbf{v}_2 + (-2)\mathbf{v}_3$, and thus $[\mathbf{x}]_S$

$= \begin{bmatrix} 3 \\ 0 \\ -2 \end{bmatrix}$. [See Example (b) of frame 18 for a similar calculation.] Then, from

(3), we get

$$[L(\mathbf{x})]_T = \begin{bmatrix} 3 & 1 & 0 \\ 1 & 2 & 5 \end{bmatrix} \begin{bmatrix} 3 \\ 0 \\ -2 \end{bmatrix} = \begin{bmatrix} 9 \\ -7 \end{bmatrix}, \tag{5}$$

and thus,

$$L(\mathbf{x}) = 9\mathbf{w}_1 - 7\mathbf{w}_2 = 9\begin{bmatrix} 1 \\ 1 \end{bmatrix} - 7\begin{bmatrix} 0 \\ 1 \end{bmatrix} = \begin{bmatrix} 9 \\ 2 \end{bmatrix}. \tag{6}$$

This checks with the preceding value in Eq. (4) for $L(\mathbf{x})$.

Example (b): Let $L: R^3 \rightarrow R^2$ be as defined in Example (a). Now let $S' = \{\mathbf{e}_1, \mathbf{e}_2, \mathbf{e}_3\}$ and $T' = \{\hat{\mathbf{e}}_1, \hat{\mathbf{e}}_2\}$ where $\mathbf{e}_1 = [1, 0, 0]^t$, $\mathbf{e}_2 = [0, 1, 0]^t$, $\mathbf{e}_3 = [0, 0, 1]^t$, $\hat{\mathbf{e}}_1 = [1, 0]^t$, and $\hat{\mathbf{e}}_2 = [0, 1]^t$. That is, S' and T' are the standard bases for R^3 and R^2, respectively. Determine the matrix A' of L with respect to the bases S' and T'.

Solution: From the defining equation for $L(\mathbf{x})$ [Eq. (1) in Example (a)], we have

$$L(\mathbf{e}_1) = \begin{bmatrix} 2 + 0 - 0 \\ 1 + 0 + 0 \end{bmatrix} = \begin{bmatrix} 2 \\ 1 \end{bmatrix} = 2\hat{\mathbf{e}}_1 + \hat{\mathbf{e}}_2.$$

Thus, $[L(\mathbf{e}_1)]_{T'} = \begin{bmatrix} 2 \\ 1 \end{bmatrix}$.

Likewise, $L(e_2) = \begin{bmatrix} 1 \\ 3 \end{bmatrix} = [L(e_2)]_{T'}$, and $L(e_3) = \begin{bmatrix} -1 \\ 2 \end{bmatrix} = [L(e_3)]_{T'}$, and

thus

$$A' = \begin{bmatrix} 2 & 1 & -1 \\ 1 & 3 & 2 \end{bmatrix}.$$

The matrix A' here is none other than the *standard matrix* for L. Actually, we have already solved this problem in the Example of frame 6.

If we change the order of the vectors in the bases S and T, then the matrix of L with respect to the bases may change.

Example (c): Let $L: R^3 \to R^2$ be as defined in Example (a). Now let

$$S'' = \{v_2, v_3, v_1\} \qquad \text{and} \qquad T'' = \{w_1, w_2\},$$

where v_1, v_2, v_3, w_1, and w_2 are as in Example (a). [Note the change in order of the v_j's in S'' as compared to S in Example (a).] Determine the matrix of L with respect to bases S'' and T''.

———————————

Solution: We find that $A'' = \begin{bmatrix} 1 & 0 & 3 \\ 2 & 5 & 1 \end{bmatrix}$. Compare this with matrix A in

Example (a).

21. It is instructive to prove Theorem 6.14. We do so now for a special case.

Proof of Theorem 6.14 where n = 3 *and* m = 2

Since V has dimension 3 and W has dimension 2, bases S and T will have the following forms: $S = \{v_1, v_2, v_3\}$ and $T = \{w_1, w_2\}$. Now for each of the basis vectors v_1, v_2, and v_3 in V, $L(v_1)$, $L(v_2)$, and $L(v_3)$ are vectors in W.

Now each $L(v_j)$ is uniquely expressed in terms of the basis $\{w_1, w_2\}$. That is,

$$L(v_j) = a_{1j}w_1 + a_{2j}w_2 \qquad \text{for } j = 1, 2, 3. \tag{1}$$

Thus, the coordinate column vector $[L(v_j)]_T$ is

$$[L(v_j)]_T = \begin{bmatrix} a_{1j} \\ a_{2j} \end{bmatrix} \qquad \text{for } j = 1, 2, 3, \tag{2}$$

where a_{1j} and a_{2j} are unique for each j. Now define the 2×3 matrix A as

$$A = \left[[L(\mathbf{v}_1)]_T \;\vdots\; [L(\mathbf{v}_2)]_T \;\vdots\; [L(\mathbf{v}_3)]_T \right] = \begin{bmatrix} a_{11} & a_{12} & a_{13} \\ a_{21} & a_{22} & a_{23} \end{bmatrix} \tag{3}$$

Now let us digress for a while and consider $\mathbf{y} = L(\mathbf{x})$ for any $\mathbf{x}$ in V. Expressing $\mathbf{x}$ in terms of S yields

$$\mathbf{x} = c_1\mathbf{v}_1 + c_2\mathbf{v}_2 + c_3\mathbf{v}_3 \tag{4}$$

and thus,

$$\begin{aligned} \mathbf{y} = L(\mathbf{x}) &= L(c_1\mathbf{v}_1 + c_2\mathbf{v}_2 + c_3\mathbf{v}_3) \\ &= c_1 L(\mathbf{v}_1) + c_2 L(\mathbf{v}_2) + c_3 L(\mathbf{v}_3). \end{aligned} \tag{5}$$

Substituting the expressions for the $L(\mathbf{v}_j)$'s from (1) into (5) yields the following equation for $\mathbf{y}$ in terms of $\{\mathbf{w}_1, \mathbf{w}_2\}$, the basis for W:

$$\mathbf{y} = (a_{11}c_1 + a_{12}c_2 + a_{13}c_3)\mathbf{w}_1 + (a_{21}c_1 + a_{22}c_2 + a_{23}c_3)\mathbf{w}_2 \tag{6}$$

Thus, the two component column vector $[\mathbf{y}]_T$ is

$$[\mathbf{y}]_T = \begin{bmatrix} a_{11}c_1 + a_{12}c_2 + a_{13}c_3 \\ a_{21}c_1 + a_{22}c_2 + a_{23}c_3 \end{bmatrix} \tag{7}$$

Now we shall show that Eq. (II) of Theorem 6.14, with A given by Eq. (3) above, yields the same expression for $[\mathbf{y}]_T$. From Eq. (4), the coordinate column vector $[\mathbf{x}]_S$ equals $[c_1, c_2, c_3]^t$. Thus, Eq. (II) yields

$$\begin{aligned} [\mathbf{y}]_T = A[\mathbf{x}]_S &= \begin{bmatrix} a_{11} & a_{12} & a_{13} \\ a_{21} & a_{22} & a_{23} \end{bmatrix} \begin{bmatrix} c_1 \\ c_2 \\ c_3 \end{bmatrix} \\ &= \begin{bmatrix} a_{11}c_1 + a_{12}c_2 + a_{13}c_3 \\ a_{21}c_1 + a_{22}c_2 + a_{23}c_3 \end{bmatrix}, \end{aligned} \tag{8}$$

as was to be shown.

Note that knowing the $m \times n$ matrix A allows us to replace the equation $\mathbf{y} = L(\mathbf{x})$ by the simple matrix–vector equation $[\mathbf{y}]_T = A[\mathbf{x}]_S$, which involves an $m \times n$ matrix and two column vectors. Researchers and scientists who use linear transformations extensively do most of their computations with the matrices associated with the linear transformations.

The next two examples involve a linear transformation from one polynomial vector space into another.

Example (a): Suppose $L: P_1 \to P_2$ is defined by $L(p(t)) = (t + 1)p(t)$.

(i) Find the matrix of L with respect to the bases $S = \{1, t\}$ and $T = \{1, t, t^2\}$ for P_1 and P_2, respectively. (Here S and T are the respective standard bases for the polynomial vector spaces P_1 and P_2. See frame 15 of Chapter Five for a review.)

(ii) For $p(t) = 2t - 1$, determine $L(p(t))$ directly, and by using the matrix from part (i) together with Theorem 6.14.

Solution: (i) Here $L(1) = (t + 1)(1) = t + 1$; thus, $[L(1)]_T = [1, 1, 0]^t$. Also, $L(t) = (t + 1)(t) = t^2 + t$; thus, $[L(t)]_T = [0, 1, 1]^t$. Thus, the matrix of L with respect to the bases S and T is

$$A = \begin{bmatrix} 1 & 0 \\ 1 & 1 \\ 0 & 1 \end{bmatrix}$$

(ii) For the direct computation of $L(p(t))$, with $p(t) = 2t - 1$, we have

$$L(p(t)) = (t + 1)(2t - 1) = 2t^2 + t - 1$$

If we use A to compute $L(p(t))$, first we observe that the coordinate column vector $[p(t)]_S$ for $p(t) = 2t - 1$ is $[p(t)]_S = [-1, 2]^t$. Thus,

$$[L(p(t))]_T = A[p(t)]_S = \begin{bmatrix} 1 & 0 \\ 1 & 1 \\ 0 & 1 \end{bmatrix} \begin{bmatrix} -1 \\ 2 \end{bmatrix} = \begin{bmatrix} -1 \\ 1 \\ 2 \end{bmatrix}$$

Thus, $L(p(t)) = (-1)(1) + 1(t) + 2(t^2) = -1 + t + 2t^2$.

Example (b): Suppose $L: P_1 \to P_2$ is the same as in Example (a) but that the bases for P_1 and P_2, respectively, are $S' = \{1 - t, 1 + t\}$ and $T' = \{1, t, t^2\}$. Note that $T' = T$ [see Example (a)].

(i) Find the matrix of L with respect to the bases $S' = \{1 - t, 1 + t\}$ and $T' = \{1, t, t^2\}$.

(ii) For $p(t) = 2t - 1$, determine $L(p(t))$ using the matrix from part (i).

Solution: (i) First, $L(1 - t) = (t + 1)(1 - t) = 1 - t^2$; thus, $[L(1 - t)]_{T'} = [1, 0, -1]^t$. Now finish part (i), and then do part (ii).

— — — — — — — — — —

(i) *Completed:* $L(1 + t) = (t + 1)(1 + t) = 1 + 2t + t^2$; thus, $[L(1 + t)]_{T'} = [1, 2, 1]^t$. Thus, the matrix A' of L with respect to S' and T' is

$$A' = \begin{bmatrix} 1 & 1 \\ 0 & 2 \\ -1 & 1 \end{bmatrix}$$

(ii)   First, we have to express $p(t) = 2t - 1$ in terms of the basis $S' = \{(1 - t), (1 + t)\}$. We proceed as in Example (b) of frame 19. First,

$$2t - 1 = c_1(1 - t) + c_2(1 + t)$$

results in $c_1 = -\frac{3}{2}$ and $c_2 = \frac{1}{2}$ after equating like powers of t on left and right. Thus, $[p(t)]_{S'} = [-\frac{3}{2}, \frac{1}{2}]^t$, and

$$[L(p(t))]_{T'} = A'[p(t)]_{S'}$$

$$= \begin{bmatrix} 1 & 1 \\ 0 & 2 \\ -1 & 1 \end{bmatrix} \begin{bmatrix} -\frac{3}{2} \\ \frac{1}{2} \end{bmatrix} = \begin{bmatrix} -1 \\ 1 \\ 2 \end{bmatrix}$$

Thus, $L(p(t)) = -1 + t + 2t^2$. This checks with Example (a).

22.   In the important special case where $V = W$ (thus, $L: V \to V$ is a linear *operator*), the *usual* policy is to choose $S = T$. The resulting matrix A in Theorem 6.14 (frame 20) is then called the *matrix of L with respect to S*. The matrix will be a square matrix, of course.

Example (a):   Let $L: R^2 \to R^2$ be the linear operator defined by

$$L\left(\begin{bmatrix} x_1 \\ x_2 \end{bmatrix}\right) = \begin{bmatrix} 2x_1 + 3x_2 \\ x_1 \end{bmatrix}$$

Find the matrix of L with respect to the basis $S = \{\mathbf{v}_1, \mathbf{v}_2\}$, where $\mathbf{v}_1 = [3, 1]^t$ and $\mathbf{v}_2 = [-1, 1]^t$.

Solution: From the definition of L,

$$L(\mathbf{v}_1) = \begin{bmatrix} 2(3) + 3(1) \\ 3 \end{bmatrix} = \begin{bmatrix} 9 \\ 3 \end{bmatrix} = 3\mathbf{v}_1 + 0\mathbf{v}_2,$$

and thus $[L(\mathbf{v}_1)]_S = \begin{bmatrix} 3 \\ 0 \end{bmatrix}$.

$$L(\mathbf{v}_2) = \begin{bmatrix} 2(-1) + 3(1) \\ -1 \end{bmatrix} = \begin{bmatrix} 1 \\ -1 \end{bmatrix} = 0\mathbf{v}_1 - \mathbf{v}_2,$$

and thus $[L(\mathbf{v}_2)]_S = \begin{bmatrix} 0 \\ -1 \end{bmatrix}$.

Therefore, the matrix of L with respect to S is

$$A = \begin{bmatrix} 3 & 0 \\ 0 & -1 \end{bmatrix}$$

Note that the matrix here is a diagonal matrix.

If, in **Example (a)**, the basis was the standard basis for R^2, namely, $\{e_1, e_2\}$, where $e_1 = [1, 0]^t$ and $e_2 = [0, 1]^t$, then the matrix of L with respect to the standard basis would be the *standard* matrix $\begin{bmatrix} 2 & 3 \\ 1 & 0 \end{bmatrix}$.

If the linear operator is the *identity* linear operator $I: V \rightarrow V$ defined by $I(\mathbf{x}) = \mathbf{x}$ for each $\mathbf{x}$ in V, then the matrix of I with respect to *any* basis S of V is the identity matrix I_n where $n = \text{Dim }(V)$.

Example (b): Demonstrate the result stated in the preceding sentence for the case $n = 3$.

Solution: Let $S = \{\mathbf{v}_1, \mathbf{v}_2, \mathbf{v}_3\}$ be *any* basis for V. Thus,

$$I(\mathbf{v}_1) = \mathbf{v}_1 = 1\mathbf{v}_1 + 0\mathbf{v}_2 + 0\mathbf{v}_3 \text{ and hence } [I(\mathbf{v}_1)]_S = \begin{bmatrix} 1 \\ 0 \\ 0 \end{bmatrix}$$

Now complete the demonstration.

— — — — — — — — — —

$$I(\mathbf{v}_2) = \mathbf{v}_2 \text{ and } [I(\mathbf{v}_2)]_S = \begin{bmatrix} 0 \\ 1 \\ 0 \end{bmatrix}$$

$$I(\mathbf{v}_3) = \mathbf{v}_3 \text{ and } [I(\mathbf{v}_3)]_S = \begin{bmatrix} 0 \\ 0 \\ 1 \end{bmatrix}$$

Thus, the matrix of I with respect to S is

$$\begin{bmatrix} 1 & 0 & 0 \\ 0 & 1 & 0 \\ 0 & 0 & 1 \end{bmatrix},$$

and this matrix is I_3.

FINAL COMMENTS

We have merely scratched the surface of the subject of linear transformations. The fundamentals covered in this chapter should prepare the interested student for further study in this area.

One useful topic pertaining to linear transformations concerns the concept of an inverse (denoted by L^{-1}, if it exists) of a linear transformation $L: R^n \rightarrow R^n$. An important fact is that if A is the standard matrix for L, then the matrix inverse A^{-1} (if it exists) will be the standard matrix of L^{-1}. For a fairly elementary discussion of this, and of the product (or composite) transformation of two or more linear transformations, refer to Shields (1980).

Another important topic related to linear transformations concerns the effect of linear transformations on geometrical regions. An important fact here is that a linear transformation $L: R^2 \rightarrow R^2$ transforms a parallellogram from the domain R^2 into a new parallelogram (in the codomain), and the ratio of the new area to the old area is equal to the absolute value of the determinant of A (the standard matrix of L). This ties in with the important equation for the *Jacobian determinant*, which occurs in calculus studies dealing with functions of several variables. The material of the current paragraph is also briefly discussed in Shields (1980).

SELF-TEST

This Self-Test will help you determine whether or not you have mastered the chapter objectives and are ready to go on to another chapter. Correct answers are given at the end of the test.

1. (a) Suppose $L(\mathbf{x})$ is the *rotation* linear transformation in R^2 (frame 3). Determine $L(\mathbf{x}) = \mathbf{x}'$ if $\mathbf{x} = [2, 3]^t$ and $\phi = 45° = \pi/4$ radians.
    (b) Given the linear transformation

$$\text{Proj}_\ell \, \mathbf{x} = \left(\frac{\mathbf{x} \cdot \mathbf{v}}{\mathbf{v} \cdot \mathbf{v}} \right) \mathbf{v} \text{ in } R^2$$

where ℓ refers to the "45 degree straight line" $x_2 = x_1$ and $\mathbf{v}$ is any nonzero vector for the line (see frame 3 for "projection" linear transformation). Determine an expression for $\text{Proj}_\ell \, \mathbf{x}$ as a two-component column vector.
    (c) Referring to part (b), determine $\text{Proj}_\ell \, \mathbf{x}$ if $\mathbf{x} = [3, 5]^t$.

2. (a) If $L(\mathbf{x})$ is a rotation linear transformation in R^2 for $\phi = 60° = \pi/3$ radians, determine the standard matrix A for L.
    (b) Determine the standard matrix A for the linear transformation $\text{Proj}_\ell \, \mathbf{x}$ of question 1b.

3. Determine whether L is one-to-one for each of the following:
    (a) $L: R^2 \rightarrow R^2$ defined by

$$L(\mathbf{x}) = \begin{bmatrix} 3x_1 + 2x_2 \\ -x_1 + x_2 \end{bmatrix}.$$

    (b) $L: R^2 \rightarrow R^2$ given in question 1b.

4. Determine Ker (L) for each of the following linear transformations. Then use Theorem 6.7 to determine whether L is one-to-one. Also, find Dim [Ker (L)].

    (a) The linear transformation of question 3a.

    (b) The linear transformation of questions 1b and 3b.

    (c) $L: R^4 \to R^3$ defined by $L(\mathbf{x}) = A\mathbf{x}$, where

$$A = \begin{bmatrix} 1 & 1 & 0 & 3 \\ 3 & -1 & -4 & 5 \\ 1 & 2 & 1 & 4 \end{bmatrix}.$$

5. For the three linear transformations of question 4, determine Dim [Range (L)]. Which linear transformations are onto?

6. (a) For $L: R^4 \to R^3$ defined by $L(\mathbf{x}) = A\mathbf{x}$, where

$$A = \begin{bmatrix} 2 & 3 & -2 & -5 \\ 3 & 2 & 1 & 7 \\ 1 & 3 & 0 & -1 \end{bmatrix},$$

find the reduced row echelon form, and Rank (A).

    (b) Determine if L is an onto linear transformation.

    (c) Find Dim [Ker (L)] by using Theorem 6.11.

7. Refer to the two linear transformations of questions 3a and 3b. Based on the answers in questions 5a and 5b as to whether each L is onto or not, and Theorem 6.12 (frame 16), determine whether each L is one-to-one.

8. (a) Let $S = \{\mathbf{v}_1, \mathbf{v}_2, \mathbf{v}_3\}$ be a basis for R^3, where $\mathbf{v}_1 = [2, 4, 0]^t$, $\mathbf{v}_2 = [-1, 1, 2]^t$, and $\mathbf{v}_3 = [2, -3, 4]^t$. For $\mathbf{x} = [10, 7, 0]^t$ determine the coordinate column vector $[\mathbf{x}]_S$.

    (b) Let $S = \{2 + t, -1 + 3t\}$ be a basis for the polynomial vector space P_1. For $p(t) = 5 + 13t$, determine $[p(t)]_S$.

9. Let $L: R^3 \to R^2$ be defined for $\mathbf{x} = [x_1, x_2, x_3]^t$ by

$$L(\mathbf{x}) = \begin{bmatrix} 3x_1 - x_2 + 2x_3 \\ 2x_1 + 4x_2 - x_3 \end{bmatrix}.$$

Let $S = \{\mathbf{v}_1, \mathbf{v}_2, \mathbf{v}_3\}$ and $T = \{\mathbf{w}_1, \mathbf{w}_2\}$, where $\mathbf{v}_1 = [1, 1, 0]^t$, $\mathbf{v}_2 = [1, 0, 1]^t$, $\mathbf{v}_3 = [1, 1, 1]^t$, $\mathbf{w}_1 = [1, 1]^t$, and $\mathbf{w}_2 = [-1, 0]^t$.

    (a) Determine the matrix A of L with respect to bases S and T.

    (b) Compute $L(\mathbf{x})$ for $\mathbf{x} = [4, 2, 1]^t$ both directly and indirectly [by using Eq. (II) of Theorem 6.14 and the matrix from part (a)].

10. $L: P_1 \to P_1$ is defined by $L(a_0 + a_1 t) = (a_0 + 3a_1) + (2a_0 + 4a_1)t$, where $p(t) = a_0 + a_1 t$ represents a typical polynomial in the domain P_1. (Remember that $L: V \to W$ acts on vectors in domain V to generate vectors in codomain W. Thus, here domain P_1 refers to the P_1 upon which L acts.)

(a) Suppose the bases are $S = \{1 - t, 1 + t\}$ for the domain P_1 and $T = \{1, 1 + t\}$ for the codomain P_1. Determine the matrix of L with respect to the bases S and T.

(b) Refer to part (a). Compute $L(q(t))$ for $q(t) = 3 - 2t$ both directly and by using Theorem 6.14 together with the matrix of part (a).

(c) Suppose that the bases for the domain P_1 and the codomain P_1 are both chosen to be the single basis $\hat{S} = \{1, t\}$. This is the standard basis for P_1. Determine the matrix of L with respect to $\hat{S}$.

(d) Use the matrix of part (c) together with Theorem 6.14 to compute $L(3 - 2t)$.

ANSWERS TO SELF-TEST

If your answers to the test questions do not agree with the ones given here, review the frames indicated in parentheses after each answer before you go on to another chapter.

1. (a) $L\left(\begin{bmatrix} 2 \\ 3 \end{bmatrix}\right) = \dfrac{1}{2}\begin{bmatrix} \sqrt{2} & -\sqrt{2} \\ \sqrt{2} & \sqrt{2} \end{bmatrix}\begin{bmatrix} 2 \\ 3 \end{bmatrix} = \dfrac{\sqrt{2}}{2}\begin{bmatrix} -1 \\ 5 \end{bmatrix}.$

(b) $\text{Proj}_\ell\ \mathbf{x} = \dfrac{1}{2}\begin{bmatrix} x_1 + x_2 \\ x_1 + x_2 \end{bmatrix}.$

(c) For $\mathbf{x} = [3, 5]^t$, $\text{Proj}_\ell\ \mathbf{x} = [4, 4]^t$. $\hspace{2cm}$ (frames 1–3)

2. (a) $A = \begin{bmatrix} \cos\phi & -\sin\phi \\ \sin\phi & \cos\phi \end{bmatrix} = \dfrac{1}{2}\begin{bmatrix} 1 & -\sqrt{3} \\ \sqrt{3} & 1 \end{bmatrix}$ for $\phi = \pi/3$ radians.

(b) A for $\text{Proj}_\ell\ \mathbf{x} = \dfrac{1}{2}\begin{bmatrix} 1 & 1 \\ 1 & 1 \end{bmatrix}.$ $\hspace{2cm}$ (frames 6–9)

3. (a) L is one-to-one.

(b) L is not one-to-one. For example, $L(\mathbf{x}) = \dfrac{1}{2}\begin{bmatrix} x_1 + x_2 \\ x_1 + x_2 \end{bmatrix} = \begin{bmatrix} 2 \\ 2 \end{bmatrix}$, for any

vector $\mathbf{x} = [x_1, x_2]^t$ such that $x_1 + x_2 = 4$. Thus, for example, $L([1, 3]^t) = L([2, 2]^t) = L([4, 0]^t) = [2, 2]^t$. Geometrically, every point on the line given by $x_1 + x_2 = 4$ (this line is perpendicular to $x_1 = x_2$) projects onto the point with coordinates $x_1 = 2$, $x_2 = 2$. (frame 10)

4. (a) Ker $(L) = \{\mathbf{0}\}$, where $\mathbf{0} = [0, 0]^t$. Dim [Ker (L)] = 0. L is one-to-one.

(b) Ker (L) consists of all vectors of the form $\mathbf{x} = r[-1, 1]^t$ and Dim[Ker (L)] = 1. L is not one-to-one.

(c) Ker (L) consists of all vectors of the form $\mathbf{x} = r[1, -1, 1, 0]^t + s[-2, -1, 0, 1]^t$ and Dim [Ker (L)] = 2. L is not one-to-one.

$\hspace{8cm}$ (frames 11–12)

5. Here Dim [Range (L)] = Rank (A).
   (a) Dim [Range (L)] = 2 and L is onto.
   (b) Dim [Range (L)] = 1 and L is not onto.
   (c) Dim [Range (L)] = 2 and L is not onto.   (frames 13–14)

6. (a) Reduced row echelon form is $\begin{bmatrix} 1 & 0 & 0 & 2 \\ 0 & 1 & 0 & -1 \\ 0 & 0 & 1 & 3 \end{bmatrix}$ and Rank (A) = 3.

   (b) L is onto since Rank (A) = m = 3.
   (c) Dim [Ker (L)] = 1.   (frames 13–15)

7. In both (a) and (b), Dim (V) = Dim (W), so Theorem 6.12 applies. In (a), L is one-to-one, since L is onto. In (b), L is not one-to-one, since L is not onto.   (frames 16, 17)

8. (a) $[\mathbf{x}]_S = \begin{bmatrix} 3 \\ -2 \\ 1 \end{bmatrix}$.

   (b) $[p(t)]_S = \begin{bmatrix} 4 \\ 3 \end{bmatrix}$.   (frames 18, 19)

9. (a) $A = \begin{bmatrix} 6 & 1 & 5 \\ 4 & -4 & 1 \end{bmatrix}$.

   (b) $L(\mathbf{x}) = \begin{bmatrix} 12 \\ 15 \end{bmatrix}$. (frames 20, 21)

10. (a) $A = \begin{bmatrix} 0 & -2 \\ -2 & 6 \end{bmatrix}$.

    (b) $L(3 - 2t) = -3 - 2t$.

    (c) $\hat{A} = \begin{bmatrix} 1 & 3 \\ 2 & 4 \end{bmatrix}$.

    (d) $L(3 - 2t) = -3 - 2t$.   (frames 21, 22)

CHAPTER SEVEN
Eigenvalues and Eigenvectors

Our work in this chapter shall concentrate on certain properties of the linear transformation $L: R^n \rightarrow R^n$, where $L(\mathbf{x})$ is specified by the product $A\mathbf{x}$ (L is called a linear operator; the matrix A is the standard matrix for L).

We shall focus on the problem of determining nonzero vectors $\mathbf{x}$ for which $A\mathbf{x}$ is a multiple of $\mathbf{x}$, that is, for which $A\mathbf{x} = \lambda\mathbf{x}$, where the quantity λ is a scalar (here a particular λ for which this equation holds is called an *eigenvalue*; associated nonzero vectors are called eigenvectors corresponding to λ). This equation, $A\mathbf{x} = \lambda\mathbf{x}$, is the starting point of our study of eigenvalue problems.

One of our main interests in eigenvalue problems is concerned with the practical question of when an $n \times n$ matrix A can be diagonalized. We say that A can be diagonalized if we can write

$$P^{-1}AP = D \qquad \text{or, equivalently,} \qquad A = PDP^{-1} \qquad \text{(1a), (1b)}$$

where D is an $n \times n$ diagonal matrix (Definition 1.7 in frame 21 of Chapter One), and P is an invertible $n \times n$ matrix. Whether it is possible to express A in the form given by Eq. (1b) is intimately related to the study of eigenvalue problems. One of the major consequences that follows if Eq. (1b) is true is that A^k, the kth power of A, can be expressed in the simple form

$$A^k = PD^kP^{-1}, \qquad (2)$$

and this is discussed later in this chapter. Equation (2) makes it easy to compute A^k since D^k is itself a diagonal matrix whose diagonal entries are easy to calculate.

The diagonalization of matrix A and the computation of A^k by means of Eq. (2) occurs in many practical applications (see, for example, Chapter Eight, Selected Applications).

OBJECTIVES

When you complete this chapter, you should be able to

- Determine eigenvalues and corresponding eigenvectors for the $n \times n$ matrix A.
- Manipulate the characteristic equation $|\lambda I_n - A| = 0$, and determine its solutions for λ (those solutions happen to be the eigenvalues of matrix A).
- Determine if matrices A and B are similar, which means that $P^{-1}AP = B$, where P is some invertible matrix.
- Determine conditions such that matrix A is similar to a diagonal matrix D (if so, A is said to be *diagonalizable*); that is, such that $P^{-1}AP = D$.
- Calculate the power A^k by using $A^k = PD^kP^{-1}$.
- Carry out the diagonalization of A by producing the diagonal matrix D whose diagonal entries are the eigenvalues of A, and the matrix P whose columns are corresponding eigenvectors.
- Diagonalize a symmetric matrix A and understand why such a matrix can always (theoretically) be diagonalized.
- Determine an *orthogonal* matrix $\hat{P}$ that diagonalizes a symmetric matrix A.
- Make use of the property that $(\hat{P})^{-1} = (\hat{P})^t$ for an orthogonal matrix $\hat{P}$.

A. INTRODUCTION TO EIGENVALUES AND EIGENVECTORS

1.   For this chapter we shall focus on linear transformations from R^n into R^n (also known as *linear operators* on R^n). We can define a linear operator L: $R^n \rightarrow R^n$ by $L(\mathbf{x}) = A\mathbf{x}$, where A is an $n \times n$ matrix and $\mathbf{x}$ is any vector in R^n. (A is the standard matrix for L; see frame 7 of Chapter Six.)

An important question for applications is to determine nonzero vectors $\mathbf{x}$ for which $A\mathbf{x}$ is a multiple of $\mathbf{x}$, that is, for which $A\mathbf{x} = \lambda\mathbf{x}$, where λ is a scalar. The multiple λ is called an eigenvalue, and the corresponding nonzero vectors are called eigenvectors. Such questions arise in many areas of application, such as mechanical engineering, chemical engineering, mechanics, biology, statistics, and differential equations.

Definition 7.1 (Eigenvalues and Eigenvectors)

For A an $n \times n$ matrix the real number (scalar) λ is called an *eigenvalue* of A if there exists a nonzero vector $\mathbf{x}$ in R^n such that

$$A\mathbf{x} = \lambda\mathbf{x} \tag{I}$$

Every nonzero vector $\mathbf{x}$ satisfying (I) for a particular λ is called an *eigenvector corresponding* to λ.

Notes: (i)   In German, *eigen* is a word that means "self," "own," or "proper." Other expressions for eigenvalues are characteristic values, proper values, or latent values, with similar names for the corresponding eigenvectors. (ii)   The zero vector (**0**) satisfies Eq. (I). Our definition requires that eigenvectors be nonzero solutions of Eq. (I) for a given λ. (iii)   Unless otherwise noted, eigenvalues will be real in this book. There are practical applications (not considered here) involving vector spaces for which complex numbers are required. In such so-called complex vector spaces, complex numbers occur in the eigenvalues and eigenvectors.

Example (a): In frame 2, we shall show that $\lambda_1 = 3$ is an eigenvalue of the matrix $A = \begin{bmatrix} 2 & 3 \\ 1 & 0 \end{bmatrix}$. Verify that $\mathbf{x}_1 = \begin{bmatrix} 3 \\ 1 \end{bmatrix}$ is a corresponding eigenvector.

Solution:

$$A\mathbf{x}_1 = \begin{bmatrix} 2 & 3 \\ 1 & 0 \end{bmatrix} \begin{bmatrix} 3 \\ 1 \end{bmatrix} = \begin{bmatrix} 6 + 3 \\ 3 + 0 \end{bmatrix} = \begin{bmatrix} 9 \\ 3 \end{bmatrix} = 3 \begin{bmatrix} 3 \\ 1 \end{bmatrix} = 3\mathbf{x}_1$$

Thus, $A\mathbf{x}_1 = 3\mathbf{x}_1$ and $\mathbf{x}_1$ is an eigenvector corresponding to $\lambda_1 = 3$.

Note: If **x** is an eigenvector corresponding to λ, then so is $k\mathbf{x}$, where k is any nonzero real number. That is, if we multiply $A\mathbf{x} = \lambda\mathbf{x}$ by k, we get $kA\mathbf{x} = k\lambda\mathbf{x}$, or $A(k\mathbf{x}) = \lambda(k\mathbf{x})$. Thus, in Example (a) other eigenvectors corresponding to $\lambda_1 = 3$ are $[6, 2]^t$, $[9, 3]^t$, $[-3, -1]^t$, and so on.

Example (b): For the matrix $A = \begin{bmatrix} 2 & 3 \\ 1 & 0 \end{bmatrix}$, another eigenvalue is $\lambda_2 = -1$. Verify that $\mathbf{x}_2 = [-1, 1]^t$ is a corresponding eigenvector.

— — — — — — — — —

$$A\mathbf{x}_2 = \begin{bmatrix} 2 & 3 \\ 1 & 0 \end{bmatrix} \begin{bmatrix} -1 \\ 1 \end{bmatrix} = \begin{bmatrix} -2 + 3 \\ -1 + 0 \end{bmatrix} = \begin{bmatrix} 1 \\ -1 \end{bmatrix} = (-1) \begin{bmatrix} -1 \\ 1 \end{bmatrix} = (-1)\mathbf{x}_2$$

2.   Though it is true that the zero vector cannot be an eigenvector, the number zero can be an eigenvalue. For example, the matrix $A = \begin{bmatrix} 2 & 3 \\ 0 & 0 \end{bmatrix}$ has an eigenvalue of zero. A corresponding eigenvector is $[3, -2]^t$. To verify this, we see that $\begin{bmatrix} 2 & 3 \\ 0 & 0 \end{bmatrix} \begin{bmatrix} 3 \\ -2 \end{bmatrix} = \begin{bmatrix} 0 \\ 0 \end{bmatrix} = \mathbf{0}$ and $\lambda \begin{bmatrix} 3 \\ -2 \end{bmatrix} = 0 \begin{bmatrix} 3 \\ -2 \end{bmatrix} = \begin{bmatrix} 0 \\ 0 \end{bmatrix} = \mathbf{0}$.

Example: For $\begin{bmatrix} 2 & 3 \\ 1 & 0 \end{bmatrix}$, find the eigenvalues and corresponding eigenvectors by working from Definition 7.1.

Solution: Equation (I) of Definition 7.1 yields

$$\begin{bmatrix} 2 & 3 \\ 1 & 0 \end{bmatrix} \begin{bmatrix} x_1 \\ x_2 \end{bmatrix} = \lambda \begin{bmatrix} x_1 \\ x_2 \end{bmatrix} \quad \text{or} \quad \begin{bmatrix} 2x_1 + 3x_2 \\ x_1 \end{bmatrix} = \begin{bmatrix} \lambda x_1 \\ \lambda x_2 \end{bmatrix} \tag{1}$$

This yields the two scalar equations

$$2x_1 + 3x_2 = \lambda x_1 \tag{2a}$$

$$x_1 \qquad = \lambda x_2, \tag{2b}$$

or

$$(\lambda - 2)x_1 - 3x_2 = 0 \tag{3a}$$

$$-x_1 + \lambda x_2 = 0, \tag{3b}$$

after subtracting from the *right* sides. Equations (3a) and (3b) constitute a homogeneous system of two equations in two unknowns. From Theorem 3.10′ (frame 13), we know that Eqs. (3a) and (3b) will have a nontrivial solution for x_1 and x_2 (i.e., a solution for which at least one of x_1 and x_2 is unequal to zero, or, equivalently, for which $\mathbf{x} \neq \mathbf{0}$) if and only if the determinant of its coefficient matrix is equal to zero. Setting this determinant equal to zero yields

$$\begin{vmatrix} (\lambda - 2) & -3 \\ -1 & \lambda \end{vmatrix} = 0, \tag{4}$$

or

$$\lambda^2 - 2\lambda - 3 = 0 \tag{5}$$

But,

$$\lambda^2 - 2\lambda - 3 = (\lambda - 3)(\lambda + 1) \tag{6}$$

Thus, Eq (5) becomes

$$(\lambda - 3)(\lambda + 1) = 0, \tag{7}$$

and thus $\lambda_1 = 3$, and $\lambda_2 = -1$ are the eigenvalues of A. We can find eigenvectors corresponding to $\lambda_1 = 3$ by substituting $\lambda_1 = 3$ into Eqs. (3a) and (3b), say. This yields

$$x_1 - 3x_2 = 0 \tag{8a}$$

$$-x_1 + 3x_2 = 0 \tag{8b}$$

We can solve this by inspection or by using the Gauss–Jordan elimination method (Method 2.1). Applying the latter to the augmented matrix of Eqs. (8a) and (8b) yields

$$\left[\begin{array}{rr|r} 1 & -3 & 0 \\ -1 & 3 & 0 \end{array}\right] \xrightarrow{\quad r_2' = r_2 + r_1 \quad} \left[\begin{array}{rr|r} 1 & -3 & 0 \\ 0 & 0 & 0 \end{array}\right] \tag{9}$$

Thus,

$$x_1 - 3x_2 = 0 \quad \text{or} \quad x_1 = 3x_2 \tag{10}$$

Thus, we have

$$x_1 = 3r \quad \text{and} \quad x_2 = r, \tag{11}$$

where r is any real number. Thus, all eigenvectors corresponding to the eigenvalue $\lambda_1 = 3$ are given by $\left[\begin{array}{c} 3r \\ r \end{array}\right] = r \left[\begin{array}{c} 3 \\ 1 \end{array}\right]$, for r a nonzero real number. Letting $r = 1$ (other values will also do) leads to the eigenvector $\mathbf{x}_1 = \left[\begin{array}{c} 3 \\ 1 \end{array}\right]$.

Now find eigenvectors corresponding to $\lambda_2 = -1$.

– – – – – – – – – – –

Putting $\lambda_2 = -1$ into Eqs. (3a) and (3b) yields

$$-3x_1 - 3x_2 = 0 \tag{12a}$$

$$-x_1 - x_2 = 0 \tag{12b}$$

This leads to $x_1 = -x_2$ and thus $\mathbf{x} = \left[\begin{array}{c} -s \\ s \end{array}\right] = s \left[\begin{array}{c} -1 \\ 1 \end{array}\right]$, with s any nonzero real number. Letting $s = 1$ leads to the eigenvector

$$\mathbf{x}_2 = \left[\begin{array}{c} -1 \\ 1 \end{array}\right]$$

corresponding to $\lambda_2 = -1$.

B.   THE CHARACTERISTIC EQUATION

3.

Definition 7.2 (The Characteristic Equation)

Suppose A is an $n \times n$ matrix and I_n is the $n \times n$ identity matrix. The equation involving the determinant $|\lambda I_n - A|$, given by

$$|\lambda I_n - A| = 0, \tag{II}$$

is defined to be the *characteristic equation* of A.

Notes: (a)   The left side of the preceding equation reduces to a polynomial of degree n, in the form

$$\lambda^n + c_1\lambda^{n-1} + c_2\lambda^{n-2} + \cdots + c_{n-1}\lambda + c_n,$$

where the c_i's are real constants. For this reason $|\lambda I_n - A|$ is called the *characteristic polynomial of A*.

(b)   Refer to frame 2. The characteristic equation for $A = \begin{vmatrix} 2 & 3 \\ 1 & 0 \end{vmatrix}$ is given by Eq. (4) [and Eq. (5)] of the solution. The characteristic polynomial, which is of degree 2, is the left side of Eq. (5).

The connection between the characteristic polynomial and the calculation of eigenvalues is indicated by the following theorem. (In Theorem 7.1, as on most occasions in this chapter, we ignore the possibility of complex eigenvalues.)

Theorem 7.1

The real values of λ that are the solutions of the characteristic equation are the eigenvalues of A. Put differently, the eigenvalues are the real roots of the characteristic polynomial.

The proof is similar in scope to the first part of the solution in frame 2.

Proof of Theorem 7.1

If λ is an eigenvalue of A with corresponding eigenvector $\mathbf{x}$, we have

$$A\mathbf{x} = \lambda\mathbf{x} \tag{i}$$

Since $I_n\mathbf{x} = \mathbf{x}$, this can be rewritten as

$$A\mathbf{x} = \lambda I_n\mathbf{x} \quad \text{or} \quad \lambda I_n\mathbf{x} - A\mathbf{x} = 0 \tag{ii}$$

or

$$(\lambda I_n - A)\mathbf{x} = 0 \tag{iii}$$

This is equivalent to the matrix–vector equation for a homogeneous system of n equations in n unknowns with coefficient matrix $B = \lambda I_n - A$. See Section C of Chapter Two. Such a system has a nonzero solution for vector $\mathbf{x}$ (i.e., $\mathbf{x} \neq 0$) if and only if the determinant of the coefficient matrix B equals zero (Theorem 3.10′). This leads to the characteristic equation

$$|\lambda I_n - A| = 0 \tag{iv) or (II}$$

Example: Determine the characteristic equation for

$$A = \begin{bmatrix} -2 & 2 & 3 \\ -2 & 3 & 2 \\ -4 & 2 & 5 \end{bmatrix}$$

Solution: First,

$$\lambda I_3 - A = \begin{bmatrix} \lambda & 0 & 0 \\ 0 & \lambda & 0 \\ 0 & 0 & \lambda \end{bmatrix} - \begin{bmatrix} -2 & 2 & 3 \\ -2 & 3 & 2 \\ -4 & 2 & 5 \end{bmatrix} \tag{1}$$

Thus,

$$\lambda I_3 - A = \begin{bmatrix} \lambda + 2 & -2 & -3 \\ 2 & \lambda - 3 & -2 \\ 4 & -2 & \lambda - 5 \end{bmatrix} \tag{2}$$

Now take the determinant of $\lambda I_3 - A$, and obtain the characteristic equation.

$$|\lambda I_3 - A| = \begin{vmatrix} \lambda + 2 & -2 & -3 \\ 2 & \lambda - 3 & -2 \\ 4 & -2 & \lambda - 5 \end{vmatrix}$$

$$= (\lambda + 2)(\lambda - 3)(\lambda - 5) - (\lambda + 2)(4) + 2(2)(\lambda - 5)$$

$$+ 2(2)(4) - 3(-4) - 3(-4)(\lambda - 3)$$

Now, after carefully collecting terms, the characteristic equation $|\lambda I_3 - A| = 0$, has the form

$$\lambda^3 - 6\lambda^2 + 11\lambda - 6 = 0 \tag{3}$$

4. Thus, the eigenvalues of A are the solutions of the preceding Eq. (3). To solve this equation we shall make use of the fact that all integer solutions (if any) of the polynomial equation

$$\lambda^n + c_1\lambda^{n-1} + \cdots + c_{n-1}\lambda + c_n = 0 \tag{4}$$

with integer coefficients (i.e., $c_1, c_2, \ldots, c_n$ are integers) must be divisors of c_n. Thus, the only possible integer solutions of Eq. (3) are the divisors of -6, namely, $\pm 1, \pm 2, \pm 3$, and ± 6. Using trial and error, we find that $\lambda = 1$ satisfies Eq. (3), and thus $\lambda_1 = 1$ is one eigenvalue. Since $\lambda = 1$ is a solution of Eq. (3) it follows, from algebra, that $(\lambda - 1)$ must be a factor of the left side of Eq. (3). Dividing $(\lambda - 1)$ into the left side of Eq. (3) yields

$$(\lambda - 1)(\lambda^2 - 5\lambda + 6) = 0 \tag{5}$$

Thus, the remaining solutions of Eq. (3) satisfy the quadratic equation

$$\lambda^2 - 5\lambda + 6 = 0 \tag{6}$$

Solve Eq. (6) for the remaining eigenvalues.

-- -- -- -- -- -- -- -- -- --

Solving by using the quadratic formula, or by using factoring, yields

$$\lambda_2 = 2, \qquad \lambda_3 = 3$$

as the remaining eigenvalues.

5. Now to find the eigenvectors for the Example of frame 3, we make use of Eq. (iii) of frame 3, which we restate here now:

$$(\lambda I_n - A)\mathbf{x} = \mathbf{0} \tag{iii}$$

Example (a): Find an eigenvector corresponding to $\lambda_1 = 1$ for the Example of frame 3.

Solution: For this example, with $\lambda_1 = 1$, Eq. (iii) becomes

$$
\begin{bmatrix} 1+2 & -2 & -3 \\ 2 & 1-3 & -2 \\ 4 & -2 & 1-5 \end{bmatrix} \begin{bmatrix} x_1 \\ x_2 \\ x_3 \end{bmatrix} = \begin{bmatrix} 0 \\ 0 \\ 0 \end{bmatrix} \tag{1}
$$

Doing the matrix multiplication on the left, and simplifying, yields the following homogeneous system of equations:

$$3x_1 - 2x_2 - 3x_3 = 0 \tag{2a}$$

$$2x_1 - 2x_2 - 2x_3 = 0 \tag{2b}$$

$$4x_1 - 2x_2 - 4x_3 = 0 \tag{2c}$$

From the augmented coefficient matrix for Eqs. (2a), (2b), and (2c), we obtain the following reduced row echelon form:

$$
\left[\begin{array}{ccc|c} 1 & 0 & -1 & 0 \\ 0 & 1 & 0 & 0 \\ 0 & 0 & 0 & 0 \end{array} \right] \tag{3}
$$

The equations corresponding to (3) are

$$x_1 - x_3 = 0 \quad \text{and} \quad x_2 = 0.$$

Thus, a solution is $x_1 = r$, $x_2 = 0$, and $x_3 = r$, or, in vector form, $\mathbf{x} = [r, 0, r]^t = r[1, 0, 1]^t$. Taking $r = 1$ yields $\mathbf{x}_1 = [1, 0, 1]^t$ as an eigenvector corresponding to $\lambda_1 = 1$.

Example (b): Now find eigenvectors corresponding to $\lambda_2 = 2$ and $\lambda_3 = 3$ for the Example of frame 3.

Solution: For $\lambda_2 = 2$, we first form Eq. (iii) for $\lambda_2 = 2$, that is, $(2I_3 - A)\mathbf{x} = 0$. The corresponding augmented coefficient matrix is,

$$
\left[\begin{array}{ccc|c} 4 & -2 & -3 & 0 \\ 2 & -1 & -2 & 0 \\ 4 & -2 & -3 & 0 \end{array} \right], \text{ and the reduced row echelon form is}
$$

$$
\left[\begin{array}{ccc|c} 1 & -\frac{1}{2} & 0 & 0 \\ 0 & 0 & 1 & 0 \\ 0 & 0 & 0 & 0 \end{array} \right]
$$

From this we obtain $x_1 = x_2/2$, $x_2 = r$, $x_3 = 0$. A vector solution is $\mathbf{x} = [r/2, r, 0]^t$, and taking $r = 2$ yields $\mathbf{x}_2 = [1, 2, 0]^t$ as an eigenvector corresponding to $\lambda_2 = 2$ (we choose $r = 2$ to cause $\mathbf{x}_2$ to have integer components.)

For $\lambda_3 = 3$, $(\lambda_3 I_3 - A)\mathbf{x} = \mathbf{0}$ leads to the augmented coefficient matrix

$$\begin{bmatrix} 5 & -2 & -3 & 0 \\ 2 & 0 & -2 & 0 \\ 4 & -2 & -2 & 0 \end{bmatrix}.$$ Now finish the work.

--- --- --- --- --- --- ---

The reduced row echelon form is

$$\begin{bmatrix} 1 & 0 & -1 & 0 \\ 0 & 1 & -1 & 0 \\ 0 & 0 & 0 & 0 \end{bmatrix}$$

and from this we obtain $\mathbf{x} = [r, r, r]^t$. Letting $r = 1$ leads to eigenvector $\mathbf{x}_3 = [1, 1, 1]^t$ corresponding to $\lambda_3 = 3$.

6. The eigenvectors of an $n \times n$ matrix A corresponding to a particular eigenvalue λ are the nonzero vectors that satisfy $A\mathbf{x} = \lambda\mathbf{x}$. Put differently, the eigenvectors corresponding to eigenvalue λ are the nonzero vectors in the solution space of $(\lambda I_n - A)\mathbf{x} = \mathbf{0}$. (This solution space is a subspace of R^n. To see this, let $B = \lambda I_n - A$, and refer to frames 6 and 18 of Chapter Five.)

Definition 7.3

The set consisting of all eigenvectors of A corresponding to a particular eigenvalue λ, as well as the zero vector, is a subspace of R^n called the *eigenspace* of A corresponding to λ. [In other words, the eigenspace is the solution space of $(\lambda I_n - A)\mathbf{x} = \mathbf{0}$.]

Note that every vector space (and an eigenspace is a vector space) must contain the zero vector. The next example illustrates how to find a basis of eigenvectors for an eigenspace.

Example: Find bases for the respective eigenspaces of

$$A = \begin{bmatrix} 3 & 0 & 0 \\ 0 & 1 & -2 \\ 0 & -2 & 1 \end{bmatrix}.$$

Solution: The characteristic equation (show) of A is $(\lambda - 3)^2(\lambda + 1) = 0$, and so the eigenvalues are $\lambda = 3$ and $\lambda = -1$. First, let us focus on the eigenspace associated with $\lambda = 3$. To find eigenvectors of A corresponding to $\lambda = 3$, we find the general vector solution of $(3I_3 - A)\mathbf{x} = \mathbf{0}$. This leads to the reduced row echelon form $\begin{bmatrix} 0 & 1 & 1 & | & 0 \\ 0 & 0 & 0 & | & 0 \\ 0 & 0 & 0 & | & 0 \end{bmatrix}$, which means that x_1 is

arbitrary (or $x_1 = r$, say), and likewise for x_3, and $x_2 = -x_3$. Put differently, we can express the vector solution as $\mathbf{x} = [r, -s, s]^t$, or, separating the "$r$ and s parts,"

$$\mathbf{x} = \begin{bmatrix} r \\ -s \\ s \end{bmatrix} = r\begin{bmatrix} 1 \\ 0 \\ 0 \end{bmatrix} + s\begin{bmatrix} 0 \\ -1 \\ 1 \end{bmatrix}$$

Thus, we obtain the eigenvectors $\mathbf{x}_1 = [1, 0, 0]^t$ and $\mathbf{x}_2 = [0, -1, 1]^t$ by respectively letting $r = 1$, $s = 0$ and then $r = 0$, $s = 1$ in the preceding vector equation. Equivalently, these eigenvectors are the vectors that multiply r and s, respectively, in the preceding equation. The vectors $\mathbf{x}_1$ and $\mathbf{x}_2$, which are independent, comprise a basis for the eigenspace corresponding to $\lambda = 3$. The dimension of this eigenspace is thus 2.

Now, find eigenvectors of A corresponding to $\lambda = -1$, and then determine a basis for the corresponding eigenspace. Remember, you have to solve the equation $(-I_3 - A)\mathbf{x} = \mathbf{0}$ for a general vector solution first.

- - - - - - - - - -

The related reduced row echelon form is $\begin{bmatrix} 1 & 0 & 0 & | & 0 \\ 0 & 1 & -1 & | & 0 \\ 0 & 0 & 0 & | & 0 \end{bmatrix}$, which means

that the general vector solution is

$$\mathbf{x} = \begin{bmatrix} 0 \\ t \\ t \end{bmatrix} = t\begin{bmatrix} 0 \\ 1 \\ 1 \end{bmatrix}.$$

This means that the single eigenvector $\mathbf{x}_3 = [0, 1, 1]^t$ (set $t = 1$) comprises a basis for the eigenspace corresponding to $\lambda = -1$. The dimension of this eigenspace hence, is 1.

7.   For a linear operator $L: V \to V$, we can also define eigenvalues and eigenvectors. A scalar λ is an *eigenvalue* of the linear operator $L: V \to V$ if there exists a nonzero vector $\mathbf{x}$ in V such that $L(\mathbf{x}) = \lambda\mathbf{x}$. Such a vector $\mathbf{x}$ is called an eigenvector corresponding to λ. (Observe that we have linear operator L here, and not matrix A.) The calculation of eigenvalues and eigenvectors for a linear operator L is made easy because of the work done in Section D of Chapter Six. We shall next illustrate the connection.

Suppose that V has dimension n and that A is the $n \times n$ matrix of L with respect to *some* basis S of V (see frame 22 of Chapter Six). Then it can be shown that

I.   The eigenvalues of L are the eigenvalues of the matrix A.
II. A vector $\mathbf{x}$ is an eigenvector of L corresponding to λ if and only if its coordinate column vector $[\mathbf{x}]_S$ is an eigenvector of A corresponding to λ.

We illustrate with an example in which the vector space V is the polynomial vector space P_1.

Example: Find the eigenvalues and corresponding eigenvectors of the linear operator $L: P_1 \to P_1$ defined as follows for a typical polynomial $p(t) = a_0 + a_1 t$ in P_1:

$$L(a_0 + a_1 t) = (2a_0 + 3a_1) + a_0 t$$

The left side here is $L(p(t))$.

Solution: The matrix of L with respect to the standard basis (for P_1) $S = \{1, t\}$ is

$$A = \begin{bmatrix} 2 & 3 \\ 1 & 0 \end{bmatrix}$$

(Convince yourself of this. Review frames 20–22 and Self-Test, question 10, of Chapter Six if necessary.) The eigenvalues of L are the eigenvalues of A, and we found these to be $\lambda_1 = 3$ and $\lambda_2 = -1$ in frame 2. Also, in frame 2 we found that corresponding eigenvectors were $\mathbf{x}_1 = [3, 1]^t$ and $\mathbf{x}_2 = [-1, 1]^t$. Now finish the problem. *Hint:* Use the preceding fact II.

— — — — — — — — — —

The column vectors just cited are *coordinate* column vectors with respect to S for the corresponding eigenvectors when expressed as polynomials in P_1. This means that $p_1(t) = 3 + t$ and $p_2(t) = -1 + t$ are eigenvectors in P_1 corresponding to $\lambda_1 = 3$ and $\lambda_2 = -1$, respectively. (Recall that the basis chosen for P_1 is $S = \{1, t\}$.)

Note: Henceforth in this chapter we shall again focus on the eigenvalue problem associated with $L: R^n \to R^n$, where we express $L(\mathbf{x})$ by $A\mathbf{x}$. Here, $n \times n$ matrix A is the standard matrix corresponding to L. The eigenvalue problem can then be posed by the equation $A\mathbf{x} = \lambda\mathbf{x}$. The results we obtain by studying $A\mathbf{x} = \lambda\mathbf{x}$ can be translated to the more general eigenvalue problem for $L: V \to V$ [and posed by the equation $L(\mathbf{x}) = \lambda\mathbf{x}$] by using ideas and techniques from the current frame.

C. DIAGONALIZATION

8. It is possible for the characteristic equation to have solutions that are not real, as the following example illustrates.

Example (a): Find the eigenvalues of the matrix $A = \begin{bmatrix} -3 & -2 \\ 4 & 1 \end{bmatrix}$.

Solution: The characteristic equation $|\lambda I_2 - A| = 0$ reduces to $\lambda^2 + 2\lambda + 5 = 0$. Solving this quadratic equation leads to the complex eigenvalues

$$\lambda_1 = -1 + 2i \quad \text{and} \quad \lambda_2 = -1 - 2i,$$

where $i = \sqrt{-1}$. We could now solve $(\lambda I_2 - A)\mathbf{x} = \mathbf{0}$ for eigenvectors corresponding to λ_1 and λ_2, respectively, in much the same way as we have done already. We would find that the eigenvectors have complex components.

In the statement of Theorem 7.1 (frame 3) we did not allow for complex eigenvalues. This will continue to be our policy in this book.

It should be noted that complex eigenvalues and eigenvectors are of considerable theoretical and of some practical importance.

Now we shall consider the concept of similar matrices. This will lead to new insights and practical tools pertaining to eigenvalue problems.

Definition 7.4 (Similar Matrices)

Suppose A and B are both $n \times n$ matrices. We say that B *is similar to A* if there is an invertible matrix P such that

$$B = P^{-1}AP. \tag{I}$$

Note that if we first premultiply $B = P^{-1}AP$ by P and then postmultiply the resulting equation by P^{-1}, we obtain $A = PBP^{-1}$. Letting $P^{-1} = Q$ means that $P = (P^{-1})^{-1} = Q^{-1}$. Thus, we can write $A = PBP^{-1}$ as

$$A = Q^{-1}BQ, \tag{Ia}$$

which means that A is similar to B. Thus, A is similar to B if and only if B is similar to A, and henceforth we will often simply say that A *and B are similar*.

Note: If A and B are similar, then $|B| = |A|$. The proof is as follows. Taking determinants in (I) yields $|B| = |P^{-1}||A||P|$ from the determinant product rule (Theorem 3.4 in frame 7). Now the product of the two numbers $|P^{-1}|$ times $|P|$—remember, determinants are numbers—equals 1 from Theorem 3.5 (frame 8). Thus, $|B| = |A|$.

Example (b): Suppose $A = \begin{bmatrix} 2 & 3 \\ 1 & 0 \end{bmatrix}$, which is the matrix of frames 1 and 2.

Consider the invertible matrix $P = \begin{bmatrix} 2 & -1 \\ 2 & 3 \end{bmatrix}$, which was picked at random.

Using a method from Chapter Two or Three, we see that
$P^{-1} = \dfrac{1}{8}\begin{bmatrix} 3 & 1 \\ -2 & 2 \end{bmatrix}$. Thus,

$$B = P^{-1}AP = \begin{bmatrix} 4 & \frac{5}{2} \\ -2 & -2 \end{bmatrix},$$

and matrix B is thus similar to matrix A. Checking, we see that $|B| = |A| = -3$.

Example (c): Let $A = \begin{bmatrix} 2 & 3 \\ 1 & 0 \end{bmatrix}$ again. Now let $\hat{P} = \begin{bmatrix} 3 & -1 \\ 1 & 1 \end{bmatrix}$. (Observe that the columns of $\hat{P}$ are eigenvectors of A.) Now find the matrix $\hat{B} = (\hat{P})^{-1}A\hat{P}$, which is also similar to matrix A.

— — — — — — — — — —

First, we calculate $(\hat{P})^{-1} = \dfrac{1}{4}\begin{bmatrix} 1 & 1 \\ -1 & 3 \end{bmatrix}$, and then

$$\hat{B} = (\hat{P})^{-1}A\hat{P} = \begin{bmatrix} 3 & 0 \\ 0 & -1 \end{bmatrix}.$$

Thus, $\hat{B}$ is also similar to A. (Checking, we see that $|\hat{B}| = |A| = -3$.)

9. Observe that $\hat{B}$ is a diagonal matrix (see Definition 1.7 in frame 21; this means the nondiagonal entries are zeros). Also, the diagonal entries of $\hat{B}$ are the eigenvalues of matrix A (see frames 1 and 2). This is no accident as we shall see later! Diagonal matrices have a strong tie-in with eigenvalue problems. In particular, one of the key results of this chapter will be to show that if the $n \times n$ matrix A has n linearly independent eigenvectors, then A is similar to a diagonal matrix whose diagonal entries are the eigenvalues of A.

Before getting back to eigenvalue problems, let us continue our digression on similar matrices. In particular, we shall now study some results that have great practical value (as will be seen in Chapter Eight, for example).

Theorem 7.2

    If A and B are similar, then the powers A^k and B^k (k is a nonnegative integer) are related by $B^k = P^{-1}A^kP$ or $A^k = PB^kP^{-1}$.

We shall demonstrate for $k = 2$. The key steps involve regrouping of terms, which is justified by the associative law of matrix multiplication [Theorem 1.3, part (a)]. First, $B = P^{-1}AP$, since B and A are similar. Then

$$B^2 = BB = (P^{-1}AP)(P^{-1}AP) = (P^{-1}A)(PP^{-1})(AP)$$
$$= (P^{-1}A)(I_n)(AP) = (P^{-1}A)(AP) = P^{-1}A^2P.$$

That is, $B^2 = P^{-1}A^2P$. Note the step $PP^{-1} = I_n$ involving the identity matrix in the preceding sequence of equations.

Thus, we have verified part of Theorem 7.2 for the case $k = 2$. Multiplying on the left by P and on the right by P^{-1} leads to the other part for $k = 2$, namely, $A^2 = PB^2P^{-1}$.

An important special case occurs if A is similar to a diagonal matrix D. The kth power of a diagonal matrix

$$D = \begin{bmatrix} d_1 & 0 & \cdots & 0 \\ 0 & d_2 & \cdots & 0 \\ \cdots & \cdots & \cdots & \cdot \\ \cdots & \cdots & \cdots & \cdot \\ 0 & 0 & \cdots & d_n \end{bmatrix} \qquad (1)$$

is given by

$$D^k = \begin{bmatrix} (d_1)^k & 0 & \cdots & 0 \\ 0 & (d_2)^2 & \cdots & 0 \\ \cdots & \cdots & \cdots & \cdots \\ \cdots & \cdots & \cdots & \cdots \\ 0 & 0 & & (d_n)^k \end{bmatrix} \qquad (2)$$

That is, D^k is itself a diagonal matrix, and the ith diagonal entry of D^k is equal to the kth power of the ith diagonal entry of D. Here is the demonstration of the calculation of D^2 for the case of a 2×2 matrix D:

$$D^2 = DD = \begin{bmatrix} d_1 & 0 \\ 0 & d_2 \end{bmatrix} \begin{bmatrix} d_1 & 0 \\ 0 & d_2 \end{bmatrix}$$
$$= \begin{bmatrix} (d_1)^2 + 0^2 & d_1(0) + 0(d_2) \\ 0(d_1) + d_2(0) & 0^2 + (d_2)^2 \end{bmatrix}$$
$$= \begin{bmatrix} (d_1)^2 & 0 \\ 0 & (d_2)^2 \end{bmatrix}.$$

Now if matrix A is similar to a diagonal matrix D, we have $A = PDP^{-1}$ for some invertible matrix P, and

$$A^k = PD^kP^{-1} \qquad (3)$$

with D^k given by Eq. (2) above. We summarize these ideas in Theorem 7.2a, which will then be illustrated with an Example.

Theorem 7.2a

Suppose matrix A is similar to diagonal matrix D; that is, $D = P^{-1}AP$ and $A = PDP^{-1}$. Then $A^k = PD^kP^{-1}$. The matrix D^k is also a diagonal matrix and its ith diagonal entry is given by $(d_i)^k$, where d_i is the ith diagonal entry of D.

Example: From Example (c) of frame 8, $A = \begin{bmatrix} 2 & 3 \\ 1 & 0 \end{bmatrix}$ is similar to diagonal

matrix $D = \begin{bmatrix} 3 & 0 \\ 0 & -1 \end{bmatrix}$ with P (formerly $\hat{P}$) given by $P = \begin{bmatrix} 3 & -1 \\ 1 & 1 \end{bmatrix}$. Compute A^5 by making use of Eqs. (2) and (3) above.

Solution: First, from Eq. (2),

$$D^5 = \begin{bmatrix} (3)^5 & 0 \\ 0 & (-1)^5 \end{bmatrix} = \begin{bmatrix} 243 & 0 \\ 0 & -1 \end{bmatrix}.$$

Thus, from Eq. (3), with $k = 5$,

$$A^5 = PD^5P^{-1} = \begin{bmatrix} 3 & -1 \\ 1 & 1 \end{bmatrix}\begin{bmatrix} 243 & 0 \\ 0 & -1 \end{bmatrix}\left(\frac{1}{4}\begin{bmatrix} 1 & 1 \\ -1 & 3 \end{bmatrix}\right)$$

Now complete the computations.

———————————

First we shift the 1/4 factor to the left. Then matrix multiplication from the right yields

$$A^5 = \frac{1}{4}\begin{bmatrix} 3 & -1 \\ 1 & 1 \end{bmatrix}\begin{bmatrix} 243 & 243 \\ 1 & -3 \end{bmatrix}$$

$$= \frac{1}{4}\begin{bmatrix} 728 & 732 \\ 244 & 240 \end{bmatrix} = \begin{bmatrix} 182 & 183 \\ 61 & 60 \end{bmatrix}$$

Note that the direct computation of A^5 by using $AAAAA$ would be very time consuming. Also, observe that little additional effort would be required if the preceding technique were used to compute a larger power of A (say A^{15}).

10.

Definition 7.5

A matrix A is *diagonalizable* if it is similar to a diagonal matrix, that is, if there is an invertible matrix P such that $P^{-1}AP$ is a diagonal matrix. In such a case we also say that A *can be diagonalized* and that P *diagonalizes* A.

The matrix $A = \begin{bmatrix} 2 & 3 \\ 1 & 0 \end{bmatrix}$ in the Example of frame 9 is diagonalizable, since it is similar to diagonal matrix $D = \begin{bmatrix} 3 & 0 \\ 0 & -1 \end{bmatrix}$. Also, $P = \begin{bmatrix} 3 & -1 \\ 1 & 1 \end{bmatrix}$ diagonalizes A.

One of the most important theorems of this chapter follows (to be partially proved shortly, for a special case):

Theorem 7.3

The $n \times n$ matrix A is diagonalizable if and only if it has n linearly independent eigenvectors. That is, $P^{-1}AP = D$, where D is diagonal matrix and P is an invertible matrix if and only if A has n linearly independent eigenvectors. Also, the diagonal entries of D are the eigenvalues of A, and P is a matrix whose columns are n linearly independent eigenvectors of A. (The order of eigenvectors in P corresponds to the order of eigenvalues in D.)

Note: This set of n linearly independent vectors of course constitutes a basis for R^n. This follows from Theorem 5.9 (frame 16) since R^n has dimension n.

Example (a): Refer to Example (c) of frame 8 and the Example of frame 9. Matrix $A = \begin{bmatrix} 2 & 3 \\ 1 & 0 \end{bmatrix}$ is similar to $D = \begin{bmatrix} 3 & 0 \\ 0 & -1 \end{bmatrix}$, with $P = \begin{bmatrix} 3 & -1 \\ 1 & 1 \end{bmatrix}$. Observe, from frame 2, that the eigenvalues are $\lambda_1 = 3$ and $\lambda_2 = -1$ and corresponding eigenvectors are $\mathbf{x}_1 = \begin{bmatrix} 3 \\ 1 \end{bmatrix}$ and $\mathbf{x}_2 = \begin{bmatrix} -1 \\ 1 \end{bmatrix}$.

To prove Theorem 7.3 one has to make use of the parts of the following theorem, which deals with algebraic properties of matrix multiplication.

Theorem 7.4

Given the $n \times n$ matrices A, P, and D, where D is a diagonal matrix whose diagonal entries are $d_1, d_2, \ldots, d_n$. Let the columns of P be denoted by $\mathbf{p}_1, \mathbf{p}_2, \ldots, \mathbf{p}_n$ (these are column vectors), where $\mathbf{p}_1 = [p_{11}, p_{21}, \ldots, p_{n1}]^t$, and so on, for $\mathbf{p}_2, \mathbf{p}_3$, etc. Thus, we can write $P = [\mathbf{p}_1 \mid \mathbf{p}_2 \mid \ldots \mid \mathbf{p}_n]$.

(a)   Then AP can be written as

$$AP = [A\mathbf{p}_1 \mid A\mathbf{p}_2 \mid \ldots \mid A\mathbf{p}_n],$$

where column j is equal to $A\mathbf{p}_j$ and $A\mathbf{p}_j$ is the column vector that results by multiplying matrix A by column vector $\mathbf{p}_j$.

(b)   Also, the matrix product PD can be written as

$$PD = \begin{bmatrix} d_1\,p_{11} & d_2\,p_{12} & \ldots & d_n\,p_{1n} \\ d_1\,p_{21} & d_2\,p_{22} & \ldots & d_n\,p_{2n} \\ \ldots\ldots\ldots\ldots\ldots\ldots\ldots \\ \ldots\ldots\ldots\ldots\ldots\ldots\ldots \\ d_1\,p_{n1} & d_2\,p_{n2} & \ldots & d_n\,p_{nn} \end{bmatrix}$$

$$= [d_1\mathbf{p}_1 \mid d_2\mathbf{p}_2 \mid \ldots \mid d_n\mathbf{p}_n]$$

That is, the columns of PD are $d_1\mathbf{p}_1$, $d_2\mathbf{p}_2$, etc.

Example (b): Demonstrate parts (a) and (b) of Theorem 7.4 for the 2×2 matrices $A = \begin{bmatrix} a_{11} & a_{12} \\ a_{21} & a_{22} \end{bmatrix}$, $P = \begin{bmatrix} p_{11} & p_{12} \\ p_{21} & p_{22} \end{bmatrix}$, $D = \begin{bmatrix} d_1 & 0 \\ 0 & d_2 \end{bmatrix}$. Note that we can write $P = [\mathbf{p}_1 \mid \mathbf{p}_2]$, where $\mathbf{p}_1 = \begin{bmatrix} p_{11} \\ p_{21} \end{bmatrix}$ and $\mathbf{p}_2 = \begin{bmatrix} p_{12} \\ p_{22} \end{bmatrix}$.

Solution: Part (a): Carrying out the matrix multiplication of A times P, we get

$$AP = \begin{bmatrix} a_{11} & a_{12} \\ a_{21} & a_{22} \end{bmatrix} \begin{bmatrix} p_{11} & p_{12} \\ p_{21} & p_{22} \end{bmatrix} \tag{1}$$

$$= \begin{bmatrix} a_{11}\,p_{11} + a_{12}\,p_{21} & a_{11}\,p_{12} + a_{12}\,p_{22} \\ a_{21}\,p_{11} + a_{22}\,p_{21} & a_{21}\,p_{12} + a_{22}\,p_{22} \end{bmatrix}$$

Also,

$$A\mathbf{p}_1 = \begin{bmatrix} a_{11}\,p_{11} + a_{12}\,p_{21} \\ a_{21}\,p_{11} + a_{22}\,p_{21} \end{bmatrix} \quad \text{and} \tag{2}$$

$$A\mathbf{p}_2 = \begin{bmatrix} a_{11}\,p_{12} + a_{12}\,p_{22} \\ a_{21}\,p_{12} + a_{22}\,p_{22} \end{bmatrix} \tag{3}$$

Here $A\mathbf{p}_1$ and $A\mathbf{p}_2$ are the two-component column vectors that result when A is multiplied by $\mathbf{p}_1$ and $\mathbf{p}_2$, respectively. Thus, from (1), (2), and (3), we see that

$$AP = [A\mathbf{p}_1 \mid A\mathbf{p}_2]. \tag{4}$$

Part (b):   Carrying out the multiplication of matrix P times matrix D,

$$PD = \begin{bmatrix} p_{11} & p_{12} \\ p_{21} & p_{22} \end{bmatrix} \begin{bmatrix} d_1 & 0 \\ 0 & d_2 \end{bmatrix}$$

$$= \begin{bmatrix} d_1\,p_{11} & d_2\,p_{12} \\ d_1\,p_{21} & d_2\,p_{22} \end{bmatrix} \tag{5}$$

Now finish the work.

— — — — — — — — — —

First, we see that

$$d_1\mathbf{p}_1 = d_1\begin{bmatrix} p_{11} \\ p_{21} \end{bmatrix} = \begin{bmatrix} d_1p_{11} \\ d_1p_{21} \end{bmatrix} \quad \text{and} \tag{6}$$

$$d_2\mathbf{p}_2 = \begin{bmatrix} d_2\,p_{12} \\ d_2\,p_{22} \end{bmatrix} \tag{7}$$

Thus, by substituting (6) and (7) into the right side of Eq. (5), we obtain

$$\begin{bmatrix} d_1\,p_{11} & d_2\,p_{12} \\ d_1\,p_{21} & d_2\,p_{22} \end{bmatrix} = [d_1\mathbf{p}_1 \mid d_2\mathbf{p}_2]. \tag{8}$$

Thus, $PD = [d_1\mathbf{p}_1 \mid d_2\mathbf{p}_2]$, and this is what we wanted to show.

11.   We are now in a position to prove Theorem 7.3. We shall do the proof for the case of $n = 3$ (the general case is no more difficult) and for the "if" direction. That is, we start by assuming that A has n linearly independent eigenvectors.

Proof of Theorem 7.3 (for n = 3 and for "If" Direction)

By hypothesis, A has three linearly independent eigenvectors $\mathbf{p}_1$, $\mathbf{p}_2$, $\mathbf{p}_3$, with corresponding eigenvalues λ_1, λ_2, and λ_3. Let P be the matrix whose columns are $\mathbf{p}_1$, $\mathbf{p}_2$, and $\mathbf{p}_3$, respectively, and label $\mathbf{p}_1 = [p_{11}, p_{21}, p_{31}]^t$, etc. Thus,

$$P = [\mathbf{p}_1 | \mathbf{p}_2 | \mathbf{p}_3] = \begin{bmatrix} p_{11} & p_{12} & p_{13} \\ p_{21} & p_{22} & p_{23} \\ p_{31} & p_{32} & p_{33} \end{bmatrix} \tag{1}$$

By part (a) of Theorem 7.4, the columns of AP are $A\mathbf{p}_1$, $A\mathbf{p}_2$, and $A\mathbf{p}_3$. That is,

$$AP = [A\mathbf{p}_1 | A\mathbf{p}_2 | A\mathbf{p}_3] \tag{2}$$

But

$$A\mathbf{p}_j = \lambda_j \mathbf{p}_j \quad \text{for } j = 1, 2, 3, \tag{3}$$

since $\mathbf{p}_j$ is an eigenvector corresponding to eigenvalue λ_j.
Thus,

$$AP = [\lambda_1 \mathbf{p}_1 | \lambda_2 \mathbf{p}_2 | \lambda_3 \mathbf{p}_3]. \tag{4}$$

Applying Theorem 7.4, part (b), to the right side of Eq. (4), we see that we can rewrite Eq. (4) as

$$AP = PD \tag{5}$$

where

$$D = \begin{bmatrix} \lambda_1 & 0 & 0 \\ 0 & \lambda_2 & 0 \\ 0 & 0 & \lambda_3 \end{bmatrix}. \tag{6}$$

Since the column vectors of P are linearly independent, P is invertible (Theorem 5.14 in frame 21); that is, P^{-1} exists. Premultiplying Eq. (5) by P^{-1} yields

$$P^{-1}AP = D. \tag{7}$$

Note: To prove the "only if" part of Theorem 7.3, we start by assuming that A is diagonalizable. The proof is similar to the proof just done, except that the order of steps is basically reversed; it appears in Anton (1977).

Example (a): Find a matrix P that diagonalizes the matrix

$$A = \begin{bmatrix} -2 & 2 & 3 \\ -2 & 3 & 2 \\ -4 & 2 & 5 \end{bmatrix} \text{ of frames 3–5.}$$

Solution: We found the following eigenvalues and corresponding eigenvectors in frames 3–5: For $\lambda_1 = 1$, $\mathbf{x}_1 = \begin{bmatrix} 1 \\ 0 \\ 1 \end{bmatrix}$; for $\lambda_2 = 2$, $\mathbf{x}_2 = \begin{bmatrix} 1 \\ 2 \\ 0 \end{bmatrix}$; for $\lambda_3 = 3$, $\mathbf{x}_3 = \begin{bmatrix} 1 \\ 1 \\ 1 \end{bmatrix}$. It is easy to check that $\{\mathbf{x}_1, \mathbf{x}_2, \mathbf{x}_3\}$ is linearly independent; for example, use Theorem 5.5 in frame 13. Thus,

$$P = [\mathbf{x}_1 \mid \mathbf{x}_2 \mid \mathbf{x}_3] = \begin{bmatrix} 1 & 1 & 1 \\ 0 & 2 & 1 \\ 1 & 0 & 1 \end{bmatrix}$$

diagonalizes A. As a check, we see that

$$P^{-1}AP = \begin{bmatrix} 2 & -1 & -1 \\ 1 & 0 & -1 \\ -2 & 1 & 2 \end{bmatrix} \begin{bmatrix} -2 & 2 & 3 \\ -2 & 3 & 2 \\ -4 & 2 & 5 \end{bmatrix} \begin{bmatrix} 1 & 1 & 1 \\ 0 & 2 & 1 \\ 1 & 0 & 1 \end{bmatrix} = \begin{bmatrix} 1 & 0 & 0 \\ 0 & 2 & 0 \\ 0 & 0 & 3 \end{bmatrix}.$$

Note that the order of the eigenvectors in P corresponds to the order of eigenvalues in the diagonal matrix. Thus, interchanging the first two columns of P would lead to $\hat{P} = \begin{bmatrix} 1 & 1 & 1 \\ 2 & 0 & 1 \\ 0 & 1 & 1 \end{bmatrix}$, and thus $(\hat{P})^{-1}A\hat{P} = \begin{bmatrix} 2 & 0 & 0 \\ 0 & 1 & 0 \\ 0 & 0 & 3 \end{bmatrix}$.

Now let us consider the question of when a square matrix is diagonalizable. Not all square matrices are diagonalizable, as we shall see in our examples. The following two theorems [proved in Anton (1977)] are helpful.

Theorem 7.5

If $\mathbf{x}_1, \mathbf{x}_2, \ldots, \mathbf{x}_r$ are eigenvectors corresponding to distinct eigenvalues $\lambda_1, \lambda_2, \ldots, \lambda_r$, then $\{\mathbf{x}_1, \mathbf{x}_2, \ldots, \mathbf{x}_r\}$ is a linearly independent set.

Theorem 7.6

The $n \times n$ matrix A is diagonalizable if it has n distinct eigenvalues.

Theorem 7.6 follows directly from Theorems 7.5 and 7.3.

Example (b): Show that $A = \begin{bmatrix} 3 & 5 \\ 1 & 3 \end{bmatrix}$ can be diagonalized. Find the appropriate diagonal matrix D.

Solution: The characteristic equation $|\lambda I_2 - A| = 0$ reduces to $\lambda^2 - 6\lambda + 4 = 0$. Solving for the eigenvalues, which are the solutions of this equation, by using the quadratic formula, yields $\lambda_1 = 3 + \sqrt{5}$ and $\lambda_2 = 3 - \sqrt{5}$. Now finish the problem.

_ _ _ _ _ _ _ _ _ _ _

Since A has two distinct eigenvalues, Theorem 7.6 tells us that A is diagonalizable. Thus,

$$P^{-1}AP = \begin{bmatrix} 3 + \sqrt{5} & 0 \\ 0 & 3 - \sqrt{5} \end{bmatrix}.$$

Note that it was not necessary to solve for the corresponding eigenvectors (and hence P and P^{-1}) in this problem.

12. We continue in our study of when a square matrix is diagonalizable. The key to the answer lies in Theorem 7.3 (frame 10) and in several other important results. Let us henceforth (unless otherwise noted) limit ourselves to the case where all the eigenvalues are real. We already know the answer if all n eigenvalues are real and distinct (Theorem 7.6).

Recall that for an $n \times n$ matrix A, the characteristic polynomial is $|\lambda I_n - A|$ and the characteristic equation is $|\lambda I_n - A| = 0$. See Definition 7.2 in frame 3. A general form for the characteristic polynomial, which displays the fact that the degree is n if A is $n \times n$, is given right after Definition 7.2. The eigenvalues are the roots of the characteristic polynomial (Theorem 7.1 in frame 3). An important fact is that if the eigenvalues are real but not all distinct, then A may or may not be diagonalizable. The characteristic polynomial of A can be written as a product of n factors, each of the form $(\lambda - \lambda_i)$ where λ_i is an eigenvalue. The following theorem is a special case of the Fundamental Theorem of Algebra.

Theorem 7.7

The characteristic polynomial (which is of degree n), namely, $|\lambda I_n - A|$, can be written in the form

$$(\lambda - \lambda_1)^{k_1}(\lambda - \lambda_2)^{k_2} \cdot \ldots \cdot (\lambda - \lambda_r)^{k_r},$$

where $\lambda_1, \lambda_2, \ldots, \lambda_r$ denote the r *distinct* eigenvalues of A ($r \leq n$, of course). The exponent k_j, which is a positive integer, is the *multiplicity* of eigenvalue λ_j for $j = 1, 2, \ldots, r$. Also, the sum of the multiplicities is n, that is,

$$k_1 + k_2 + \cdots + k_r = n.$$

Again, note that the characteristic polynomial has degree n.

Notes: (i)   In the first statement of the preceding theorem we have ordinary multiplication. (ii)   For the special case where there are n distinct eigenvalues, then each k_j is equal to 1, and $|\lambda I_n - A| = (\lambda - \lambda_1)(\lambda - \lambda_2) \cdot \ldots \cdot (\lambda - \lambda_n)$.

Now let us focus on the eigenvalue λ_j, which has multiplicity k_j. The eigenspace of λ_j is the solution space of $(\lambda_j I_n - A)\mathbf{x} = \mathbf{0}$ [see Definition 7.3 in frame 6]. The dimension of the eigenspace is the number of linearly independent (eigen)vectors in a basis for the eigenspace. Denote this dimension by Dim (λ_j). It can be shown that Dim (λ_j) is a positive integer between 1 and k_j inclusive. That is,

$$1 \leq \text{Dim } (\lambda_j) \leq k_j \qquad \text{for } j = 1, 2, \ldots, r. \tag{I}$$

Now if each Dim (λ_j) equals k_j, then the total number of such eigenvectors is n, since $k_1 + k_2 + \cdots + k_r = n$ (here we choose k_j linearly independent eigenvectors from each eigenspace). Moreover, it can be shown that this collection of n eigenvectors is a linearly independent set. Thus, by Theorem 7.3 it follows that A can be diagonalized.

On the other hand, if for some eigenvalue the number of linearly independent eigenvectors is less than the multiplicity, then A cannot be diagonalized. That is, if Dim $(\lambda_j) < k_j$ for some eigenvalue λ_j, then A cannot be diagonalized.

Example: Determine if $A = \begin{bmatrix} 3 & 0 & 0 \\ 4 & 3 & 0 \\ 3 & 0 & -1 \end{bmatrix}$ can be diagonalized.

Solution: The characteristic polynomial (which is of degree 3) is $|\lambda I_3 - A|$ $= (\lambda - 3)^2(\lambda + 1)$, as you should verify. Thus, the eigenvalues of A are

$\lambda = 3$, with multiplicity 2, and $\lambda = -1$, with multiplicity 1. Now consider the eigenspace associated with the eigenvalue $\lambda = 3$. (We wish to find the dimension of this eigenspace.) This consists of the vector solutions of $(3I_3 - A)\mathbf{x} = \mathbf{0}$, that is, of

$$\begin{bmatrix} 0 & 0 & 0 \\ -4 & 0 & 0 \\ -3 & 0 & 4 \end{bmatrix}\begin{bmatrix} x_1 \\ x_2 \\ x_3 \end{bmatrix} = \begin{bmatrix} 0 \\ 0 \\ 0 \end{bmatrix} \quad \text{or} \quad \begin{array}{l} 0x_1 + 0x_2 + 0x_3 = 0 \\ -4x_1 + 0x_2 + 0x_3 = 0 \ . \\ -3x_1 + 0x_2 + 4x_3 = 0 \end{array}$$

The reduced row echelon form is $\left[\begin{array}{ccc|c} 1 & 0 & 0 & 0 \\ 0 & 0 & 1 & 0 \\ 0 & 0 & 0 & 0 \end{array}\right]$, which indicates that $x_1 = 0$,

$x_3 = 0$, and x_2 is *arbitrary*. This means that $\mathbf{x} = [0, r, 0]^t = r[0, 1, 0]^t$ is the form for the general vector solution. Thus, a basis consists of the single eigenvector $[0, 1, 0]^t$. Now finish the problem.

— — — — — — — — — —

Thus, the dimension of the solution space of $(3I_3 - A)\mathbf{x} = \mathbf{0}$ is 1. In other words, the eigenspace for the eigenvalue $\lambda = 3$ has dimension 1. Thus, A cannot be diagonalized, since the eigenvalue $\lambda = 3$ has multiplicity 2.

13.

Example (a): Determine if $A = \begin{bmatrix} 3 & 0 & 3 \\ 0 & 3 & 0 \\ 0 & 0 & -1 \end{bmatrix}$ can be diagonalized.

Solution: The characteristic polynomial here is $(\lambda - 3)^2(\lambda + 1)$, as in the Example of frame 12. Thus, the eigenvalue $\lambda = 3$ has multiplicity 2 and the eigenvalue $\lambda = -1$ has multiplicity 1.

For $\lambda = 3$, the eigenspace consists of vector solutions of $(3I_3 - A)\mathbf{x} = 0$.

From this equation we obtain the reduced row echelon form $\left[\begin{array}{ccc|c} 0 & 0 & 1 & 0 \\ 0 & 0 & 0 & 0 \\ 0 & 0 & 0 & 0 \end{array}\right]$,

which means that $x_3 = 0$ with x_1 and x_2 both arbitrary. Thus, a general vector solution is

$$\mathbf{x} = \begin{bmatrix} r \\ s \\ 0 \end{bmatrix} = r\begin{bmatrix} 1 \\ 0 \\ 0 \end{bmatrix} + s\begin{bmatrix} 0 \\ 1 \\ 0 \end{bmatrix}.$$

Thus, $[1, 0, 0]^t$ and $[0, 1, 0]^t$ are linearly independent eigenvectors corresponding to $\lambda = 3$, and these constitute a basis for the eigenspace of $\lambda = 3$. Thus, the dimension and multiplicity associated with $\lambda = 3$ both equal 2.

For $\lambda = -1$, Eq. (I) of frame 12 reduces to $1 \leq \text{Dim}\ (\lambda_j) \leq 1$, which means that the dimension of the eigenspace associated with $\lambda = -1$ *has to* equal 1.

Just for practice, let us verify this. For $\lambda = -1$, the eigenspace consists of vector solutions of $(-I_3 - A)\mathbf{x} = \mathbf{0}$. This leads to the reduced row echelon

form $\begin{bmatrix} 1 & 0 & \frac{3}{4} & \big| & 0 \\ 0 & 1 & 0 & \big| & 0 \\ 0 & 0 & 0 & \big| & 0 \end{bmatrix}$, which means that $\mathbf{x} = [-3r/4,\ 0,\ r]^t$. Setting $r = 4$ means

that $\mathbf{x} = [-3,\ 0,\ 4]^t$ is an eigenvector associated with $\lambda = -1$. Thus, a basis consists of $[-3,\ 0,\ 4]^t$, and the eigenspace has dimension equal to 1.

In summary, the comments of frame 12 lead us to conclude that A can be diagonalized. Our analysis of the matrix A of this example will continue later in this frame, in Example (b).

Let us consider again the comments of frame 12, which follow Theorem 7.7. Now, Theorem 7.5 of frame 11 is a special case of a more general theorem, which we shall discuss now. Suppose $\lambda_1, \lambda_2, \ldots, \lambda_r$ are distinct eigenvalues, and we determine a linearly independent set of eigenvectors in each of the corresponding eigenspaces (such a set is also a *basis* set for the corresponding eigenspace). If we then combine all these vectors into an overall set, this resulting set will still be linearly independent. (For example, if we combine two linearly independent vectors from one eigenspace with three linearly independent vectors from another eigenspace, we obtain a linearly independent set of five vectors.)

Now, if for each λ_j the total number of linearly independent eigenvectors equals the multiplicity [in symbols, $\text{Dim}\ (\lambda_j) = k_j$ for $j = 1, 2, \ldots, r$], then an overall set containing all such vectors will contain n vectors, since $k_1 + k_2 + \cdots + k_r = n$. By the preceding comments, this overall set will be independent. Thus, in this case, matrix A is diagonalizable.

Another point of interest concerns the concept of rank of a matrix. Remember that the rank of a matrix is the number of leading entries in the reduced row echelon form of that matrix (Section D of Chapter Two). For eigenvalue λ_j, the eigenspace consists of vector solutions of $(\lambda_j I_n - A)\mathbf{x} = \mathbf{0}$. The dimension of this eigenspace is equal to the number of nonleading variables among the variables $x_1, x_2, \ldots, x_n$. We obtain the number of nonleading variables by inspecting the reduced row echelon form of matrix $(\lambda_j I_n - A)$. The number of nonleading variables equals n minus the number of leading variables, and the latter number equals the rank of matrix $(\lambda_j I_n - A)$. Thus, the dimension of the eigenspace corresponding to eigenvalue λ_j is given by

$$\text{Dim}\ (\lambda_j) = n - \text{Rank}\ (\lambda_j I_n - A) \tag{II}$$

Let us illustrate. In Example (a), for $\lambda = 3$, we had $(3I_3 - A) = \begin{bmatrix} 0 & 0 & -3 \\ 0 & 0 & 0 \\ 0 & 0 & 4 \end{bmatrix}$. The corresponding reduced row echelon form is $\begin{bmatrix} 0 & 0 & 1 \\ 0 & 0 & 0 \\ 0 & 0 & 0 \end{bmatrix}$. [Refer to Example (a); delete the last column of the reduced row echelon form of $(3I_3 - A)|0$.] Thus, Rank $(3I_3 - A) = 1$, and from (II), we see that the dimension of the eigenspace for $\lambda = 3$ is 2.

Example (b): For Example (a) we established that matrix A can be diagonalized. Determine the diagonal matrix D to which A is similar. Also, find a form for matrix P in $P^{-1}AP = D$.

Solution: We relabel the *three* eigenvalues $\lambda_1 = 3$, $\lambda_2 = 3$, and $\lambda_3 = -1$. Here $\lambda = 3$ is an eigenvalue of multiplicity 2. Observe that the labeling here differs from that in Theorem 7.7, and in the prior discussion of this frame. Thus, a form for D is

$$D = \begin{bmatrix} 3 & 0 & 0 \\ 0 & 3 & 0 \\ 0 & 0 & -1 \end{bmatrix}.$$

Now determine an appropriate form for P.

- - - - - - - - - - -

A form for P is $P = \begin{bmatrix} 1 & 0 & -3 \\ 0 & 1 & 0 \\ 0 & 0 & 4 \end{bmatrix}$, since the first two columns of P are

linearly independent eigenvectors corresponding to $\lambda = 3$, and the third column is an eigenvector corresponding to $\lambda = -1$ [see Example (a) of this frame].

14. The reader should note that practically all the topics studied in the previous chapters have been used in solving eigenvalue–eigenvector and associated diagonalization problems. Some of these topics are matrix products, reduced row echelon form, rank, matrix inverses, determinants, linear independence, and bases for subspaces.

The key point of the previous two frames is the following (we assume all eigenvalues of A are real here):

A can be diagonalized if and only if, for each eigenvalue of multiplicity k (where $k > 1$), the corresponding eigenspace has dimension equal to k.

Example (a): Determine if matrix $A = \begin{bmatrix} 2 & 2 \\ 0 & 2 \end{bmatrix}$ can be diagonalized.

Solution: The characteristic polynomial of A is $(\lambda - 2)^2$. This means that $\lambda = 2$ is an eigenvalue of multiplicity 2. The eigenspace consists of vector solutions of $(2I_2 - A)\mathbf{x} = \mathbf{0}$, and this has reduced row echelon form $\begin{bmatrix} 0 & 1 & 0 \\ 0 & 0 & 0 \end{bmatrix}$. Thus, eigenvectors are of the form $\begin{bmatrix} r \\ 0 \end{bmatrix}$, which means that a basis consists of $\begin{bmatrix} 1 \\ 0 \end{bmatrix}$, say, and the eigenspace has dimension 1. [From another point of view, using Eq. (II) of frame 13, Rank $(2I_2 - A) = 1$ and Dim (λ) $= 2 - 1 = 1$ for $\lambda = 2$.]

Thus, matrix A cannot be diagonalized. Reviewing, we observe that two linearly independent eigenvectors for A do not exist.

An interesting special case occurs if a matrix is upper triangular or lower triangular.

Definition 7.6

(a) An $n \times n$ matrix A is upper triangular if all entries below the main diagonal are zeros, that is, if $a_{ij} = 0$ for $i > j$.

(b) An $n \times n$ matrix A is lower triangular if all entries above the main diagonal are zeros, that is, $a_{ij} = 0$ for $j > i$.

Example (b): The following matrices are upper triangular:

$$\begin{bmatrix} 3 & 4 \\ 0 & 2 \end{bmatrix}, \quad \begin{bmatrix} 3 & 0 & 3 \\ 0 & 3 & 0 \\ 0 & 0 & -1 \end{bmatrix}, \quad \begin{bmatrix} 7 & 2 & -3 \\ 0 & 6 & 5 \\ 0 & 0 & 4 \end{bmatrix}, \quad \begin{bmatrix} 4 & 0 & 0 \\ 0 & -6 & 0 \\ 0 & 0 & 7 \end{bmatrix}$$

The following matrices are lower triangular:

$$\begin{bmatrix} 7 & 0 \\ 6 & -2 \end{bmatrix}, \quad \begin{bmatrix} 7 & 0 & 0 \\ -2 & 6 & 0 \\ 4 & 0 & 4 \end{bmatrix}, \quad \begin{bmatrix} 11 & 0 & 0 \\ 4 & 0 & 0 \\ -6 & 5 & 9 \end{bmatrix}, \quad \begin{bmatrix} 4 & 0 & 0 \\ 0 & -6 & 0 \\ 0 & 0 & 7 \end{bmatrix}$$

Any diagonal matrix is both upper triangular and lower triangular.

Theorem 7.8

If a matrix is upper or lower triangular, then its eigenvalues are its diagonal entries.

Example (c): Determine if the matrix $A = \begin{bmatrix} 1 & 0 & 0 \\ 0 & 4 & 0 \\ 4 & 0 & 4 \end{bmatrix}$ is diagonalizable. If it is, determine the diagonal matrix D and an appropriate form for matrix P.

Solution: Since the matrix is lower triangular, the eigenvalues are $\lambda_1 = 1$, $\lambda_2 = 4$, and $\lambda_3 = 4$. Thus, $\lambda = 4$ is an eigenvalue of multiplicity 2. The eigenspace for $\lambda = 4$ consists of vector solutions of $(4I_3 - A)\mathbf{x} = \mathbf{0}$, for which

the reduced row echelon form is $\begin{bmatrix} \underline{1} & 0 & 0 & | & 0 \\ 0 & 0 & 0 & | & 0 \\ 0 & 0 & 0 & | & 0 \end{bmatrix}$. This means that eigenvectors

have the form $\mathbf{x} = [0, r, s]^t = r[0, 1, 0]^t + s[0, 0, 1]^t$. Thus, $\mathbf{x}_2 = [0, 1, 0]^t$ and $\mathbf{x}_3 = [0, 0, 1]^t$ are linearly independent eigenvectors, and the eigenspace for $\lambda = 4$ has dimension 2. Now finish the problem.

— — — — — — — — — —

The eigenspace for $\lambda = 1$ consists of vector solutions of $(I_3 - A)\mathbf{x} = \mathbf{0}$,

for which the reduced row echelon form is $\begin{bmatrix} \underline{1} & 0 & \frac{3}{4} & | & 0 \\ 0 & \underline{1} & 0 & | & 0 \\ 0 & 0 & 0 & | & 0 \end{bmatrix}$. Thus, the vector

solutions (eigenvectors) have the form $\mathbf{x} = [-3t/4, 0, t]^t$, and so a particular eigenvector (choose $t = 4$) associated with $\lambda_1 = 1$ is $\mathbf{x}_1 = [-3, 0, 4]^t$.

Thus, A can be diagonalized since the eigenvectors $\mathbf{x}_1$, $\mathbf{x}_2$, and $\mathbf{x}_3$ are linearly independent. [Actually, we knew this after finding two linearly independent eigenvectors corresponding to $\lambda = 4$, the eigenvalue of multiplicity 2. See comments in frame 13, in the solution to Example (a).] Thus, a diagonal matrix is

$$D = \begin{bmatrix} \lambda_1 & 0 & 0 \\ 0 & \lambda_2 & 0 \\ 0 & 0 & \lambda_3 \end{bmatrix} = \begin{bmatrix} 1 & 0 & 0 \\ 0 & 4 & 0 \\ 0 & 0 & 4 \end{bmatrix},$$

and an associated P matrix (such that $P^{-1}AP = D$ — we say P "diagonalizes" matrix A) is

$$P = [\mathbf{x}_1 \mid \mathbf{x}_2 \mid \mathbf{x}_3] = \begin{bmatrix} -3 & 0 & 0 \\ 0 & 1 & 0 \\ 4 & 0 & 1 \end{bmatrix}.$$

D. DIAGONALIZATION OF SYMMETRIC MATRICES

15. Throughout this chapter we have been studying the eigenvalue properties of an $n \times n$ matrix A, all of whose entries were real numbers. We found that, on occasion, some eigenvalues of such a matrix can have complex form [see Example (a) of frame 8]. Actually, if complex eigenvalues do occur at all, they occur in pairs. There is one very important case for which an $n \times n$ matrix A with real entries will have only real eigenvalues. This is so when A is a symmetric matrix! Symmetric matrices arise in many applications.

From Definition 1.6 (frame 21), an $n \times n$ matrix A is symmetric if $a_{ij} = a_{ji}$ for all i and j. In other words, $n \times n$ matrix A is symmetric if $A^t = A$.

Let us list some of the major theorems and results dealing with the eigenvalue and diagonalization properties of symmetric matrices.

Theorem 7.9

All the eigenvalues of a symmetric matrix A are real numbers.

It is useful at this point to recall the discussion of frame 12, which of course applies to symmetric matrices, since it applies to all real-entried matrices. (See, in particular, Theorem 7.7.)

There we indicated that for an $n \times n$ matrix A, the characteristic polynomial (which is of degree n) could be written as

$$(\lambda - \lambda_1)^{k_1}(\lambda - \lambda_2)^{k_2} \cdot \ldots \cdot (\lambda - \lambda_r)^{k_r},$$

where the λ_j's denote the r distinct eigenvalues of A and the positive integer exponent k_j denotes the *multiplicity* of eigenvalue λ_j for $j = 1, 2, \ldots, r$. Also,

$$k_1 + k_2 + \ldots + k_r = n.$$

Recall that in frame 12, we indicated that

$$1 \le \text{Dim}\,(\lambda_j) \le k_j \qquad \text{for } j = 1, 2, \ldots, r. \tag{I}$$

In particular, this says that the dimension of the eigenspace associated with eigenvalue λ_j is less than or equal to k_j. In the case of a symmetric matrix, we have

$$\text{Dim}\,(\lambda_j) = k_j \qquad \text{for } j = 1, 2, \ldots, r. \tag{I$'$}$$

Stating this in theorem form, we have

Theorem 7.10

If eigenvalue λ_j of a symmetric matrix A has multiplicity k_j, then the eigenspace corresponding to λ_j has dimension equal to k_j also.

Now we know that if we combine together all the basis eigenvectors from the different λ_j eigenspaces into a single set, this resulting set will still be linearly independent. Now for the case of a symmetric matrix, the total number of such eigenvectors is n [k_j linearly independent vectors from a basis set for each λ_j eigenspace, and a total of $k_1 + k_2 + \ldots + k_r = n$ linearly independent eigenvectors]. Thus, from master Theorem 7.3 (frame 10), we draw the following conclusion.

Theorem 7.11

The $n \times n$ symmetric matrix A is diagonalizable, since it has n linearly independent eigenvectors. Thus, $P^{-1}AP = D$, which says that A is similar to a diagonal matrix D. The diagonal entries of D are the eigenvalues of A, and P is a matrix whose columns are n linearly independent eigenvectors of A.

Example: For the symmetric matrix $A = \begin{bmatrix} 3 & 2 & 2 \\ 2 & 3 & 2 \\ 2 & 2 & 3 \end{bmatrix}$, determine the eigenvalues and corresponding eigenvectors. Show that A can be diagonalized.

Solution: The characteristic polynomial is (verify) $|\lambda I_3 - A| = (\lambda - 1)^2(\lambda - 7)$. Thus, eigenvalue $\lambda = 1$ has multiplicity 2. By solving $(I_3 - A)\mathbf{x} = \mathbf{0}$ we find, as expected, two linearly independent eigenvectors $\mathbf{x}_1 = [-1, 1, 0]^t$ and $\mathbf{x}_2 = [-1, 0, 1]^t$. This indicates that a basis for the eigenspace of $\lambda = 1$ is $\{\mathbf{x}_1, \mathbf{x}_2\}$ and that the dimension of the eigenspace is 2.
Finish the problem.

— — — — — — — — — — — — —

For $\lambda = 7$ we solve $(7I_3 - A)\mathbf{x} = \mathbf{0}$ and obtain the single eigenvector $\mathbf{x}_3 = [1, 1, 1]^t$. (Thus, a basis for the $\lambda = 7$ eigenspace is $\{\mathbf{x}_3\}$, and the dimension of the eigenspace is 1.)
Let us label the eigenvalues as follows: $\lambda_1 = \lambda_2 = 1$ and $\lambda_3 = 7$. Now A is diagonalizable. It is similar to

$$ D = \begin{bmatrix} \lambda_1 & 0 & 0 \\ 0 & \lambda_2 & 0 \\ 0 & 0 & \lambda_3 \end{bmatrix} = \begin{bmatrix} 1 & 0 & 0 \\ 0 & 1 & 0 \\ 0 & 0 & 7 \end{bmatrix}, $$

with $P = [\mathbf{x}_1 \mid \mathbf{x}_2 \mid \mathbf{x}_3] = \begin{bmatrix} -1 & -1 & 1 \\ 1 & 0 & 1 \\ 0 & 1 & 1 \end{bmatrix}.$

16. An important property associated with symmetric matrices is indicated in the following theorem.

Theorem 7.12

If A is a symmetric matrix, then eigenvectors corresponding to different eigenvalues of A are orthogonal.

We know from Chapters Four (frame 7) and Five (frame 23), that column vectors **u** and **v** are orthogonal (perpendicular) if $\mathbf{u} \cdot \mathbf{v} = 0$. [Here, of course, we have the dot product of vectors **u** and **v**.]

Theorem 7.12 can easily be proved by making use of the following result.

If **x** and **y** are any column vectors in R^n, and A is any $n \times n$ matrix, then

$$(A\mathbf{x}) \cdot \mathbf{y} = \mathbf{x} \cdot (A^t \mathbf{y}) \tag{I}$$

Thus, we can "move the matrix A across the dot product," provided we change it into its transpose. The proof of (I) is both easy and instructive. First, we recall that

$$\mathbf{u} \cdot \mathbf{v} = \mathbf{u}^t \mathbf{v}, \tag{II}$$

as noted in frame 4 of Chapter Four. (Suppose **u** and **v** are n component column vectors. On the right side, we have matrix multiplication of the $1 \times n$ matrix **u** transpose times the $n \times 1$ matrix **v**.) Either side is equal to the sum $u_1 v_1 + u_2 v_2 + \cdots + u_n v_n$. The proof of (I) follows, where the steps (to be explained below) are indicated by letters above the equals signs:

$$
\overset{\text{(a)}}{(A\mathbf{x}) \cdot \mathbf{y}} = \overset{\text{(b)}}{(A\mathbf{x})^t \mathbf{y}} = \overset{\text{(c)}}{(\mathbf{x}^t A^t) \mathbf{y}} = \overset{\text{(d)}}{\mathbf{x}^t (A^t \mathbf{y})} = \mathbf{x} \cdot (A^t \mathbf{y})
$$

(a) First, we use Eq. (II). (b) Then we use the transpose product rule $(AB)^t = B^t A^t$; see Theorem 1.4 in frame 19. (c) Next, apply the associative rule of Theorem 1.3 in frame 13. (d) Then use Eq. (II) again.

Before proving Theorem 7.12, we illustrate it.

Example (a): Show how Theorem 7.12 is verified from inspecting the results of the Example of frame 15.

Solution: The two linearly independent eigenvectors $\mathbf{x}_1 = [-1, 1, 0]^t$ and $\mathbf{x}_2 = [-1, 0, 1]^t$ correspond to $\lambda = 1$ and eigenvector $\mathbf{x}_3 = [1, 1, 1]^t$ corresponds to $\lambda = 7$.

We see that $\mathbf{x}_1 \cdot \mathbf{x}_3 = \mathbf{x}_2 \cdot \mathbf{x}_3 = 0$.

Now let us prove Theorem 7.12.

Proof of Theorem 7.12

Let $\mathbf{x}_1$ and $\mathbf{x}_2$ be eigenvectors of A corresponding to the distinct eigenvalues λ_1 and λ_2 of A. Thus,

$$A\mathbf{x}_1 = \lambda_1 \mathbf{x}_1 \quad \text{and} \quad A\mathbf{x}_2 = \lambda_2 \mathbf{x}_2. \tag{1a), (1b}$$

Now

$$(A\mathbf{x}_1) \cdot \mathbf{x}_2 = (\lambda_1\mathbf{x}_1) \cdot \mathbf{x}_2 = \lambda_1(\mathbf{x}_1 \cdot \mathbf{x}_2). \tag{2}$$

Also,

$$\mathbf{x}_1 \cdot (A\mathbf{x}_2) = \mathbf{x}_1 \cdot (\lambda_2\mathbf{x}_2) = \lambda_2(\mathbf{x}_1 \cdot \mathbf{x}_2). \tag{3}$$

Now, for the left side of (2),

$$(A\mathbf{x}_1) \cdot \mathbf{x}_2 = \mathbf{x}_1 \cdot (A^t\mathbf{x}_2) = \mathbf{x}_1 \cdot (A\mathbf{x}_2), \tag{4}$$

where first we have used (I), and then the condition that $A = A^t$ (since A is symmetric). Thus, the left sides of (2) and (3) are equal, and so the right sides are also. This means that

$$(\lambda_1 - \lambda_2)(\mathbf{x}_1 \cdot \mathbf{x}_2) = 0, \tag{5}$$

and since $\lambda_1 \neq \lambda_2$, we conclude that $\mathbf{x}_1 \cdot \mathbf{x}_2 = 0$.

Example (b): For symmetric matrix $A = \begin{bmatrix} 4 & 0 & 0 \\ 0 & 2 & 3 \\ 0 & 3 & 2 \end{bmatrix}$, find the eigenvalues and corresponding eigenvectors. Display the diagonal matrix D to which A is similar. Also, verify Theorem 7.12.

Solution: The characteristic polynomial is $(\lambda - 4)(\lambda - 5)(\lambda + 1)$, which means that the eigenvalues are $\lambda_1 = -1$, $\lambda_2 = 4$, and $\lambda_3 = 5$. Solving $(\lambda I_3 - A)\mathbf{x} = \mathbf{0}$ after substituting the respective values for the three preceding eigenvalues, we find (verify) corresponding eigenvectors to be

$$\mathbf{x}_1 = [0, -1, 1]^t, \quad \mathbf{x}_2 = [1, 0, 0]^t, \quad \mathbf{x}_3 = [0, 1, 1]^t.$$

Now finish the problem.

— — — — — — — — — —

A is similar to

$$D = \begin{bmatrix} \lambda_1 & 0 & 0 \\ 0 & \lambda_2 & 0 \\ 0 & 0 & \lambda_3 \end{bmatrix} = \begin{bmatrix} -1 & 0 & 0 \\ 0 & 4 & 0 \\ 0 & 0 & 5 \end{bmatrix}$$

We see that $\mathbf{x}_1 \cdot \mathbf{x}_2 = \mathbf{x}_1 \cdot \mathbf{x}_3 = \mathbf{x}_2 \cdot \mathbf{x}_3 = 0$, thus verifying Theorem 7.12.

17.   Now Theorem 7.12 tells us that eigenvectors corresponding to different eigenvalues are perpendicular. We know we can produce a set of n linearly independent eigenvectors if A is a symmetric matrix (Theorem 7.11 of frame 15). Let us focus on the particular eigenspace associated with the eigenvalue λ_j, which has multiplicity k_j. From Theorem 7.10 (frame 15) we know we can determine k_j linearly independent eigenvectors in this eigenspace (the dimension of the eigenspace is k_j). Such a set of k_j linearly independent eigenvectors is a basis for the eigenspace. From Theorem 5.18 (Gram–Schmidt process) in frame 27, we know that for each such basis of k_j eigenvectors, an orthogonal basis of eigenvectors can be constructed. Thus, we can form an orthogonal set containing n eigenvectors by combining the separate bases of orthogonal eigenvectors from the different eigenspaces into a single overall set. (The total number of eigenvectors is n, since $k_1 + k_2 + \cdots + k_r = n$.) If we normalize these vectors (frame 23 of Chapter Five), we obtain an orthonormal set of n eigenvectors. (Such a set is an orthonormal *basis* for R^n, since R^n has dimension n.)

Thus, we can construct an $n \times n$ matrix $\hat{P}$ whose columns are the n eigenvectors from this orthonormal set. Such a matrix is called an *orthogonal* matrix (*not* an orthonormal matrix, which would seem more plausible; we shall have more to say about orthogonal matrices later). Clearly, $(\hat{P})^{-1}A\hat{P} = D$, as master Theorem 7.3 of frame 10 indicates. Matrix D is a diagonal matrix with eigenvalues along the main diagonal.

To make things systematic, the following method (algorithm) indicates steps for determining such a matrix $\hat{P}$.

Method 7.1 (For Determining an Orthogonal Matrix $\hat{P}$)

Step 1:   Determine a basis of linearly independent eigenvectors for each eigenspace of A.

Step 2:   Obtain an orthonormal basis of eigenvectors for each eigenspace by applying the Gram–Schmidt process (when necessary), and then normalizing.

Step 3:   Combine all the bases of Step 2 into an orthonormal set of n eigenvectors. Then form the $n \times n$ matrix $\hat{P}$ whose columns are the individual vectors from this orthonormal set.

Example: Refer to the Example of frame 15. Determine an orthogonal matrix $\hat{P}$ by using Method 7.1.

Solution: For the eigenvectors $\mathbf{x}_1 = [-1, 1, 0]^t$ and $\mathbf{x}_2 = [-1, 0, 1]^t$ corresponding to $\lambda = 1$, we apply the Gram–Schmidt process (frame 27 of Chapter Five):

$$\mathbf{v}_1 = \mathbf{x}_1 = [-1, 1, 0]^t$$

$$\mathbf{v}_2 = \mathbf{x}_2 - \text{Proj}_{W_1}\mathbf{x}_2 = \mathbf{x}_2 - \frac{\mathbf{x}_2 \cdot \mathbf{v}_1}{\mathbf{v}_1 \cdot \mathbf{v}_1}\mathbf{v}_1$$

$$= \mathbf{x}_2 - \tfrac{1}{2}\mathbf{v}_1 = [-\tfrac{1}{2}, -\tfrac{1}{2}, 1]^t.$$

The set $\{v_1, v_2\}$ is an orthogonal basis for the eigenspace corresponding to $\lambda = 1$. Normalizing these eigenvectors leads to

$$y_1 = \frac{v_1}{\|v_1\|} = [-1/\sqrt{2}, 1/\sqrt{2}, 0]^t,$$

$$y_2 = [-1/\sqrt{6}, -1/\sqrt{6}, 2/\sqrt{6}]^t,$$

and $\{y_1, y_2\}$ is an orthonormal basis for the eigenspace corresponding to $\lambda = 1$. Corresponding to $\lambda = 7$, we have the single eigenvector $x_3 = [1, 1, 1]^t$. Now finish the calculations.

— — — — — — — — — —

Normalizing yields $y_3 = [1/\sqrt{3}, 1/\sqrt{3}, 1/\sqrt{3}]^t$. The set $\{y_1, y_2, y_3\}$ is an orthonormal basis for R^3, and the matrix $\hat{P}$ is

$$\hat{P} = [y_1 \mid y_2 \mid y_3] = \begin{bmatrix} -1/\sqrt{2} & -1/\sqrt{6} & 1/\sqrt{3} \\ 1/\sqrt{2} & -1/\sqrt{6} & 1/\sqrt{3} \\ 0 & 2/\sqrt{6} & 1/\sqrt{3} \end{bmatrix}.$$

18. Next, we turn to a discussion of the usual definition and properties of an orthogonal matrix.

Definition 7.7 (Orthogonal Matrix)

An $n \times n$ invertible matrix A is said to be orthogonal if $A^{-1} = A^t$.

Note: Equivalently, A is orthogonal if $A^t A = I_n$. (Remember, if $BA = I_n$, then it follows that $AB = I_n$. See frame 27 of Chapter Two. Thus, above it is unnecessary to also require that $AA^t = I_n$.)

Example (a): The matrix $A = \begin{bmatrix} \frac{3}{5} & -\frac{4}{5} \\ \frac{4}{5} & \frac{3}{5} \end{bmatrix}$ is orthogonal, since $A^t = \begin{bmatrix} \frac{3}{5} & \frac{4}{5} \\ -\frac{4}{5} & \frac{3}{5} \end{bmatrix}$ and $AA^t = A^t A = I_2$. Also, the matrix $\hat{P}$ in the answer of the Example of frame 17 is orthogonal, since $(\hat{P})^t \hat{P} = I_3$.

The following theorem is useful.

Theorem 7.13

The $n \times n$ matrix A is orthogonal if and only if the columns of A form an orthonormal set of vectors in R^n.

Notes: (i)   This theorem thus justifies our use of the word *orthogonal* to describe the matrix $\hat{P}$ of frame 17. (ii)   The word *columns* above can be replaced by the word *rows*, and the theorem would still be true.

Let us do a proof of the "if" part of the theorem for the case of a 2×2 matrix.

Partial Proof of Theorem 7.13

Given matrix $A = \begin{bmatrix} a_1 & b_1 \\ a_2 & b_2 \end{bmatrix}$ for which the columns, identified as $\mathbf{a} = \begin{bmatrix} a_1 \\ a_2 \end{bmatrix}$ and $\mathbf{b} = \begin{bmatrix} b_1 \\ b_2 \end{bmatrix}$, form an orthonormal set. Thus,

$$\|\mathbf{a}\|^2 = \mathbf{a} \cdot \mathbf{a} = (a_1)^2 + (a_2)^2 = 1 \tag{1}$$

and

$$\|\mathbf{b}\|^2 = \mathbf{b} \cdot \mathbf{b} = (b_1)^2 + (b_2)^2 = 1 \tag{2}$$

and

$$\mathbf{a} \cdot \mathbf{b} = a_1 b_1 + a_2 b_2 = 0 \tag{3}$$

Now $A^t = \begin{bmatrix} a_1 & a_2 \\ b_1 & b_2 \end{bmatrix}$ and hence

$$A^t A = \begin{bmatrix} (a_1)^2 + (a_2)^2 & a_1 b_1 + a_2 b_2 \\ b_1 a_1 + b_2 a_2 & (b_1)^2 + (b_2)^2 \end{bmatrix} = \begin{bmatrix} 1 & 0 \\ 0 & 1 \end{bmatrix} = I_2 \tag{4}$$

Here we used Eqs. (1), (2), and (3) to show the components of $A^t A$ were 1's and 0's.

The proof of the converse involves the same steps, but in reverse.

Note the inconsistency in terminology. An orthogonal matrix has the property that its columns are orthonormal vectors.

Example (b):   Show that the standard matrix for a rotation linear transformation is an orthogonal matrix by using the "if" part of Theorem 7.13. Refer to frame 3 of Chapter Six. Then compute the inverse of the matrix.

Solution: Here $A = \begin{bmatrix} \cos \phi & -\sin \phi \\ \sin \phi & \cos \phi \end{bmatrix}$, where ϕ denotes the particular angle of rotation. Identifying the columns as $\mathbf{a} = \begin{bmatrix} \cos \phi \\ \sin \phi \end{bmatrix}$ and $\mathbf{b} = \begin{bmatrix} -\sin \phi \\ \cos \phi \end{bmatrix}$, we see that

$$\mathbf{a} \cdot \mathbf{a} = (\cos \phi)^2 + (\sin \phi)^2 = 1,$$

$$\mathbf{b} \cdot \mathbf{b} = (\sin \phi)^2 + (\cos \phi)^2 = 1.$$

Also,

$$\mathbf{a} \cdot \mathbf{b} = -(\cos \phi)(\sin \phi) + (\sin \phi)(\cos \phi) = 0.$$

Thus, the column vectors form an orthonormal set, and hence A is an orthogonal matrix, by Theorem 7.13.

Then $A^{-1} = A^t = \begin{bmatrix} \cos \phi & \sin \phi \\ -\sin \phi & \cos \phi \end{bmatrix}$.

The following theorem summarizes the work of the past few frames. The matrices D and $\hat{P}$ are both $n \times n$ matrices.

Theorem 7.14

If A is a symmetric $n \times n$ matrix, then there exists an orthogonal matrix $\hat{P}$ such that $(\hat{P})^{-1}A\hat{P} = D$, a diagonal matrix. (In short, A is similar to a diagonal matrix.) The diagonal entries of D are the eigenvalues of A, and the columns of $\hat{P}$ are n orthonormal eigenvectors of A. Also, the $(\hat{P})^{-1}$ in the equation can be replaced by $(\hat{P})^t$, to which it is equal.

Notes: (i) The converse of the first sentence of the theorem is valid also.

(ii) Essentially, Theorem 7.14 follows directly from Theorem 7.11 (frame 15) and from applying Method 7.1 (frame 17) to transform linearly independent sets of vectors to orthonormal sets.

(iii) The main value of working with an orthogonal matrix $\hat{P}$ to diagonalize a symmetric matrix A is that it is so easy to compute the inverse of $\hat{P}$ [since $(\hat{P})^{-1} = (\hat{P})^t$].

Example (c): For the examples of frames 15 and 17, we found the orthogonal matrix

$$\hat{P} = \begin{bmatrix} -1/\sqrt{2} & -1/\sqrt{6} & 1/\sqrt{3} \\ 1/\sqrt{2} & -1/\sqrt{6} & 1/\sqrt{3} \\ 0 & 2/\sqrt{6} & 1/\sqrt{3} \end{bmatrix}.$$

One can verify that

$$(\hat{P})^{-1}A\hat{P} = (\hat{P})^t A\hat{P} = \begin{bmatrix} \lambda_1 & 0 & 0 \\ 0 & \lambda_2 & 0 \\ 0 & 0 & \lambda_3 \end{bmatrix} = \begin{bmatrix} 1 & 0 & 0 \\ 0 & 1 & 0 \\ 0 & 0 & 7 \end{bmatrix}.$$

Example (d): Refer to Example (b) of frame 16. Determine an orthogonal matrix $\hat{P}$ such that $(\hat{P})^t A\hat{P} = D$.

Solution: All we have to do is normalize the eigenvectors $\mathbf{x}_1 = [0, -1, 1]^t$, $\mathbf{x}_2 = [1, 0, 0]^t$, and $\mathbf{x}_3 = [0, 1, 1]^t$, since the $\mathbf{x}_j$'s correspond to distinct eigenvalues. (Thus, in applying Method 7.1, there is no need to use the Gram–Schmidt process.) Now finish the work.

– – – – – – – – – –

$$\hat{P} = \begin{bmatrix} 0 & 1 & 0 \\ -1/\sqrt{2} & 0 & 1/\sqrt{2} \\ 1/\sqrt{2} & 0 & 1/\sqrt{2} \end{bmatrix}$$

It follows that $(\hat{P})^t A \hat{P} = D = \begin{bmatrix} -1 & 0 & 0 \\ 0 & 4 & 0 \\ 0 & 0 & 5 \end{bmatrix}$.

19. The following example reviews many of the topics covered in this section.

Example: For the symmetric matrix $A = \begin{bmatrix} 5 & 3 & 0 & 0 \\ 3 & 5 & 0 & 0 \\ 0 & 0 & 5 & 3 \\ 0 & 0 & 3 & 5 \end{bmatrix}$, determine the ei-

genvalues and corresponding eigenvectors. Then determine an orthogonal matrix $\hat{P}$ such that $(\hat{P})^t A \hat{P} = D$.

Solution: The characteristic polynomial is $(\lambda - 2)^2(\lambda - 8)^2$, which means the eigenvalues of A are

$$\lambda_1 = 2, \qquad \lambda_2 = 2, \qquad \lambda_3 = 8, \qquad \lambda_4 = 8.$$

For the eigenspace corresponding to $\lambda = 2$, we solve $(2I_4 - A)\mathbf{x} = \mathbf{0}$. From this we obtain the two linearly independent eigenvectors $\mathbf{x}_1 = [1, -1, 0, 0]^t$ and $\mathbf{x}_2 = [0, 0, 1, -1]^t$.
Now finish the problem.

– – – – – – – – –

For the eigenspace corresponding to $\lambda = 8$, we solve $(8I_4 - A)\mathbf{x} = \mathbf{0}$. From this we obtain the two linearly independent eigenvectors $\mathbf{x}_3 = [1, 1, 0, 0]^t$ and $\mathbf{x}_4 = [0, 0, 1, 1]^t$.

Since $\mathbf{x}_1$ and $\mathbf{x}_2$ are already orthogonal ($\mathbf{x}_1 \cdot \mathbf{x}_2 = 0$), all we have to do is normalize them, and likewise for $\mathbf{x}_3$ and $\mathbf{x}_4$. Thus, we obtain the set of 4 orthonormal eigenvectors $S = \{\mathbf{y}_1, \mathbf{y}_2, \mathbf{y}_3, \mathbf{y}_4\}$, where each $\mathbf{y}_j = \mathbf{x}_j/\|\mathbf{x}_j\| = \mathbf{x}_j/\sqrt{2}$, and thus,

$$\mathbf{y}_1 = [1/\sqrt{2}, -1/\sqrt{2}, 0, 0]^t, \qquad \mathbf{y}_2 = [0, 0, 1/\sqrt{2}, -1/\sqrt{2}]^t,$$

$$\mathbf{y}_3 = [1/\sqrt{2}, 1/\sqrt{2}, 0, 0]^t, \qquad \mathbf{y}_4 = [0, 0, 1/\sqrt{2}, 1/\sqrt{2}]^t$$

By the way, the set S is also an orthonormal basis for R^4. Thus, we obtain orthogonal matrix $\hat{P}$, in which the scalar $1/\sqrt{2}$ has been factored out.

$$\hat{P} = [\mathbf{y}_1 \mid \mathbf{y}_2 \mid \mathbf{y}_3 \mid \mathbf{y}_4] = \frac{1}{\sqrt{2}} \begin{bmatrix} 1 & 0 & 1 & 0 \\ -1 & 0 & 1 & 0 \\ 0 & 1 & 0 & 1 \\ 0 & -1 & 0 & 1 \end{bmatrix}$$

The diagonal matrix to which A is similar is

$$D = \begin{bmatrix} \lambda_1 & 0 & 0 & 0 \\ 0 & \lambda_2 & 0 & 0 \\ 0 & 0 & \lambda_3 & 0 \\ 0 & 0 & 0 & \lambda_4 \end{bmatrix} = \begin{bmatrix} 2 & 0 & 0 & 0 \\ 0 & 2 & 0 & 0 \\ 0 & 0 & 8 & 0 \\ 0 & 0 & 0 & 8 \end{bmatrix}.$$

FINAL COMMENTS

The material covered in the current chapter provides a brief introduction to the following topics: the eigenvalue problem, the diagonalization of a matrix, properties of orthogonal matrices. Several books that deal more extensively with these topics will be listed below.

We learned that every symmetric matrix can be diagonalized but that some nonsymmetric matrices cannot be. An important fact is that even if matrix A cannot be diagonalized (this means that there is no diagonal matrix to which A is *similar*), there is definitely a particular type of matrix, known as a *Jordan canonical form*, to which A is similar. This Jordan canonical form matrix, call it J, is "nearly diagonal" in its form. To illustrate, we saw that the 3×3 matrix $A = \begin{bmatrix} 3 & 0 & 0 \\ 4 & 3 & 0 \\ 3 & 0 & -1 \end{bmatrix}$ in the Example of frame 12 cannot be diagonalized.

However, A is similar to the Jordan canonical form matrix $J = \begin{bmatrix} 3 & 1 & 0 \\ 0 & 3 & 0 \\ 0 & 0 & -1 \end{bmatrix}$. Observe that the eigenvalues $\lambda_1 = 3$, $\lambda_2 = 3$, and $\lambda_3 = -1$ occur on the main diagonal as for a diagonal matrix, but note the presence of the entry $j_{12} = 1$.

Another important fact is that if a matrix is diagonalizable, then its Jordan canonical form matrix is precisely the diagonal matrix D to which A is similar (and which has the eigenvalues of A along the main diagonal). A more extensive discussion of the Jordan canonical form may be found in Grossman (1980a), Lipschutz (1968), and Hoffman and Kunze (1971).

There are many interesting and important properties of orthogonal matrices. For example, if a linear operator $L: R^n \rightarrow R^n$ is defined by $L(\mathbf{x}) = A\mathbf{x}$, where A is an orthogonal matrix, then L has the property of preserving the dot product;

that is, $L(\mathbf{x}) \cdot L(\mathbf{y}) = \mathbf{x} \cdot \mathbf{y}$ for any vectors $\mathbf{x}$ and $\mathbf{y}$ in the domain R^n. This further implies that linear operator L preserves length [i.e., $\|L(\mathbf{x})\| = \|\mathbf{x}\|$] and angle. Another interesting fact is that if A is a 2×2 orthogonal matrix, then the linear operator L given by $L(\mathbf{x}) = A\mathbf{x}$ is either a reflection or rotation. For a further discussion of the properties of orthogonal matrices, refer to Bloch and Michaels (1977) and Kolman (1980).

An elementary and interesting treatment of some of the geometrical aspects of eigenvalue problems may be found in Shields (1980) and Bloch and Michaels (1977).

SELF-TEST

This Self-Test will help you determine whether or not you have mastered the chapter objectives and are ready to go on to another chapter. Correct answers are given at the end of the test.

1. For each of the following matrices find the characteristic polynomial, eigenvalues, and a set of linearly independent eigenvectors.

(a) $\begin{bmatrix} 5 & -4 \\ 7 & -6 \end{bmatrix}$,   (b) $\begin{bmatrix} 6 & -2 & -5 \\ 6 & -1 & -6 \\ 4 & -2 & -3 \end{bmatrix}$.

2. Determine which of the following matrices is diagonalizable. If a matrix is not, indicate why not. If a matrix can be diagonalized, list an invertible matrix P such that $P^{-1}AP$ is a diagonal matrix, and list D.

   (a) Matrix of question 1a.

   (b) Matrix of question 1b.

   (c) $\begin{bmatrix} 2 & 0 & 3 \\ 0 & 3 & 5 \\ 0 & 0 & 3 \end{bmatrix}$.   (d) $\begin{bmatrix} 2 & 0 & 3 \\ 0 & 3 & 0 \\ 0 & 0 & 3 \end{bmatrix}$.   (e) $\begin{bmatrix} 4 & -5 \\ 0 & 4 \end{bmatrix}$.

3. For $A = \begin{bmatrix} -1 & 0 & 0 \\ 0 & 3 & 0 \\ 3 & 0 & 3 \end{bmatrix}$,

   (a) Indicate why A is lower triangular.

   (b) Determine if A can be diagonalized. If so, determine the diagonal matrix D, and an appropriate form for matrix P.

   (c) Verify that $A\mathbf{x} = \lambda\mathbf{x}$ for the eigenvalue of multiplicity 2, and for a pair of linearly independent eigenvectors in the corresponding eigenspace; do the same for the eigenvalue of multiplicity 1 and an eigenvector corresponding to that eigenvalue.

4. For each of the following symmetric matrices, find a matrix P such that $P^{-1}AP$ is diagonal. List the diagonal matrix D.

   (a) $\begin{bmatrix} 5 & 3 \\ 3 & 5 \end{bmatrix}$.   (b) $\begin{bmatrix} 1 & 0 & 0 \\ 0 & -2 & 3 \\ 0 & 3 & -2 \end{bmatrix}$.   (c) $\begin{bmatrix} 3 & -1 & -1 \\ -1 & 3 & -1 \\ -1 & -1 & 3 \end{bmatrix}$.

5.   For parts (a), (b), and (c) of question 4, determine an orthogonal matrix $\hat{P}$ such that $(\hat{P})^{-1}A\hat{P}$ is diagonal. [Remember, $(\hat{P})^{-1} = (\hat{P})^t$, since $\hat{P}$ is an orthogonal matrix.]

6.   Refer to the matrix in question 4a. Use Theorem 7.2a (frame 9) to compute A^4.

7.   (a)   Given that matrix $B = \begin{bmatrix} -\frac{1}{3} & -\frac{2}{3} & \frac{2}{3} \\ \frac{2}{3} & -\frac{2}{3} & -\frac{1}{3} \\ -\frac{2}{3} & -\frac{1}{3} & -\frac{2}{3} \end{bmatrix}$, use the "if" part of Theorem

7.13 (frame 18) to show that B is an orthogonal matrix.

(b)   Then compute the inverse of B.

ANSWERS TO SELF-TEST

If your answers to the test questions do not agree with the following ones, review the frames indicated in parentheses after each answer before you go on to another chapter.

1.   (a) $\lambda^2 + \lambda - 2$; $\lambda_1 = -2$, $\lambda_2 = 1$; $\mathbf{x}_1 = [4, 7]^t$, $\mathbf{x}_2 = [1, 1]^t$.

(b) $\lambda^3 - 2\lambda^2 - \lambda + 2$; $\lambda_1 = -1$, $\lambda_2 = 1$, $\lambda_3 = 2$; $\mathbf{x}_1 = [1, 0, 1]^t$, $\mathbf{x}_2 = [1, 2, 0]^t$, $\mathbf{x}_3 = [1, 1, 1]^t$.   (frames 1–5)

2.   (a)   Can be diagonalized. $P = [\mathbf{x}_1 \mid \mathbf{x}_2] = \begin{bmatrix} 4 & 1 \\ 7 & 1 \end{bmatrix}$, $D = \begin{bmatrix} -2 & 0 \\ 0 & 1 \end{bmatrix}$.

(b)   Can be diagonalized. $P = \begin{bmatrix} 1 & 1 & 1 \\ 0 & 2 & 1 \\ 1 & 0 & 1 \end{bmatrix}$, $D = \begin{bmatrix} -1 & 0 & 0 \\ 0 & 1 & 0 \\ 0 & 0 & 2 \end{bmatrix}$.

(c)   Cannot be diagonalized, since $\lambda = 3$ is of multiplicity 2 but eigenspace for $\lambda = 3$ has dimension 1 (i.e., for $\lambda = 3$, two linearly independent eigenvectors don't exist).

(d)   Can be diagonalized. $P = \begin{bmatrix} 0 & 3 & 1 \\ 1 & 0 & 0 \\ 0 & 1 & 0 \end{bmatrix}$, $D = \begin{bmatrix} 3 & 0 & 0 \\ 0 & 3 & 0 \\ 0 & 0 & 2 \end{bmatrix}$.

(e)   Cannot be diagonalized since $\lambda = 4$ is of multiplicity 2, but eigenspace for $\lambda = 4$ has dimension 1.   (frames 3–13)

3.   (a)   It is lower triangular because $a_{12} = a_{13} = a_{23} = 0$.

(b) A can be diagonalized. $D = \begin{bmatrix} -1 & 0 & 0 \\ 0 & 3 & 0 \\ 0 & 0 & 3 \end{bmatrix}$, $P = [\mathbf{x}_1 \mid \mathbf{x}_2 \mid \mathbf{x}_3] = \begin{bmatrix} -4 & 0 & 0 \\ 0 & 1 & 0 \\ 3 & 0 & 1 \end{bmatrix}$.

(c) $A\mathbf{x}_1 = A[-4, 0, 3]^t = [4, 0, -3]^t = (-1)\mathbf{x}_1$; $A\mathbf{x}_2 = A[0, 1, 0]^t = 3\mathbf{x}_2$; $A\mathbf{x}_3 = A[0, 0, 1]^t = 3\mathbf{x}_3$.   (frames 1–14)

4. (a) $P = [\mathbf{x}_1 \mid \mathbf{x}_2] = \begin{bmatrix} -1 & 1 \\ 1 & 1 \end{bmatrix}, D = \begin{bmatrix} \lambda_1 & 0 \\ 0 & \lambda_2 \end{bmatrix} = \begin{bmatrix} 2 & 0 \\ 0 & 8 \end{bmatrix}.$

(b) $P = [\mathbf{x}_1 \mid \mathbf{x}_2 \mid \mathbf{x}_3] = \begin{bmatrix} 0 & 1 & 0 \\ -1 & 0 & 1 \\ 1 & 0 & 1 \end{bmatrix}, D = \begin{bmatrix} \lambda_1 & 0 & 0 \\ 0 & \lambda_2 & 0 \\ 0 & 0 & \lambda_3 \end{bmatrix} = \begin{bmatrix} -5 & 0 & 0 \\ 0 & 1 & 0 \\ 0 & 0 & 1 \end{bmatrix}.$

(c) $P = [\mathbf{x}_1 \mid \mathbf{x}_2 \mid \mathbf{x}_3] = \begin{bmatrix} 1 & -1 & -1 \\ 1 & 0 & 1 \\ 1 & 1 & 0 \end{bmatrix}, D = \begin{bmatrix} \lambda_1 & 0 & 0 \\ 0 & \lambda_2 & 0 \\ 0 & 0 & \lambda_3 \end{bmatrix} = \begin{bmatrix} 1 & 0 & 0 \\ 0 & 4 & 0 \\ 0 & 0 & 4 \end{bmatrix}.$

(frames 15–17)

5. (a) $\hat{P} = [\mathbf{x}_1/\sqrt{2} \mid \mathbf{x}_2/\sqrt{2}] = \dfrac{1}{\sqrt{2}}\begin{bmatrix} -1 & 1 \\ 1 & 1 \end{bmatrix}.$

(b) $\hat{P} = \begin{bmatrix} 0 & 1 & 0 \\ -1/\sqrt{2} & 0 & 1/\sqrt{2} \\ 1/\sqrt{2} & 0 & 1/\sqrt{2} \end{bmatrix}.$

(c) Use Gram–Schmidt process on $\{\mathbf{x}_2, \mathbf{x}_3\}$:

$\mathbf{v}_2 = \mathbf{x}_2; \mathbf{v}_3 = \mathbf{x}_3 - \text{Proj}_{W_2}\mathbf{x}_3 = \mathbf{x}_3 - \dfrac{\mathbf{x}_3 \cdot \mathbf{v}_2}{\mathbf{v}_2 \cdot \mathbf{v}_2}\mathbf{v}_2 = [-\tfrac{1}{2}, 1, -\tfrac{1}{2}]^t.$ Normalize

$\mathbf{v}_2$ and $\mathbf{v}_3$ to $\mathbf{y}_2 = [-1/\sqrt{2}, 0, 1/\sqrt{2}]^t$ and $\mathbf{y}_3 = [-1/\sqrt{6}, 2/\sqrt{6}, -1/\sqrt{6}]^t.$ Also $\mathbf{y}_1 = \mathbf{x}_1/\|\mathbf{x}_1\| = \mathbf{x}_1/\sqrt{3} = [1/\sqrt{3}, 1/\sqrt{3}, 1/\sqrt{3}]^t.$ Then

$$\hat{P} = [\mathbf{y}_1 \mid \mathbf{y}_2 \mid \mathbf{y}_3] = \begin{bmatrix} 1/\sqrt{3} & -1/\sqrt{2} & -1/\sqrt{6} \\ 1/\sqrt{3} & 0 & 2/\sqrt{6} \\ 1/\sqrt{3} & 1/\sqrt{2} & -1/\sqrt{6} \end{bmatrix}$$ (frames 15–19)

6. First, $D^4 = \begin{bmatrix} 2^4 & 0 \\ 0 & 8^4 \end{bmatrix} = \begin{bmatrix} 16 & 0 \\ 0 & 4096 \end{bmatrix},$ and $P^{-1} = \left(-\dfrac{1}{2}\right)\begin{bmatrix} 1 & -1 \\ -1 & -1 \end{bmatrix}.$ Then,

$A^4 = \begin{bmatrix} 2056 & 2040 \\ 2040 & 2056 \end{bmatrix}.$ (frame 9)

7. (a) Labeling the columns of B as $\mathbf{b}_1$, $\mathbf{b}_2$, and $\mathbf{b}_3$, we see that $\mathbf{b}_1 \cdot \mathbf{b}_2 = \mathbf{b}_1 \cdot \mathbf{b}_3 = \mathbf{b}_2 \cdot \mathbf{b}_3 = 0$ and $\mathbf{b}_1 \cdot \mathbf{b}_1 = \mathbf{b}_2 \cdot \mathbf{b}_2 = \mathbf{b}_3 \cdot \mathbf{b}_3 = 1$, thus indicating that the columns of B form an orthonormal set of vectors in R^3.

(b) $B^{-1} = B^t = \begin{bmatrix} -\frac{1}{3} & \frac{2}{3} & -\frac{2}{3} \\ -\frac{2}{3} & -\frac{2}{3} & -\frac{1}{3} \\ \frac{2}{3} & -\frac{1}{3} & -\frac{2}{3} \end{bmatrix}$

(frame 18)

CHAPTER EIGHT
Selected Applications

There are many areas of application of the forms and techniques of linear and matrix algebra. For example, linear and matrix algebra are used to a great degree in physics, engineering, quantitative business analysis, economics, computer science, and the life sciences. Also, linear algebra is intimately involved with the formulation of the branch of calculus concerned with functions of several variables. In statistics, matrix methods are employed in analyzing linear statistical models (where, for example, a dependent variable is a linear function of two or more independent variables).

In this chapter we present a brief collection of applications involving linear and matrix algebra. Of course, other applications were studied in prior chapters (see, for example, Section G of Chapter Two and Section D of Chapter Three). The references at the back of the book contain exhaustive treatments in many areas of application. In particular, the interested reader is referred to Searle and Hausman (1970) for business and economic applications, to Searle (1966) for biological and life sciences applications, and to Wylie (1975) for engineering and natural science applications.

Also, Rorres and Anton (1979) and Grossman (1980b) are totally concerned with applications in many areas of the social, natural, and computer sciences.

The very important subject, *linear programming*, is based to a large extent on linear algebra. It is treated in detail in Cooper and Steinberg (1974), Kolman and Beck (1980), and Rothenberg (1979). A condensed treatment of linear programming may be found in Grossman (1980b), Kolman (1980), Rorres and Anton (1979), Rothenberg (1980), and Shields (1980).

OBJECTIVES

When you complete this chapter you should be able to

- Represent and analyze stepwise phenomena that occur in the business, marketing, and scientific fields by means of the Markov process model.
- Determine the behavior of certain Markov processes after a long period of time.

- Use diagonalization techniques for the analysis and calculation of numerical results in Markov processes.
- Set up and use the Leslie model of population growth for analyzing population changes with time in animal and human populations.
- Do calculations with the Leslie matrix and age distribution vectors.
- Use the Leslie model to determine approximate population growth behavior after a long period of time.

A. MARKOV PROCESSES

Note: We mention at the start that the material in this section involves some elementary probability ideas and a simplified understanding of the limit concept (frame 8). However, an attempt has been made to proceed in an elementary fashion so that one only slightly familiar with these ideas will be able, with effort, to follow the development. Also, later in this section (frame 8), we make use of some of the diagonalization of matrices material from Section C of Chapter Seven.

1. Consider a system that, at any particular time, will be in only one of a finite number of states. For example, a particular person is either rich, middle class, or poor at any given time. A computer is either working or not working (known as "down") at a particular time. The sales of a store are either better than, worse than, or roughly the same as the sales of the same store during the preceding year.

As time goes on, the system may move from one state to another. In many applications, we know (or can estimate) probabilities associated with being in given states at some point in time, and we wish to know the probabilities associated with those states at some later time. Often we can predict such probabilities with relative ease. If a system has the properties of a Markov process, this will be the case. [A. A. Markov was a great Russian mathematician (1856–1922) who developed the theory of stochastic processes; a Markov process is a particular kind of stochastic process. The latter consists of a sequence of probability experiments in which each experiment has a finite number of states. Stochastic is derived from the Greek word *stochos,* which means "guess."]

The key ideas and other conventions associated with Markov processes are best illustrated with some examples. We now introduce two examples, which will serve as models for several of our later illustrative computations dealing with Markov processes.

Example (a): A particular brand of soap (brand I) is being promoted by a company. A market survey analysis in several test communities reveals that 70% of those who use brand I will again purchase brand I next month; the remaining 30% will switch to another brand. Of those people using other brands of soap (referred to collectively as II), 40% will change over to brand I in the next month.

The preceding information can be summarized conveniently in the following *transition* matrix, where the percents have been converted to decimals.

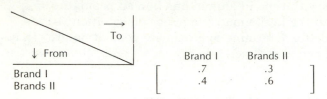

The numbers in each row may be regarded as probabilities, denoted by p_{ij}'s. As is the convention with matrices, the i refers to the row and the j to the column. Note that the probabilities in each row add up to 1. The entire transition matrix is denoted by P. As an example of how to read the entries of the matrix, observe that $p_{12} = .3$; this indicates that there is a .3 (or 30%) probability that a person will switch from brand I to another brand in one month. Also, $p_{21} = .4$ reveals that there is a 40% probability that a person will switch from another brand over to brand I. In general, p_{ij} is the probability of switching from state i to state j in the next step. In the example, a step is a one-month interval.

Example (b): A long-range study of the operation of a computer over hourly periods reveals that if the computer is working at any point in time, there is a 90% probability if will be working one hour later; thus, there is a 10% probability that it will not be working one hour later. Also, if the computer is not working (i.e., it is "down") at any point in time, there is a 40% probability that it will not be working one hour later. Let the two states of the machine be labeled 1 for working and 2 for "down." Show the matrix of transition probabilities. Let row 1 and column 1 refer to the working state.

———————————

Solution: The transition matrix is as follows:

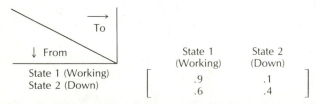

For example, $p_{21} = .6$ reveals there is a 60% probability the computer will be working one hour after a point in time at which it is down.

2. After these two examples, we are now ready for some definitions and terminology pertaining to Markov processes.

Suppose a sequence of experiments (or trials) is performed in which there are m possible outcomes or *states* 1, 2, . . ., m for each trial. In Example (a) of frame 1, m equals 2, since there are two states in each trial, namely purchasing brand I or purchasing another brand (brands II). Likewise, in Example (b) of frame 1, m equals 2; the two states here are the "working" and "down" states of the computer. A sequence of trials in the latter example refers to observations of the state of the computer (working or down) at points in time that are one hour apart (say, at 1:00 P.M., 2:00 P.M., 3:00 P.M., etc.). On occasion we will use the term *observation times* or just *times* in place of *trials*.

Definition 8.1

A Markov process is a process in which the probability of a system being in a specific state at a given trial depends only on its state at the immediately preceding trial. The transition from one trial to the next trial is called a *step* or *interval*.

Suppose a system has m possible states 1, 2, . . . , m. The symbol p_{ij} (for $i = 1, 2, . . . , m; j = 1, 2, . . . , m$) denotes the probability that if the system is in state i at a given trial, it will be in state j at the next trial. The term p_{ij} is called the *transition probability* from state i to state j.

In a Markov process, each p_{ij} depends only on i and j; that is, each p_{ij} is independent of how many trials have occurred. Put another way, each p_{ij} does not change with time (here we associate a sequence of trials with the passage of time).

Notes: (i)   To illustrate the last paragraph of Definition 8.1, refer to Example (b) of frame 1. Suppose the state of the computer is observed at one-hour intervals beginning at 1:00 P.M. Thus, for example, consider the meaning of $p_{12} = .1$. This means there is a .1 probability that the computer will be down one hour later if it is working at, say, 2:00 P.M. Likewise, there will be a .1 probability the computer will be down one hour later if it is working at 6:00 P.M. This same statement holds for any one-hour time interval whatsoever that begins at the start of an hour. (ii)   The transition probability p_{ij} is a special case of a *conditional probability*. As a conditional probability, the symbol for p_{ij} would be $P(j|i)$, where the dividing line (|) is read as "given that" or "on condition that." Thus, we can read p_{ij}, or its equivalent $P(j|i)$, as the "probability the system goes to state j in the next trial given that it is in state i at some particular trial." (iii)   By definition, any probability is a number between 0 and 1 inclusive.

The $m \times m$ matrix P consisting of the p_{ij} entries is called the *transition matrix*. The general transition matrix for a two-state system ($m = 2$) is given as follows:

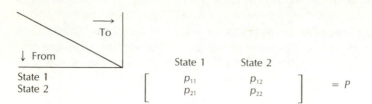

The transition matrices displayed in Examples (a) and (b) of frame 1 were both 2×2.

Each p_{ij} lies between 0 and 1 inclusive, since each p_{ij} is a probability. Also, the sum of the probabilities in each row must equal 1, that is,

$$p_{i1} + p_{i2} + \cdots + p_{im} = 1 \qquad \text{for } i = 1, 2, \ldots, m$$

Note: A vector $\mathbf{v}$ with m components $v_1, v_2, \ldots, v_m$ such that each $v_i \geq 0$, and such that $v_1 + v_2 + \cdots + v_m = 1$ is called a *probability vector*. Thus, each row of the transition matrix P is a probability row vector.

Example: The following is a Markov model for the sales characteristics of a store over one-year periods. The states are the following: plus, if sales increase by more than 3% over the preceding year; small change, if sales fluctuate by less than 3%; minus, if sales decrease by more than 3% of the sales of the preceding year. Suppose the three-state transition matrix P is as follows:

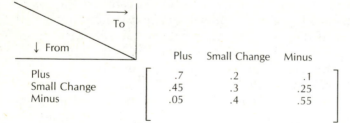

The entries in row 1 are $p_{11} = .7$, $p_{12} = .2$, and $p_{13} = .1$. These indicate the probabilities of plus, small change, and minus, respectively, during a year following a year of plus change. Note that the probabilities in each row add up to 1.

Now interpret the entries in rows 2 and 3.

The entries in row 2, namely, $p_{21} = .45$, $p_{22} = .30$, and $p_{23} = .25$ indicate the probabilities of plus, small change, and minus, respectively, during a year following a year of small change. The entry $p_{31} = .05$ indicates there is a .05 probability of plus (i.e., a sales increase by more than 3%) for a year following a minus year.

3.   Suppose we have a typical Markov process for which the transition matrix P is known. Suppose that at some particular trial (or time) the probability that the system is in state i is x_i. Let us denote the proabilities associated with all the states by the m component probability row vector **x**:

$$\mathbf{x} = [x_1, x_2, \ldots, x_m]$$

This vector is called the *probability distribution* of the system at some particular trial. (Note that each $x_i \geq 0$, and $x_1 + x_2 + \cdots + x_m = 1$.) It is important to be able to determine the probability distribution one step later, that is, on the next trial. For a Markov process the latter probability distribution depends on the x_i's (the probability distribution before the transition) and on the p_{ij}'s (the transition probabilities).

To make matters simple, suppose we know the probability distribution $\mathbf{x} = [x_1, x_2]$ for a two-state system at some given trial. The general 2×2 transition matrix is given right after Definition 8.1. Let the probability distribution on the next trial be denoted by $\mathbf{x}' = [x_1', x_2']$.

We shall now derive the equation for x_1'. This derivation employs several basic probability concepts, which should be at least intuitively meaningful to students with limited backgrounds in probability. [Elementary treatments of probability may be found in Anton and Kolman (1978) and Rothenberg (1980).] Henceforth, in the following discussion, $P(A)$ will be used as a symbol for the probability of the event A.

The term x_i is the probability of being in state i on a given trial, or, in our new symbols, P(state i on given trial). The term x_i' is P(state i on the next trial). There are two so-called mutually exclusive routes to state 1 on the next trial, namely, one that starts in state 1 and makes a transition to state 1 and the other that starts in state 2 and makes a transition to state 1. Thus, using what is sometimes referred to as the *total probability rule,* we have

P(state 1 on next trial) = P(state 1 on given trial *and* transition to state 1 on next trial) + P(state 2 on given trial *and* transition to state 1 on next trial).   (1)

Using the *multiplication rule* on the two right-hand probabilities, we have

P(state 1 on next trial) = P(state 1 on given trial) · P(state 1 on next trial | state 1 on given trial) + P(state 2 on given trial) · P(state 1 on next trial | state 2 on given trial).   (2)

The two terms on the right that contain the dividing line are conditional probabilities; the dividing line is read "given that" or "on condition that." For example, we read P (state 1 on next trial | state 2 on given trial) as "the probability the system will go to state 1 on the next trial on condition that it is in state 2 on the given trial."

Now, the two conditional probabilities on the right of Eq. (2) are none other than the transition probabilities p_{11} and p_{21}, respectively. [See Note (ii) after Definition 8.1 in frame 2 for a discussion of the p_{ij} symbol.] Employing the x_i (probability for given trial) and x_i' (probability for next trial) symbols in Eq. (2), we obtain

$$x_1' = x_1 p_{11} + x_2 p_{21} \tag{3}$$

In the same way we can derive the following equation for x_2', the probability of being in state 2 on the next trial:

$$x_2' = x_1 p_{12} + x_2 p_{22} \tag{4}$$

We can combine Eqs. (3) and (4) into the following row vector equation:

$$[x_1', x_2'] = [x_1 p_{11} + x_2 p_{21}, \ x_1 p_{12} + x_2 p_{22}] \tag{5}$$

Now $\mathbf{x}'$, the probability distribution on the next trial, equals the left side of Eq. (5), and the right side is equal to the product of $\mathbf{x}$, the probability distribution on the given trial (here, equal to $[x_1, x_2]$), multiplied on the right by P, the 2×2 transition matrix (the equation for the 2×2 transition matrix is given in frame 2 after the notes that follow Definition 8.1). Thus, Eq. (5) may be rewritten as the following single matrix–vector equation:

$$\mathbf{x}' = \mathbf{x}P \tag{6}$$

Equation (6) holds in general for any m state system. The key equation needed for deriving Eq. (6), in the general case, is the following, which is an extension of Eqs. (3) and (4):

$$x_j' = x_1 p_{1j} + x_2 p_{2j} + \ldots + x_m p_{mj} \tag{7}$$

Eq. (7) represents m equations, namely one for each of $j = 1, 2, \ldots,$ and m. We restate Eq. (6) in the following theorem.

Theorem 8.1

If $\mathbf{x}$ is the probability distribution at some given trial, then $\mathbf{x}'$, the probability distribution at the next trial (i.e., after one step), is given by

$$\mathbf{x}' = \mathbf{x}P$$

Suppose the probability distribution at some initial time or trial (say, on trial 0) is given by

$$\mathbf{x}^{(0)} = [x_1^{(0)}, \ x_2^{(0)}, \ \ldots, \ x_m^{(0)}] \qquad \text{(Initial Probability Distribution)}$$

Let $\mathbf{x}^{(1)}$ and $\mathbf{x}^{(2)}$ denote the probability distributions on trials 1 and 2, respectively; that is, one and two steps later, respectively. Then, from Theorem 8.1, we have

$$\mathbf{x}^{(1)} = \mathbf{x}^{(0)}P, \qquad \text{and} \qquad \mathbf{x}^{(2)} = \mathbf{x}^{(1)}P. \qquad \text{(8),} \quad \text{(9)}$$

We can continue in this way to find $\mathbf{x}^{(n)}$, the probability distribution on trial n (or, equivalently, after n steps) in terms of $\mathbf{x}^{(n-1)}$, the probability distribution on trial $(n-1)$:

$$\mathbf{x}^{(n)} = \mathbf{x}^{(n-1)}P \qquad (10)$$

If we replace $\mathbf{x}^{(1)}$ in Eq. (9) by its value as given in Eq. (8), we obtain

$$\mathbf{x}^{(2)} = (\mathbf{x}^{(0)}P)P, \qquad \text{or} \qquad \mathbf{x}^{(2)} = \mathbf{x}^{(0)}P^2. \qquad (11)$$

Here P^2 is the matrix product of P times P. The generalization of Eq. (11) is stated in the following theorem:

Theorem 8.2

The probability distribution on trial n (or, equivalently, after n steps) is given in terms of the initial probability distribution $\mathbf{x}^{(0)}$ and the nth power of the transition matrix P by the following equation:

$$\mathbf{x}^{(n)} = \mathbf{x}^{(0)}P^n$$

Note: Here P^n denotes the nth power of the matrix P (see frame 17 of Chapter One for a discussion of powers of a matrix). The term $\mathbf{x}^{(n)}$ does *not* denote a power of the row vector $\mathbf{x}$. The superscript (n) in $\mathbf{x}^{(n)}$ is merely a label for trial n.

In our initial calculations we shall find it easier to use the scheme suggested by Eqs. (8), (9), and (10) to calculate probability distributions after a certain number of steps (provided the number of steps is small).

Example: Refer to Example (b) of frame 1, which deals with the operation of a computer. Suppose that at some time the computer is working. Let this time be the initial time. This means that the initial probability distribution is

$$\mathbf{x}^{(0)} = [1, 0],$$

which indicates the computer is in state 1 (the working state) with probability 1. To find the probability distribution after one hour, we have to compute the matrix product $\mathbf{x}^{(0)}P$. Thus, after one hour, the probability distribution is given by

$$[1, 0]\begin{bmatrix} .9 & .1 \\ .6 & .4 \end{bmatrix} = [1(.9) + 0(.6), \ 1(.1) + 0(.4)] = [.9, .1]$$

Thus, $\mathbf{x}^{(1)} = [.9, .1]$, which indicates the probability is .9 that the computer is working after one hour and .1 that it is not.

Find the probability distribution after two hours, that is, after two steps. This is $\mathbf{x}^{(2)}$. The easiest approach here is to use Eq. (9).

_____

$$\mathbf{x}^{(2)} = \mathbf{x}^{(1)}P = [.9, .1]\begin{bmatrix} .9 & .1 \\ .6 & .4 \end{bmatrix} = [.81 + .06, \ .09 + .04] = [.87, .13]$$

This indicates that after two hours there is a probability of .87 that the computer is working.

4.

Example: Refer to Example (a) of frame 1, which deals with the tendencies of people to use soap brand I or some other brand. Suppose that in some town 40% of the people use brand I at the beginning of January. In probability terms this is equivalent to stating that the probability is 40% (or .40) that a person picked at random at the beginning of January uses brand I. Assume the transition matrix of Example (a) applies. Determine the percentage of people that use brand I after one month and after two months. Here the initial probability distribution is $\mathbf{x}^{(0)} = [.4, .6]$, corresponding to the data applying at the beginning of January.

Solution: After one month, the probability distribution is given by $\mathbf{x}^{(0)}P$, that is, by

$$[.4, .6]\begin{bmatrix} .7 & .3 \\ .4 & .6 \end{bmatrix} = [(.4)(.7) + (.6)(.4), \ (.4)(.3) + (.6)(.6)]$$

$$= [.52, .48].$$

Thus, $\mathbf{x}^{(1)} = [.52, .48]$, which indicates that after one month, 52% of the people use brand I and 48% use some other brand.

Now finish the problem. That is, calculate $\mathbf{x}^{(2)}$.

_ _ _ _ _ _ _ _ _

After two months, the probability distribution is given by $\mathbf{x}^{(2)} = \mathbf{x}^{(1)}P$. That is,

$$\mathbf{x}^{(2)} = [.52, .48]\begin{bmatrix} .7 & .3 \\ .4 & .6 \end{bmatrix} = [.364 + .192, \ .156 + .288] = [.556, .444]$$

This indicates that after two months 55.6% of the people use brand I and 44.4% use some other brand.

5.   The long-run behavior of Markov processes is often of interest. The reason for this is that for certain Markov processes, the probability distribution approaches some fixed (constant) vector after a number of steps, regardless of what the initial distribution is. For example, consider a continuation of the calculations of the Example of frame 3. These calculations deal with the functioning of a computer that initially was working; each step is of one hour's duration.

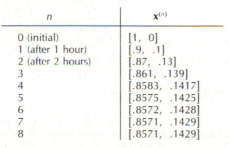

n	$\mathbf{x}^{(n)}$
0 (initial)	[1, 0]
1 (after 1 hour)	[.9, .1]
2 (after 2 hours)	[.87, .13]
3	[.861, .139]
4	[.8583, .1417]
5	[.8575, .1425]
6	[.8572, .1428]
7	[.8571, .1429]
8	[.8571, .1429]

Thus, after about six hours the probability of the computer working stabilizes to about .857 (to three decimal places). In fact, it can be shown that regardless of the initial probability distribution, the successive probability distributions will approach the fixed (or constant) probability distribution, [.857, .143], if we round off probabilities to three decimal places.

This is not accidental. The following definition and theorem indicate that this stabilization occurs in a certain type of Markov process.

Definition 8.2

A transition matrix P is called *regular* if some power of P has all positive entries.

Each of the transition matrices of Example (a) and (b) of frame 1, and the Example of frame 2 is regular. For each, P itself (i.e., the first power of P) has all positive entries.

The transition matrix $P = \begin{bmatrix} \frac{3}{4} & \frac{1}{4} \\ 1 & 0 \end{bmatrix}$ is regular, since the matrix $P^2 = P \cdot P =$

$\begin{bmatrix} \frac{13}{16} & \frac{3}{16} \\ \frac{3}{4} & \frac{1}{4} \end{bmatrix}$ has only positive entries. The transition matrix $P = \begin{bmatrix} \frac{1}{2} & \frac{1}{2} \\ 0 & 1 \end{bmatrix}$ is not

regular, since every power of P will have the row 2, column 1 entry equal to zero.

The following very important theorem is proved in Kemeny and Snell (1976).

Theorem 8.3

Suppose that P, the transition matrix for a Markov process, is regular. Then

(a) Regardless of the initial probability distribution $\mathbf{x}^{(0)}$, the successive probability distributions will approach the same fixed probability distribution $\mathbf{t}$. Here, $\mathbf{t} = [t_1, t_2, \ldots, t_m]$, say. The row vector $\mathbf{t}$ is called the *steady-state* or *stationary* probability distribution of the Markov process.

(b) As the trial label n increases without bound (in symbols, as $n \to \infty$, where ∞ means infinity), the nth power of P, namely P^n, approaches a fixed matrix in which each row is given by the steady-state probability distribution $\mathbf{t}$. Put differently, $P^n \to T$ as $n \to \infty$, where the $m \times m$ matrix T is given as follows:

$$T = \begin{bmatrix} t_1 & t_2 & t_3 & \cdots & t_m \\ t_1 & t_2 & t_3 & \cdots & t_m \\ \cdots & \cdots & \cdots & \cdots & \cdots \\ \cdots & \cdots & \cdots & \cdots & \cdots \\ t_1 & t_2 & t_3 & \cdots & t_m \end{bmatrix}.$$

(c) The vector $\mathbf{t}$ is the unique probability vector solution of the equation $\mathbf{t}P = \mathbf{t}$.

Notes: (i) A consequence of part (a) of Theorem 8.3 is that for a regular transition matrix, after many trials the probability that state i occurs is approximately equal to component t_i of the row vector $\mathbf{t}$. (ii) Each component of $\mathbf{t}$ is positive and, naturally, all the components add up to 1 (since $\mathbf{t}$ is a probability row vector). (iii) According to part (a) of Theorem 8.3, the effect of the initial probability distribution tends to disappear as the number of steps increases.

At the beginning of this frame we saw for the computer situation [introduced in Example (b) of frame 1] that the probability distribution seemed to approach the row vector [.857, .143], to three decimal places if the initial probability distribution was $\mathbf{x}^{(0)} = [1, 0]$.

Example: Show that the steady-state probability distribution for the computer situation is equal to [.857, .143], to three decimal places. Use part (c) of Theorem 8.3.

Solution: Here we seek a probability row vector $\mathbf{t} = [t_1, t_2]$ that satisfies the matrix–vector equation $\mathbf{t}P = \mathbf{t}$. Thus, we have

$$[t_1, t_2]\begin{bmatrix} .9 & .1 \\ .6 & .4 \end{bmatrix} = [t_1, t_2], \tag{1}$$

using the transition matrix P from Example (b) of frame 1. Doing the matrix multiplication of the left side of Eq. (1) leads to

$$[.9t_1 + .6t_2, \quad .1t_1 + .4t_2] = [t_1, t_2]. \tag{2}$$

Equating respective components on the left and right yields

$$.9t_1 + .6t_2 = t_1 \tag{3a}$$

$$.1t_1 + .4t_2 = t_2 \tag{3b}$$

Also,

$$t_1 + t_2 = 1, \tag{4}$$

since $\mathbf{t}$ is a probability distribution. Now only one of (3a) and (3b) is useful, since if we add them, we get the identity $t_1 + t_2 = t_1 + t_2$. Thus, using (3a) and (4), we can solve for *unique* values of t_1 and t_2.

Now finish the calculations.

— — — — — — — — — —

From Eq. (3a), we get

$$.1t_1 - .6t_2 = 0 \qquad \text{or} \qquad t_1 - 6t_2 = 0. \tag{5}$$

Applying the Gauss–Jordan elimination method (Method 2.1) to Eqs. (4) and (5), we obtain

$$\begin{bmatrix} 1 & 1 & 1 \\ 1 & -6 & 0 \end{bmatrix} \xrightarrow{\ r_2' = r_2 - r_1\ } \begin{bmatrix} 1 & 1 & 1 \\ 0 & -7 & -1 \end{bmatrix}$$

$$\xrightarrow{\ r_2' = r_2/(-7)\ } \begin{bmatrix} 1 & 1 & 1 \\ 0 & 1 & \frac{1}{7} \end{bmatrix}$$

$$\xrightarrow{\ r_1' = r_1 - r_2\ } \begin{bmatrix} 1 & 0 & \frac{6}{7} \\ 0 & 1 & \frac{1}{7} \end{bmatrix},$$

which yields $t_1 = \frac{6}{7}$, and $t_2 = \frac{1}{7}$. Expressing the components of $\mathbf{t}$ to three decimal places leads to $\mathbf{t} = [.857, .143]$.

6. Refer to Example (a) of frame 1 and the calculations of the Example of frame 4. The probability distributions $\mathbf{x}^{(n)}$ give the proportions of people in a town who use and don't use soap brand I. Suppose we continue the calculations of the Example of frame 4, where the initial probability distribution was $\mathbf{x}^{(0)} = [.4, .6]$. Each step in the table below is of a one-month duration.

n	$\mathbf{x}^{(n)}$
0 (initial)	$[.4, .6]$
1 (after 1 month)	$[.52, .48]$
2 (after 2 months)	$[.556, .444]$
3	$[.5668, .4332]$
4	$[.5700, .4300]$
5	$[.5710, .4290]$
6	$[.5713, .4287]$

Thus, after about five months, the proportion of people in the town who use brand I stabilizes at about .571, to three decimal places. If the initial probability distribution were $\mathbf{x}^{(0)} = [.5, .5]$, we have the following table. Remember that the transition matrix P is still equal to $\begin{bmatrix} .7 & .3 \\ .4 & .6 \end{bmatrix}$.

n	$\mathbf{x}^{(n)}$
0	$[.5, .5]$
1	$[.55, .45]$
2	$[.565, .435]$
3	$[.5695, .4305]$
4	$[.5709, .4292]$
5	$[.5713, .4287]$

Thus, once again the probability distribution stabilizes to $[.571, .429]$, to three decimal places. These calculations merely reaffirm part (a) of Theorem 8.3.

Example: Determine the steady-state probability distribution for the transition matrix of Example (a) of frame 1 by using Theorem 8.3, part (c). The resulting **t** should be close to [.571, .429] if we round off to three decimal places.

_ _ _ _ _ _ _ _ _ _

Solution: From $\mathbf{t}P = \mathbf{t}$, we have

$$[t_1, t_2]\begin{bmatrix} .7 & .3 \\ .4 & .6 \end{bmatrix} = [t_1, t_2] \tag{1}$$

Multiplying on the left side of Eq. (1) yields

$$[.7t_1 + .4t_2, \ .3t_1 + .6t_2] = [t_1, t_2] \tag{2}$$

Equating the first components on the left and right yields

$$.7t_1 + .4t_2 = t_1 \tag{3}$$

We also have

$$t_1 + t_2 = 1 \tag{4}$$

Solving (3) and (4) simultaneously leads to $t_1 = \frac{4}{7}$ and $t_2 = \frac{3}{7}$. That is, $\mathbf{t} = [.571, .429]$ to three decimal places, and this result agrees with the long-run $\mathbf{x}^{(n)}$'s obtained previously.

7.

Example: Refer to the transition matrix for the Example of frame 2. Determine the steady-state probability distribution **t**. This will reveal to us the long-run probabilities relating to the increase, small change, or decrease of sales for the store.

Solution: The equation we have to solve is $\mathbf{t}P = \mathbf{t}$, which is as follows for the given transition matrix:

$$[t_1, t_2, t_3]\begin{bmatrix} .7 & .2 & .1 \\ .45 & .3 & .25 \\ .05 & .4 & .55 \end{bmatrix} = [t_1, t_2, t_3] \tag{1}$$

Doing the matrix multiplication on the left side of (1) leads to

$$[.7t_1 + .45t_2 + .05t_3, \ .2t_1 + .3t_2 + .4t_3, \ .1t_1 + .25t_2 + .55t_3] = [t_1, t_2, t_3] \tag{2}$$

Equating the first and second components, respectively, on the left and right, leads to

$$.7t_1 + .45t_2 + .05t_3 = t_1 \tag{3}$$

$$.2t_1 + .3t_2 + .4t_3 = t_2 \tag{4}$$

Equating the third components would not be useful. The following equation, which pertains to the fact that **t** is a probability vector, is useful:

$$t_1 + t_2 + t_3 = 1 \tag{5}$$

We can now solve Eqs. (3), (4), and (5) simultaneously to obtain unique values for t_1, t_2, and t_3. Let us employ the substitution method. From Eq. (5), we have $t_3 = 1 - t_1 - t_2$, and substituting this into Eqs. (3) and (4) yields

$$-.35t_1 + .4t_2 = -.05 \tag{6}$$

and

$$-.2t_1 - 1.1t_2 = -.4 \tag{7}$$

Solving (7) for t_1 in terms of t_2 yields

$$t_1 = 2 - 5.5t_2 \tag{8}$$

Substituting t_1 from (8) back into (6) leads to the following single equation in the unknown t_2.

$$-.7 + 1.195t_2 + .4t_2 = -.05 \quad \text{or} \quad 2.325t_2 = .65$$

This leads to

$$t_2 = .280 \tag{9}$$

Now complete the calculations. [Make use of Eqs. (8) and (5), in that order.]

– – – – – – – – – –

Substituting the t_2 value from (9) into (8) yields

$$t_1 = .460 \tag{10}$$

From Eq. (5), we have $t_3 = 1 - t_1 - t_2$, and after substituting from Eqs. (9) and (10), we obtain

$$t_3 = .260 \tag{11}$$

Thus, the steady-state probability distribution is $\mathbf{t} = [.460, .280, .260]$. This indicates that, after a long period of time, there is a .460 probability of plus change, a .280 probability of small change, and a .260 probability of a minus change in any year, regardless of the initial probability distribution.

8. Note: The work of this frame will involve procedures pertaining to the diagonalizing of a matrix. See Section (C)—in particular, frame 9—of Chapter Seven.

From Theorem 8.2 (frame 3), we saw that another way to compute $\mathbf{x}^{(n)}$ is from

$$\mathbf{x}^{(n)} = \mathbf{x}^{(0)}P^n \tag{1}$$

This equation would be practical if we had a simple way for computing the matrix power P^n as a function of n. Fortunately, a simple way does exist, provided that P can be diagonalized. If P can be diagonalized, then we know from Section (C) of Chapter Seven that the $m \times m$ matrix P can be expressed as

$$P = RDR^{-1}, \tag{2}$$

where all the matrices on the right are $m \times m$. Here D is a diagonal matrix whose diagonal entries are the eigenvalues $\lambda_1, \lambda_2, \ldots, \lambda_m$ of matrix P. Thus, if P is 2×2 and its eigenvalues are λ_1 and λ_2, then

$$D = \begin{bmatrix} \lambda_1 & 0 \\ 0 & \lambda_2 \end{bmatrix} \tag{3}$$

In matrix R the columns are a sequence of eigenvectors $\mathbf{r}_1, \mathbf{r}_2, \ldots, \mathbf{r}_m$, which correspond to the eigenvalues $\lambda_1, \lambda_2, \ldots, \lambda_m$ of matrix D. Thus, if R is 2×2, we have

$$R = [\mathbf{r}_1 \mid \mathbf{r}_2] \tag{4}$$

where the two-component column vector $\mathbf{r}_1$ is an eigenvector corresponding to eigenvalue λ_1, and $\mathbf{r}_2$ is an eigenvector corresponding to λ_2.

Now, from Eq. (2) and Theorem 7.2a (frame 9), we have the following equation for P^n, the nth power of P:

$$P^n = RD^nR^{-1} \tag{5}$$

Here D^n is a diagonal matrix whose diagonal entries are $(\lambda_1)^n$, $(\lambda_2)^n$, ..., and $(\lambda_m)^n$, respectively.

Example: Consider the situation pertaining to the functioning of a computer [Example (b) of frame 1, and the Examples of frames 3 and 5].

(a) Determine P^n by using Eq. (5) of the previous discussion. (b) Then use the P^n equation from part (a) to compute $\mathbf{x}^{(2)}$ if $\mathbf{x}^{(0)} = [1, 0]$. Repeat if $\mathbf{x}^{(0)} = [.6, .4]$. (c) Determine the limiting form of P^n as n increases without bound (in other words, as n becomes infinite, or, in symbols, as $n \to \infty$), and from this determine the steady-state probability distribution t. Compare the latter result with the result already calculated in the Example of frame 5.

Solution: (a)   First, we calculate the eigenvalues and eigenvectors of the transition matrix P, where

$$P = \begin{bmatrix} .9 & .1 \\ .6 & .4 \end{bmatrix}, \tag{6}$$

by using the method of Section (B) of Chapter Seven. From the characteristic equation,

$$\begin{vmatrix} \lambda - .9 & -.1 \\ -.6 & \lambda - .4 \end{vmatrix} = \lambda^2 - 1.3\lambda + .3 = 0, \tag{7}$$

we find that

$$\lambda_1 = 1 \quad \text{and} \quad \lambda_2 = .3. \tag{8}$$

Corresponding eigenvectors are

$$\mathbf{r}_1 = \begin{bmatrix} 1 \\ 1 \end{bmatrix} \quad \text{and} \quad \mathbf{r}_2 = \begin{bmatrix} -1 \\ 6 \end{bmatrix}. \tag{9}$$

Thus,

$$R = \begin{bmatrix} 1 & -1 \\ 1 & 6 \end{bmatrix} \tag{10}$$

Next, we find the inverse of R [see Section (F) of Chapter Two, for example] to be

$$R^{-1} = \frac{1}{7} \begin{bmatrix} 6 & 1 \\ -1 & 1 \end{bmatrix}. \tag{11}$$

Thus, substituting into Eq. (5) yields the following equation for P^n:

$$P^n = \frac{1}{7}\begin{bmatrix} 1 & -1 \\ 1 & 6 \end{bmatrix}\begin{bmatrix} 1 & 0 \\ 0 & (.3)^n \end{bmatrix}\begin{bmatrix} 6 & 1 \\ -1 & 1 \end{bmatrix}. \tag{12}$$

Note that the diagonal entries of D^n are $(1)^n$ [which equals 1] and $(.3)^n$, respectively. Now Eq. (12) can be reduced to

$$P^n = \frac{1}{7}\begin{bmatrix} 6 + (.3)^n & 1 - (.3)^n \\ 6 - 6(.3)^n & 1 + 6(.3)^n \end{bmatrix}, \tag{13}$$

after multiplying the three matrices on the right. We can check Eq. (13) by substituting $n = 1$. Doing this, we get $P = P^1 = \begin{bmatrix} .9 & .1 \\ .6 & .4 \end{bmatrix}$, as expected.

(b)   From Theorem 8.2, $\mathbf{x}^{(n)} = \mathbf{x}^{(0)}P^n$, and, thus, $\mathbf{x}^{(2)} = \mathbf{x}^{(0)}P^2$. From Eq. (13),

$$P^2 = \begin{bmatrix} .87 & .13 \\ .78 & .22 \end{bmatrix} \tag{14}$$

Thus, for $\mathbf{x}^{(0)} = [1, 0]$, we have

$$\mathbf{x}^{(2)} = [1, 0]\begin{bmatrix} .87 & .13 \\ .78 & .22 \end{bmatrix} = [.87, .13]$$

This checks out with the result for $\mathbf{x}^{(2)}$ calculated in the Example of frame 3. If $\mathbf{x}^{(0)} = [.6, .4]$, we obtain

$$\mathbf{x}^{(2)} = [.6, .4]\begin{bmatrix} .87 & .13 \\ .78 & .22 \end{bmatrix} = [.834, .166].$$

(c)   Refer to Eq. (13). We wish to find the limiting form that P^n approaches as n increases without bound. We use an intuitively clear result on limits (proved in calculus), which indicates that

$$a^n \to 0 \qquad \text{as} \quad n \to \infty \qquad \text{if} \quad -1 < a < 1 \tag{15}$$

Expression (15) indicates that a^n approaches zero as n increases without bound provided that a is strictly between -1 and 1. [The symbolic form $n \to \infty$ can also be read "as n becomes infinite"; here ∞ is the symbol for infinity.] Thus, in Eq. (13), $(.3)^n \to 0$ as n increases without bound. Using this fact, now express the limiting form of P^n as n increases without bound. Work with Eq. (13).

From Eq. (13), and from the fact that $(.3)^n \to 0$ as $n \to \infty$, we see that the limiting form of P^n is

$$\begin{bmatrix} \frac{6}{7} & \frac{1}{7} \\ \frac{6}{7} & \frac{1}{7} \end{bmatrix}$$

as n increases without bound, Thus, for our current situation, this is the matrix T cited in Theorem 8.3 (frame 5). From an inspection of the rows [Theorem 8.3, Part (b)], we see that the steady-state probability distribution vector t is given by $t = \begin{bmatrix} \frac{6}{7}, & \frac{1}{7} \end{bmatrix}$. This is in agreement with the result calculated in the Example of frame 5.

Note: It can be shown that the results of parts (a) and (c) of Theorem 8.3 are consequences of part (b) of the theorem. That is, as n becomes infinite, if P^n approaches a matrix each of whose rows is the same probability vector, then that probability vector is the steady-state probability distribution t, and, moreover, t satisfies $tP = t$.

B. THE LESLIE MODEL OF POPULATION GROWTH

Note: The development in parts of this section use concepts dealing with eigenvalues and eigenvectors, diagonalization, and the limit concept. The first two topics are covered in Chapter Seven, and the limit concept was discussed briefly in frame 8.

9. The *Leslie model of population growth*, developed in the 1940s, describes the growth of the female portion of a human or animal population. If one assumes the male population is always in the same proportion to the female population regardless of age (say, equal to the female population), then one has a good idea of the growth behavior of the population as a whole. For a detailed discussion of the Leslie model, refer to Rorres and Anton (1979), and Grossman (1980b).

In the Leslie model, we have a division of ages into n age classes, each of equal duration. If the maximum age attained by a female is M years, then each class is M/n years in duration, as in Table 8.1.

TABLE 8.1   Table of Age Classes for Leslie Model

Age Class	Age Interval
C_1	$[0, \ M/n)$
C_2	$[M/n, \ 2M/n)$
C_3	$[2M/n, \ 3M/n)$
$\vdots$	$\vdots$
C_{n-1}	$[(n-2)M/n, \ (n-1)M/n)$
C_n	$[(n-1)M/n, \ M]$

Example (a): For example, if $M = 45$ years and $n = 3$, we have the three age classes C_1, C_2, and C_3. Since $M/n = 45/3 = 15$, each age interval has a 15-year duration. The intervals corresponding to the classes are $[0, 15)$, $[15, 30)$, and $[30, 45]$. The interval $[15, 30)$ means all ages between 15 and 30 years, including 15 years—this is the meaning of $[15$—but not including 30 years.

Suppose we know the number of females in each of the n classes at time $t = 0$. Let $x_1^{(0)}$ be the number of females in the class C_1, $x_2^{(0)}$ be the number of females in class C_2, and so on for all n age classes. We define the *initial age distribution vector* $\mathbf{x}^{(0)}$ as follows:

$$\mathbf{x}^{(0)} = \begin{bmatrix} x_1^{(0)} \\ x_2^{(0)} \\ \cdot \\ \cdot \\ \cdot \\ x_n^{(0)} \end{bmatrix}$$

Note: In the development of this section, the vectors will be column vectors.

As time goes on, the number of females within each age class changes because of birth, death, and aging. The goal of the Leslie model is to describe this process in a definite way.

In the Leslie model we observe the number of females in each age class at discrete times t_0, t_1, t_2, $\ldots$, t_k, $\ldots$, where the duration between any two successive observation times is equal to the width of the age intervals; from Table 8.1, the latter value is M/n. Thus, the observation times are as follows: $t_0 = 0$; $t_1 = M/n$; $t_2 = 2M/n$; $t_3 = 3M/n$; $\ldots$; $t_k = kM/n$; and so on.

The *age distribution vector* $\mathbf{x}^{(k)}$ *at observation time* t_k is defined to be

$$\mathbf{x}^{(k)} = \begin{bmatrix} x_1^{(k)} \\ x_2^{(k)} \\ \cdot \\ \cdot \\ \cdot \\ x_n^{(k)} \end{bmatrix} \qquad \text{for } k = 0, 1, 2, \ldots \\ \text{(i.e., for } t_0 = 0, t_1 = M/n, t_2 = 2M/n, \text{ etc.)} \tag{I}$$

Hence, $x_i^{(k)}$ denotes the number of females in the ith class at time t_k.

Example (b): See Example (a). The number of females in the three age classes at $t_0 = 0$ and $t_1 = 15$ are given as follows:

Age Class	Age Interval	No. Females at $t_0 = 0$	No. Females at $t_1 = 15$
C_1	[0, 15)	50	165
C_2	[15, 30)	30	25
C_3	[30, 45]	20	15

Determine the age distribution vectors $\mathbf{x}^{(0)}$ and $\mathbf{x}^{(1)}$.

Solution:

$$\mathbf{x}^{(0)} = \begin{bmatrix} x_1^{(0)} \\ x_2^{(0)} \\ x_3^{(0)} \end{bmatrix} = \begin{bmatrix} 50 \\ 30 \\ 20 \end{bmatrix}$$

Now find $\mathbf{x}^{(1)}$. This is $\mathbf{x}^{(k)}$ for $k = 1$. Note that $k = 1$ refers to the time $t_1 = M/n = 45/3 = 15$.

- - - - - - - - - -

$$\mathbf{x}^{(1)} = \begin{bmatrix} x_1^{(1)} \\ x_2^{(1)} \\ x_3^{(1)} \end{bmatrix} = \begin{bmatrix} 165 \\ 25 \\ 15 \end{bmatrix}$$

10. Now we define the birth and death parameters for the Leslie model.

Definition 8.3

The quantity a_i (for $i = 1, 2, \ldots, n$) is the average number of females born per female during the time a female is in the ith age class.

The quantity b_i (for $i = 1, 2, \ldots, n-1$) is the fraction of females in the ith age class (i.e., in class C_i) who survive to pass into the $(i + 1)$st age class (i.e., into age class C_{i+1}).

It should be clear that

$$a_i \geqslant 0 \qquad \text{for } i = 1, 2, \ldots, n \tag{1}$$
$$\text{and} \quad 0 < b_i \leqslant 1 \qquad \text{for } i = 1, 2, \ldots, n-1 \tag{2}$$

We assume that each $b_i > 0$, for otherwise there would be no females in the $(i + 1)$st age class.

Example (a): Refer to the Examples of frame 9. Suppose that 50% of the females in age class C_1 survive to pass into age class C_2, and likewise for those in age class C_2 who survive to pass into age class C_3. Suppose that the average number of females born per female in age classes C_1, C_2, and C_3 are 0, 3.5, and 3, respectively. Determine the b_i's and a_i's for this situation.

Solution: $b_1 = b_2 = \frac{1}{2}$ (or 0.5); $a_1 = 0$, $a_2 = 3.5$, $a_3 = 3.0$.

Refer again to the description of $x_i^{(k)}$ in frame 9. We would like to relate the $x_i^{(k)}$ terms [$x_i^{(k)}$ is the number of females in the ith age class at time t_k] to the $x_i^{(k-1)}$ terms [$x_i^{(k-1)}$ is the number of females in the ith age class at time t_{k-1}].

The females in the first age class (i.e., in class C_1) at time t_k are those females *born* between time t_{k-1} and t_k. Thus,

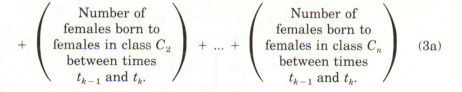

$$\begin{pmatrix} \text{Number of females} \\ \text{in class } C_1 \\ \text{at time } t_k \end{pmatrix} = \begin{pmatrix} \text{Number of} \\ \text{females born to} \\ \text{females in class } C_1 \\ \text{between times} \\ t_{k-1} \text{ and } t_k. \end{pmatrix}$$

$$+ \begin{pmatrix} \text{Number of} \\ \text{females born to} \\ \text{females in class } C_2 \\ \text{between times} \\ t_{k-1} \text{ and } t_k. \end{pmatrix} + \dots + \begin{pmatrix} \text{Number of} \\ \text{females born to} \\ \text{females in class } C_n \\ \text{between times} \\ t_{k-1} \text{ and } t_k. \end{pmatrix} \qquad (3a)$$

Symbolically, the left side is $x_1^{(k)}$. For the right side, let us focus on the second term, for example. It is equal to $x_2^{(k-1)}$, the number of females in class C_2 who are alive at t_{k-1} multiplied by a_2, the average number of females born to a female in age class C_2; the product $a_2 x_2^{(k-1)}$ is the result. A similar discussion holds for the other $(n-1)$ terms on the right side of Eq. (3a). Thus, we can write Eq. (3a) mathematically as

$$x_1^{(k)} = a_1 x_1^{(k-1)} + a_2 x_2^{(k-1)} + \dots + a_{n-1} x_{n-1}^{(k-1)} + a_n x_n^{(k-1)} \qquad (3b)$$

Let us focus on the equations for the other age classes, that is, on equations for $x_2^{(k)}$, $x_3^{(k)}$, and so on.

The number $x_i^{(k)}$ of females in age class C_i (for $i = 2, 3, \dots, n$) at time t_k is equal to the number of females from the previous age class at time t_{k-1} who survive to be alive at time t_k. Thus, $x_i^{(k)}$ equals $x_{i-1}^{(k-1)}$ times b_{i-1} (see definition for the b_i terms at the beginning of this frame). Thus,

$$x_i^{(k)} = b_{i-1} x_{i-1}^{(k-1)} \qquad \text{for } i = 2, 3, \dots, n. \qquad (4)$$

If we write out the equations from (4) for $i = 2, 3, \ldots, n$, in succession below Eq. (3b), and then form column vectors on left and right, we get the following column vector equation:

$$
\begin{bmatrix} x_1^{(k)} \\ x_2^{(k)} \\ x_3^{(k)} \\ \cdot \\ \cdot \\ \cdot \\ x_n^{(k)} \end{bmatrix} = \begin{bmatrix} a_1 x_1^{(k-1)} + a_2 x_2^{(k-1)} + \ldots + a_{n-1} x_{n-1}^{(k-1)} + a_n x_n^{(k-1)} \\ b_1 x_1^{(k-1)} \\ b_2 x_2^{(k-1)} \\ \cdots\cdots\cdots\cdots\cdots\cdots\cdots\cdots\cdots\cdots\cdots\cdots\cdots\cdots \\ b_{n-1} x_{n-1}^{(k-1)} \end{bmatrix}
\tag{5}
$$

Now, the left side is the vector $\mathbf{x}^{(k)}$ [see Eq. (I) in frame 9].

Definition 8.4 (Leslie Matrix)

The $n \times n$ Leslie matrix A is defined to be

$$
A = \begin{bmatrix}
a_1 & a_2 & \ldots & a_{n-1} & a_n \\
b_1 & 0 & \ldots & 0 & 0 \\
0 & b_2 & \ldots & 0 & 0 \\
\cdot & \cdot & \ldots & \cdot & \cdot \\
\cdot & \cdot & \ldots & \cdot & \cdot \\
0 & 0 & \ldots & b_{n-1} & 0
\end{bmatrix}
\tag{6}
$$

Recall that n is the total number of age classes, each of which is of the same duration (M/n years).

Example (b): Express the right side of Eq. (5) as a product involving the Leslie matrix A.

– – – – – – – – – –

Right side $= A\mathbf{x}^{(k-1)}$,
$\tag{7}$

since $\mathbf{x}^{(k-1)} = [x_1^{(k-1)}, x_2^{(k-1)}, \ldots, x_{n-1}^{(k-1)}, x_n^{(k-1)}]^t$.

11. Because of the preceding manipulations, we can rewrite Eq. (5) compactly as $\mathbf{x}^{(k)} = A\mathbf{x}^{(k-1)}$. We summarize in the following theorem.

Theorem 8.4

The age distribution vector at time t_k, namely $\mathbf{x}^{(k)}$, is related to the age distribution vector at time t_{k-1}, namely $\mathbf{x}^{(k-1)}$, by Eq. (II) below, where A is the Leslie matrix:

$$\mathbf{x}^{(k)} = A\mathbf{x}^{(k-1)} \qquad \text{for } k = 1, 2, 3, \ldots \tag{II}$$

From Eq. (II) it follows that

$$\mathbf{x}^{(1)} = A\mathbf{x}^{(0)} \tag{1}$$

$$\mathbf{x}^{(2)} = A\mathbf{x}^{(1)} = A(A\mathbf{x}^{(0)}) = A^2\mathbf{x}^{(0)} \tag{2}$$

$$\mathbf{x}^{(3)} = A\mathbf{x}^{(2)} = A(A^2\mathbf{x}^{(0)}) = A^3\mathbf{x}^{(0)} \tag{3}$$

We have a similar result for any observation time t_k.

Theorem 8.5

The age distribution vector at time t_k, namely $\mathbf{x}^{(k)}$, is related to the initial age distribution vector $\mathbf{x}^{(0)}$ by the equation:

$$\mathbf{x}^{(k)} = A^k\mathbf{x}^{(0)} \qquad \text{for } k = 1, 2, 3, \ldots \tag{III}$$

where A^k is the kth power of Leslie matrix A.

Example: Refer to the Examples of the previous frames in this section, and, in particular, to Example (a) in frame 10. (a) Determine the Leslie matrix A. (b) If the initial age distribution is given by $x_1^{(0)} = 50$, $x_2^{(0)} = 30$, and $x_3^{(0)} = 20$, that is, $\mathbf{x}^{(0)} = [50, 30, 20]^t$, determine $\mathbf{x}^{(1)}$. (c) Determine $\mathbf{x}^{(2)}$ and $\mathbf{x}^{(3)}$.

Solution: It is useful to recall that each time interval is of a 15-year duration (see Examples in frame 9). Thus, the observation times are $t_0 = 0$, $t_1 = 15$, $t_2 = 30$, $t_3 = 45$, etc.
   (a)   From Example (a) of frame 10,

$$A = \begin{bmatrix} a_1 & a_2 & a_3 \\ b_1 & 0 & 0 \\ 0 & b_2 & 0 \end{bmatrix} = \begin{bmatrix} 0 & 3.5 & 3 \\ \frac{1}{2} & 0 & 0 \\ 0 & \frac{1}{2} & 0 \end{bmatrix}$$

(b)   From Theorem 8.4 or 8.5,

$$\mathbf{x}^{(1)} = A\mathbf{x}^{(0)} = \begin{bmatrix} 0 & 3.5 & 3 \\ \frac{1}{2} & 0 & 0 \\ 0 & \frac{1}{2} & 0 \end{bmatrix} \begin{bmatrix} 50 \\ 30 \\ 20 \end{bmatrix} = \begin{bmatrix} 165 \\ 25 \\ 15 \end{bmatrix}$$

Thus, after 15 years, there are, respectively, 165, 25, and 15 females in age classes C_1, C_2, and C_3.

Now do part (c).

— — — — — — — — — —

Since we have just calculated $\mathbf{x}^{(1)}$ it is easiest to calculate $\mathbf{x}^{(2)}$ from $\mathbf{x}^{(2)} = A\mathbf{x}^{(1)}$. Thus,

$$\mathbf{x}^{(2)} = A\mathbf{x}^{(1)} = \begin{bmatrix} 0 & 3.5 & 3 \\ \frac{1}{2} & 0 & 0 \\ 0 & \frac{1}{2} & 0 \end{bmatrix} \begin{bmatrix} 165 \\ 25 \\ 15 \end{bmatrix} = \begin{bmatrix} 132.5 \\ 82.5 \\ 12.5 \end{bmatrix}$$

Computing $\mathbf{x}^{(3)}$ in similar fashion,

$$\mathbf{x}^{(3)} = A\mathbf{x}^{(2)} = \begin{bmatrix} 0 & 3.5 & 3 \\ \frac{1}{2} & 0 & 0 \\ 0 & \frac{1}{2} & 0 \end{bmatrix} \begin{bmatrix} 132.5 \\ 82.5 \\ 12.5 \end{bmatrix} = \begin{bmatrix} 326.25 \\ 66.25 \\ 41.25 \end{bmatrix}$$

Thus, after 45 years, there are (after rounding off) 326 females between 0 and 15 years of age, 66 between 15 and 30 years of age, and 41 between 30 and 45 years of age.

12.   **Note:**   The next three frames involve material from Chapter Seven dealing with eigenvalue–eigenvector ideas and the diagonalization properties of Leslie matrix A.

Observe the similarity between the preceding Example and several of the Examples from Section A; see, for instance, frames 3 and 4.

There are some other similarities as well. Suppose the Leslie matrix A can be diagonalized. Thus, we can write A as

$$A = PDP^{-1}, \tag{IV}$$

where D is a diagonal matrix whose main diagonal entries are the eigenvalues of matrix A. The columns of matrix P are eigenvectors corresponding to the eigenvalues. Further, we have the following simple equation for the kth power of A, namely, A^k:

$$A^k = PD^kP^{-1} \tag{V}$$

Here D^k is a diagonal matrix whose diagonal entries are the powers of the corresponding diagonal entries of D. (See frame 9 of Chapter Seven, especially Theorem 7.2a.)

If we use Eq. (V) in conjunction with $\mathbf{x}^{(k)} = A^k\mathbf{x}^{(0)}$ [Eq. (III) of Theorem 8.5], we will have an easy way of calculating $\mathbf{x}^{(k)}$, the age distribution vector at time t_k.

Example: (a) For the Example in frame 11, determine D, P, and P^{-1} for the Leslie matrix A. (b) Determine A^3 from Eq. (V) with $k = 3$, and then substitute the result into Eq. (III) of frame 11 to compute $\mathbf{x}^{(3)}$. Take the initial age distribution to be the same as that cited in the Example of frame 11.

Solution: (a) You should verify that the characteristic polynomial is $\lambda^3 - \frac{7}{4}\lambda - \frac{3}{4}$, and that the eigenvalues are

$$\lambda_1 = \tfrac{3}{2}, \qquad \lambda_2 = -\tfrac{1}{2}, \qquad \lambda_3 = -1.$$

Corresponding to these eigenvalues, we have the following eigenvectors:

$$\mathbf{p}_1 = \begin{bmatrix} 9 \\ 3 \\ 1 \end{bmatrix}, \qquad \mathbf{p}_2 = \begin{bmatrix} 1 \\ -1 \\ 1 \end{bmatrix}, \qquad \mathbf{p}_3 = \begin{bmatrix} 4 \\ -2 \\ 1 \end{bmatrix}$$

Thus, the diagonal matrix D and a suitable P matrix are given by

$$D = \begin{bmatrix} \tfrac{3}{2} & 0 & 0 \\ 0 & -\tfrac{1}{2} & 0 \\ 0 & 0 & -1 \end{bmatrix}, \qquad P = [\mathbf{p}_1 \mid \mathbf{p}_2 \mid \mathbf{p}_3] = \begin{bmatrix} 9 & 1 & 4 \\ 3 & -1 & -2 \\ 1 & 1 & 1 \end{bmatrix}.$$

Further, you should verify that

$$P^{-1} = \left(\frac{1}{20}\right)\begin{bmatrix} 1 & 3 & 2 \\ -5 & 5 & 30 \\ 4 & -8 & -12 \end{bmatrix}.$$

(b) Now calculate A^3 and then $\mathbf{x}^{(3)}$.

_ _ _ _ _ _ _ _ _ _ _

From $A^3 = PD^3P^{-1}$, we have

$$A^3 = \left(\frac{1}{20}\right)\begin{bmatrix} 9 & 1 & 4 \\ 3 & -1 & -2 \\ 1 & 1 & 1 \end{bmatrix}\begin{bmatrix} \frac{27}{8} & 0 & 0 \\ 0 & -\frac{1}{8} & 0 \\ 0 & 0 & -1 \end{bmatrix}\begin{bmatrix} 1 & 3 & 2 \\ -5 & 5 & 30 \\ 4 & -8 & -12 \end{bmatrix}$$

$$= \frac{1}{8}\begin{bmatrix} 6 & 49 & 42 \\ 7 & 6 & 0 \\ 0 & 7 & 6 \end{bmatrix}.$$

Thus, $\mathbf{x}^{(3)}$ is given by

$$\mathbf{x}^{(3)} = A^3\mathbf{x}^{(0)} = \frac{1}{8}\begin{bmatrix} 6 & 49 & 42 \\ 7 & 6 & 0 \\ 0 & 7 & 6 \end{bmatrix}\begin{bmatrix} 50 \\ 30 \\ 20 \end{bmatrix} = \begin{bmatrix} 326.25 \\ 66.25 \\ 41.25 \end{bmatrix},$$

which agrees with the answer in frame 11.

Note: The amount of effort required to calculate A^k from PD^kP^{-1} would not be much greater if k were considerably larger than 3 (say, $k = 10$).

13. Now we shall study the limiting behavior for the Leslie model of population growth. That is, we will study growth patterns as the observation time t_k (or, equivalently, as the index k) increases without bound; in symbols, we indicate this increase without bound by writing $t_k \to \infty$ or $k \to \infty$.

The following theorem, stated without proof, will guide us in our analysis of the Leslie model as t_k increases without bound.

Theorem 8.6

Given the Leslie matrix A [see Definition 8.4 in frame 10]. Suppose that

(i) $a_i \geqslant 0$ for $i = 1, 2, 3, \ldots, n$.
(ii) At least two successive a_i entries are positive; that is, for some integer j, both $a_j > 0$ and $a_{j+1} > 0$.
(iii) $0 < b_i \leqslant 1$ for $i = 1, 2, \ldots, n-1$.

Then,

(A) The matrix A has a unique positive eigenvalue λ_1.
(B) The eigenvalue λ_1 has multiplicity 1. [See frame 12 of Chapter Seven for a review of multiplicity.]
(C) Corresponding to λ_1 is an eigenvector all of whose components are positive.
(D) Any other eigenvalue λ_i of A satisfies $|\lambda_i| < \lambda_1$.

Notes: (a)   Hypothesis (ii) is needed in order to conclude (D). (b)   In (D) the symbol $|\lambda_i|$ means the *absolute value* of λ_i. It should be noted that complex eigenvalues can occur for Leslie matrices. The absolute value of complex eigenvalue $\hat{\lambda} = c + di$, where $i = \sqrt{-1}$, is $|\hat{\lambda}| = \sqrt{c^2 + d^2}$.

Example: Verify Theorem 8.6 with respect to our model Leslie matrix problem. Refer to the Examples in frames 11 and 12 for matrix A and eigenvalues and eigenvectors for A.

Solution: For Leslie matrix $A = \begin{bmatrix} 0 & 3.5 & 3 \\ \frac{1}{2} & 0 & 0 \\ 0 & \frac{1}{2} & 0 \end{bmatrix}$, we first check hypotheses (i)

through (iii). We see that all a_i's are nonnegative, and the succeeding entries a_2 and a_3 are both positive. Since $b_1 = b_2 = \frac{1}{2}$, hypothesis (iii) is satisfied. Thus, all the hypotheses are satisfied.

Now show that conclusions (A) through D are all true.

- - - - - - - - - - -

The unique positive eigenvalue is $\lambda_1 = 1.5$, and it has multiplicity 1, since the characteristic polynomial $\lambda^3 - \frac{7}{4}\lambda - \frac{3}{4}$ can be factored to

$$\left(\lambda - \frac{3}{2}\right)\left(\lambda + \frac{1}{2}\right)\left(\lambda + 1\right).$$

[This is the proper form, since the characteristic polynomial has degree 3 and there are three distinct eigenvalues. Each factor is of the form $(\lambda - \lambda_i)$, where λ_i is an eigenvalue.]

The components of eigenvector $\mathbf{p}_1 = [9, 3, 1]^t$ are all positive.

Since $|\lambda_2| = |-\frac{1}{2}| = \frac{1}{2}$ and $|\lambda_3| = |-1| = 1$, we see that conclusion (D) is true.

14.   For the rest of our analysis, we assume that A is diagonalizable, or equivalently, that A has n linearly independent eigenvectors. (This will be the case if, as in our model Leslie matrix problem, there are n distinct eigenvalues.) The following theorem is very useful.

Theorem 8.7

Suppose all the hypotheses of Theorem 8.6 are satisfied and, moreover, A has n linearly independent eigenvectors. For the unique positive eigenvalue λ_1, let $\mathbf{p}_1$ be a corresponding eigenvector for which all components are positive.

Then, for large values of k (theoretically, as $k \to \infty$), an *approximate* equation for the age distribution vector $\mathbf{x}^{(k)}$ is

$$\mathbf{x}^{(k)} \cong c(\lambda_1)^k \mathbf{p}_1. \tag{A}$$

(Here $\cong$ means "approximately equal to," and c is a proportionality constant.) In particular, this states that for large values of k, the age distribution vector $\mathbf{x}^{(k)}$ *stabilizes* and is *approximately* proportional to $\mathbf{p}_1$. Furthermore, from (A), it follows that

$$\mathbf{x}^{(k)} \cong \lambda_1 \mathbf{x}^{(k-1)} \tag{B}$$

for large values of k. This means that for large values of k (or, equivalently, of time), the age distribution vector is (approximately) a scalar multiple of the preceding age distribution vector, where the scalar is the unique positive eigenvalue λ_1.

The theorem tells us that, for large values of time, the *proportion* of females in each of the age classes becomes (approximately) constant, and that the number of females in each age class will change by a factor equal to λ_1 during each time period. (That is, from t_{k-1} to t_k. As we know from frame 9, the duration of a time period is equal to M/n years.) Thus, we have the following criterion:

If $\lambda_1 > 1$, the total population eventually increases in size.
If $\lambda_1 = 1$, the total population eventually becomes constant in size.
If $\lambda_1 < 1$, the total population eventually decreases in size.

Proof of Theorem 8.7

Since we hypothesize that A has n linearly independent eigenvectors, we let $S = \{\mathbf{p}_1, \mathbf{p}_2, \mathbf{p}_3, \ldots, \mathbf{p}_n\}$ be a set of such eigenvectors. Also, let $\mathbf{p}_1$ be an eigenvector corresponding to unique positive eigenvalue λ_1, for which all components are positive. Now S forms a basis for R^n [part (a) of Theorem 5.9 in frame 16], and so we can express $\mathbf{x}^{(0)}$, the initial age distribution vector, as

$$\mathbf{x}^{(0)} = c_1 \mathbf{p}_1 + c_2 \mathbf{p}_2 + \ldots + c_n \mathbf{p}_n. \tag{1}$$

Here the c_i's form a unique collection of constants. Now, from Eq. (III) of Theorem 8.5, we have

$$\mathbf{x}^{(k)} = A^k \mathbf{x}^{(0)} \tag{2}$$

(Our goal is to develop an approximate equation for $\mathbf{x}^{(k)}$ as k increases without bound.) Now it can be shown that

$$A^k \mathbf{p}_i = (\lambda_i)^k \mathbf{p}_i, \tag{3}$$

if $\mathbf{p}_i$ is an eigenvector corresponding to eigenvalue λ_i. [For example, for $k = 2$, we have $A^2\mathbf{p}_i = A(A\mathbf{p}_i) = A(\lambda_i\mathbf{p}_i) = \lambda_i(A\mathbf{p}_i) = \lambda_i(\lambda_i\mathbf{p}_i) = (\lambda_i)^2\mathbf{p}_i$, where we twice used the fact that $A\mathbf{p}_i = \lambda_i\mathbf{p}_i$. A similar derivation applies to any integer k.] If we substitute (1) into (2), and then make use of (3), we get

$$\mathbf{x}^{(k)} = A^k(c_1\mathbf{p}_1 + c_2\mathbf{p}_2 + \ldots + c_n\mathbf{p}_n) \tag{4a}$$
$$= c_1(\lambda_1)^k\mathbf{p}_1 + c_2(\lambda_2)^k\mathbf{p}_2 + \ldots + c_n(\lambda_n)^k\mathbf{p}_n \tag{4b}$$

Let us factor out the $(\lambda_1)^k$ from the right side of (4b):

$$\mathbf{x}^{(k)} = (\lambda_1)^k\left[c_1\mathbf{p}_1 + c_2\left(\frac{\lambda_2}{\lambda_1}\right)^k\mathbf{p}_2 + \ldots + c_n\left(\frac{\lambda_n}{\lambda_1}\right)^k\mathbf{p}_n\right] \tag{5}$$

From Theorem 8.6, part (D), we have for $i \neq 1$, $\left|\dfrac{\lambda_i}{\lambda_1}\right| < 1$, and from this it follows that $\left(\dfrac{\lambda_i}{\lambda_1}\right)^k$ approaches zero as k increases without bound. Thus, for k very large, Eq. (5) becomes

$$\mathbf{x}^{(k)} \cong c_1(\lambda_1)^k\mathbf{p}_1, \tag{6 or (A)}$$

and this is Eq. (A) in Theorem 8.7. Now if k is very large, so too is $(k - 1)$, and thus Eq. (6) applies also for index $k - 1$. Thus,

$$\mathbf{x}^{(k-1)} \cong c_1(\lambda_1)^{k-1}\mathbf{p}_1. \tag{7}$$

A comparison of Eqs. (6) and (7) yields

$$\mathbf{x}^{(k)} \cong \lambda_1\mathbf{x}^{(k-1)}, \tag{8 or (B)}$$

since the right side of (6) is λ_1 times the right side of (7). Eq. (8) is none other than Eq. (B) in Theorem 8.7.

Example: For our model Leslie matrix problem, determine the approximate population growth behavior after many years. Make use of the results from the Examples of frames 12 and 13.

Solution: The unique positive eigenvalue is $\lambda_1 = \frac{3}{2}$, and corresponding to this is the eigenvector with all positive entries $\mathbf{p}_1 = \begin{bmatrix} 9 \\ 3 \\ 1 \end{bmatrix}$. Thus, the females will eventually be distributed among the three age classes in the ratios 9:3:1. This corresponds to 69.2% of the females in the youngest class (from the fraction 9/13), 23.1% in the second age class, and 7.7% in the oldest age class. (Recall that the youngest age class, C_1, is from 0 to 15 years of age, and the oldest age class, C_3, is from 30 to 45 years of age.)

Now determine the approximate percentage increase in total population every 15 years, after many years. (Remember the time period for both the age classes and the observation times is 15 years. See the Example in frame 9.) *Hint:* Use Eq. (B) of Theorem 8.7.

— — — — — — — — — —

From Eq. (B) of Theorem 8.7, we have

$$\mathbf{x}^{(k)} \cong \left(\tfrac{3}{2}\right)\mathbf{x}^{(k-1)}$$

after many years. This means that every 15 years the number of females in each of the three classes will increase by about 50% (since $\tfrac{3}{2} = 1.50$). Thus, the total number of females in the population will also increase by about 50% every 15 years.

SELF-TEST

This Self-Test will help you determine whether or not you have mastered the chapter objectives and are ready to go on to another chapter. Correct answers are given at the end of the test.

1. A certain isolated town has two supermarket stores, which we shall label I and II. Within the town (which has a constant population) there always exists a shift of customers from one store to the other. During each month store I retains 80% of its customers, and loses 20% to store II, while store II retains 60% of its customers and loses 40% to store I. Display the transition matrix for this situation. Each step has a one-month duration. *Hint:* $p_{12} = .2$ since 20% of the customers switch from store I to II.

2. Refer to question 1. Suppose that it is known that during January 55% of the customers in the town shopped at store I and 45% at store II. Determine the distribution of customers during

    (a) The month of February.

    (b) The month of March.

3. A long-range study of a brand of photocopying machine (e.g., a Xerox machine) over one-day periods revealed that if the machine is working during any one day, there is an 80% probability it will be working, and a 20% probability it won't be working the next day. If the machine is not working during a day, there is a 55% probability it won't be working and a 45% probability it will be working the next day. If the states are labeled 1 for working and 2 for not working, determine the transition matrix.

4. Refer to question 3. Suppose that during a particular day, the photocopier is *not* working. This can be denoted by the *initial* probability distribution $\mathbf{x}^{(0)} = [0, 1]$. Determine the probability distribution:

    (a) One day later.

    (b) Two days later.

5. Determine the steady-state (i.e., long-run) probability distribution for the Markov process described in question 1. Interpret.

6. Determine the steady-state probability distribution for the Markov process described in question 3. Interpret.

7. Suppose the females of a particular animal population are divided into two age classes and the Leslie matrix is $A = \begin{bmatrix} \frac{3}{4} & \frac{1}{3} \\ \frac{3}{4} & 0 \end{bmatrix}$. Beginning with the initial age distribution vector $\mathbf{x}^{(0)} = \begin{bmatrix} 3000 \\ 7000 \end{bmatrix}$, use the scheme of the Example of frame 11 to calculate $\mathbf{x}^{(1)}$, $\mathbf{x}^{(2)}$, $\mathbf{x}^{(3)}$, and $\mathbf{x}^{(4)}$.

8. Refer to question 7.

    (a) Calculate the eigenvalues, eigenvectors, and the matrixes D, P, and P^{-1} for matrix A. Note that $A = PDP^{-1}$.

    (b) If $\mathbf{x}^{(0)}$ is $\begin{bmatrix} 3000 \\ 7000 \end{bmatrix}$ as in question 7, calculate $\mathbf{x}^{(5)}$ using the exact equation $\mathbf{x}^{(5)} = A\mathbf{x}^{(4)}$ [see answers to question 7], and the approximate equation $\mathbf{x}^{(5)} \cong \lambda_1 \mathbf{x}^{(4)}$.

    (c) Calculate A^5 from PD^5P^{-1}, and then use $\mathbf{x}^{(5)} = A^5\mathbf{x}^{(0)}$ to calculate $\mathbf{x}^{(5)}$ if $\mathbf{x}^{(0)} = \begin{bmatrix} 3000 \\ 7000 \end{bmatrix}$.

    (d) Determine the approximate population growth behavior after many years.

ANSWERS TO SELF-TEST

If your answers to the test questions do not agree with the following answers, review the frames indicated in parentheses after each answer before you go on to another chapter.

1.
	I	II
I	.8	.2
II	.4	.6

(frames 1, 2)

2. (a) $\mathbf{x}^{(1)} = [.62, .38]$, meaning 62% shopped at store I and 38% at store II during February.

    (b) $\mathbf{x}^{(2)} = [.648, .352]$.

(frames 3, 4)

3.

	State 1 (Working)	State 2 (Not Working)	
State 1 (Working)	.8	.2	(frames 1, 2)
State 2 (Not Working)	.45	.55	

4. (a) $\mathbf{x}^{(1)} = [.45, .55]$, meaning there is a .45 probability the machine will be working *one* day later and a .55 probability that it won't be.

(b) $\mathbf{x}^{(2)} = [.6075, .3925]$.   (frames 3, 4)

5. $\mathbf{t} = [\frac{2}{3}, \frac{1}{3}]$. This implies that, in the long run, $\frac{2}{3}$, or 66.7%, of the customers

will be shopping at store I and $\frac{1}{3}$, or 33.3%, will be shopping at store II.

(frames 5, 6)

6. $\mathbf{t} = [\frac{9}{13}, \frac{4}{13}]$. This implies that, in the long run, the photocopying machine

will be found working $\frac{9}{13}$, or 69.2%, of the time and not working $\frac{4}{13}$, or 30.8%, of the time.   (frames 5, 6)

7. $\mathbf{x}^{(1)} = \begin{bmatrix} 4,583.33 \\ 2,250.00 \end{bmatrix}$, $\mathbf{x}^{(2)} = \begin{bmatrix} 4,187.50 \\ 3,437.50 \end{bmatrix}$,

(frames 9–11)

$\mathbf{x}^{(3)} = \begin{bmatrix} 4,286.46 \\ 3,140.62 \end{bmatrix}$, $\mathbf{x}^{(4)} = \begin{bmatrix} 4,261.72 \\ 3,214.85 \end{bmatrix}$

8. (a) $\lambda_1 = 1$, $\lambda_2 = -\frac{1}{4}$; $\mathbf{p}_1 = \begin{bmatrix} 4 \\ 3 \end{bmatrix}$, $\mathbf{p}_2 = \begin{bmatrix} -1 \\ 3 \end{bmatrix}$; $D = \begin{bmatrix} 1 & 0 \\ 0 & -\frac{1}{4} \end{bmatrix}$;

$P = \begin{bmatrix} 4 & -1 \\ 3 & 3 \end{bmatrix}$; $P^{-1} = \left(\frac{1}{15} \right) \begin{bmatrix} 3 & 1 \\ -3 & 4 \end{bmatrix}$

(b) *Exact:* $\mathbf{x}^{(5)} = \begin{bmatrix} 4,267.91 \\ 3,196.29 \end{bmatrix}$.   *Approx.:* $\mathbf{x}^{(5)} \cong \begin{bmatrix} 4,261.72 \\ 3,214.85 \end{bmatrix}$

(c) First, $A^5 = \left(\frac{1}{15} \right) \begin{bmatrix} 11.9971 & 4.0039 \\ 9.0088 & 2.9883 \end{bmatrix}$. Then, $\mathbf{x}^{(5)} = \begin{bmatrix} 4,267.9 \\ 3,196.3 \end{bmatrix}$ as in 8b.

Thus, approximately 4,268 and 3,196 females in each of the two age classes, respectively, after five time periods.

(d) Since $\mathbf{p}_1 = \begin{bmatrix} 4 \\ 3 \end{bmatrix}$, the females will eventually be distributed among the

two age classes in the ratio 4:3. That is, 57.14% (from $\frac{4}{7}$) in the first class and 42.86% in the second class. Since $\lambda_1 = 1$, eventually the number of females in each of the two age classes, and hence in the total population, will become constant.   (frames 11–14)

CHAPTER NINE

Numerical, Computer, and Calculator Methods

The computations for so many real-world and classroom linear algebra (matrix algebra) problems are accomplished by means of computers and advanced calculators (such as programmable calculators). One reason for this is that the matrices encountered in practice often are large in size (many rows and/or many columns). Also, the entries are often not simple in their forms. Associated with the extensive use of computers for solving and analyzing linear algebra problems are the numerical methods that form the basis for the computer programs that have been written.

In this chapter we shall briefly study some numerical methods for solving linear algebra problems. Also, at the end of Section A we shall list several very popular and frequently used numerical methods and indicate references in which such techniques are discussed in detail.

Section B of this chapter is concerned with computer methods. We shall briefly study how the FORTRAN programming language can be used for solving linear algebra problems. Our main attention here will be focused on using the already available International Mathematical and Statistical Libraries (IMSL) FORTRAN subroutines to do linear algebra calculations. Then we shall study how the MAT statements, which are available in some versions of the BASIC programming language, can be used to do linear algebra calculations.

We shall also make a brief survey of some of the software available for doing linear algebra calculations on personal computers.

In Section C we shall illustrate how programmable calculators in conjunction with software modules can be used to do routine linear algebra calculations.

OBJECTIVES

When you complete this chapter you should be able to

- Use the Gaussian elimination method to transform a matrix to row echelon form.

- Apply the back-substitution process to a row echelon form, and find the solution to a linear system of equations.
- Use Gaussian elimination with pivoting, and rounding off to a prescribed number of significant digits, to find an approximate solution to a linear system of equations.
- Use a FORTRAN program that calls upon IMSL subroutines to do linear algebra calculations.*
- Use a BASIC program incorporating MAT statements to do linear algebra calculations.*
- Use software available for personal computers to do linear algebra calculations on such computers.*
- Use a TI 58C (or 59) programmable calculator together with a Master Library Module to do linear algebra calculations.*

A. NUMERICAL METHODS (GAUSSIAN ELIMINATION)

1. In Section B of Chapter Two we discussed the Gauss–Jordan elimination method (Method 2.1 in frame 14) for transforming a matrix (augmented or otherwise) to reduced row echelon form. In this section we present a very similar method (Gaussian elimination method), which is more efficient from a computational point of view for solving a system of linear equations (on the average, fewer total calculations are required). The reader is advised to review material from Chapter Two, in particular the three elementary row operations (Definition 2.3 in frame 8), the reduced row echelon form (Definition 2.4 in frame 11), and Method 2.1.

A matrix is in *row echelon form* if the first three properties associated with the reduced row echelon form definition are satisfied.

Definition 9.1

A matrix is in row echelon form if it satisfies the following three properties:

1. All rows, consisting entirely of zeros, if any, are at the bottom of the matrix.
2. For a row that does not consist entirely of zeros the first nonzero entry of the row is a 1. We call this entry the *leading entry* of the row. Often we shall underline the leading entry of the row (as in $\underline{1}$). Also, for

---

*Of course, we realize that a person's skill in a particular area relating to computers or calculators (e.g., using a specific computer programming language) depends on the person's prior knowledge in that particular area.

the leading entry of row i, we identify the column location (or label) of the leading entry as column J_i.

3.   If row i and $(i + 1)$ are two successive rows that do not consist entirely of zeros, then the leading entry of row $(i + 1)$ is to the right of the leading entry of row i. That is, $J_{(i+1)} > J_i$. [Recall that row $(i + 1)$ is positionally lower than row i.]

The row echelon form differs from the reduced row echelon form in the following way: If a column contains the leading entry of some row, the entries above the leading entry are not necessarily zeros. The entries below the leading entry are zeros, as in the reduced row echelon form. We shall maintain our policy of usually underlining the leading entry of a row.

Example (a):   The following matrices are in row echelon form.

$$\text{(i)} \begin{bmatrix} \underline{1} & 7 & -2 & 3 \\ 0 & \underline{1} & 7 & 2 \\ 0 & 0 & \underline{1} & 4 \end{bmatrix} \qquad \text{(ii)} \begin{bmatrix} \underline{1} & 2 & -4 & 2 \\ 0 & 0 & \underline{1} & 6 \\ 0 & 0 & 0 & 0 \end{bmatrix} \qquad \text{(iii)} \begin{bmatrix} \underline{1} & 6 & 5 & 0 \\ 0 & \underline{1} & 2 & -3 \\ 0 & 0 & 0 & \underline{1} \end{bmatrix}$$

Note that if a matrix is in reduced row echelon form, it is automatically in row echelon form.

We will now illustrate the Gaussian elimination method, which is used to transform any matrix (augmented or not) to row echelon form. Then, if the starting matrix is the augmented coefficient matrix for a linear system of equations, a final process known as *back-substitution* is used to find the solution or solutions for the system (if such exist).

We will apply the method to the same augmented matrix as in frame 14 of Chapter Two. That is, the augmented matrix for the system of three equations in four unknowns is

$$A|\mathbf{b} = \begin{bmatrix} 0 & 0 & 7 & 14 & -7 \\ 2 & -8 & 4 & 18 & 0 \\ 3 & -12 & 1 & 13 & 7 \end{bmatrix} \tag{I}$$

As with Method 2.1, there is one stage for each row until the method is applied to all rows, or until the remaining rows consist entirely of zeros. Here, as in Method 2.1, four steps comprise each stage. The first three steps are identical in both methods; the only difference is in the fourth step. Here the fourth step is labeled *Step 4'*. In Step 4' we *clear out* the portion of a column which is *below* a given row.

In the Gaussian elimination method, Method 9.1, the *general* steps are bracketed (as in [ ]), and to the right of the central dividing line.

Recall that we indicate the interchange of rows i and j by $r_i \leftrightarrow r_j$.

Method 9.1 (Gaussian Elimination Method)

$$\begin{bmatrix} 0 & 0 & 7 & 14 & -7 \\ 2 & -8 & 4 & 18 & 0 \\ 3 & -12 & -1 & 13 & 7 \end{bmatrix}$$
↑
Left-most nonzero column.

Stage 1 (for Row 1)

$\begin{bmatrix} \text{Step 1:} & \text{Locate the left-most column} \\ \text{that does not contain all zeros. This} \\ \text{locates column } J_1. \end{bmatrix}$

Here the left-most nonzero column is column 1. Thus, $J_1 = 1$.

$$\begin{bmatrix} 2 & -8 & 4 & 18 & 0 \\ 0 & 0 & 7 & 14 & -7 \\ 3 & -12 & -1 & 13 & 7 \end{bmatrix}$$

$\begin{bmatrix} \text{Step 2:} & \text{If necessary, interchange the} \\ \text{top row with another row such that} \\ \text{the entry at top of column } J_1 \text{ is un-} \\ \text{equal to zero.} \end{bmatrix}$

Here we interchange rows 1 and 2 [type (a) row operation]. Symbolically, $r_1 \leftrightarrow r_2$.

$$\begin{bmatrix} \underline{1} & -4 & 2 & 9 & 0 \\ 0 & 0 & 7 & 14 & -7 \\ 3 & -12 & -1 & 13 & 7 \end{bmatrix}$$

$\begin{bmatrix} \text{Step 3:} & \text{Let the entry now at the top} \\ \text{of the column found at step 1 be } a. \text{ If} \\ a \neq 1 \text{ multiply the first row by } 1/a \\ \text{in order to generate a leading entry} \end{bmatrix}$

for row 1. $\left[\text{Type (b) row operation:} \right.$
$r_1' = \left(\dfrac{1}{a}\right)r_1.$ $\left. \right]$ If $a = 1$, no multiplication is needed. Underline leading entry in row 1.

Here $r_1' = \left(\frac{1}{2}\right) r_1 = r_1/2$, since we multiply row 1 by $\frac{1}{2}$.

$$\begin{bmatrix} \underline{1} & -4 & 2 & 9 & 0 \\ 0 & 0 & 7 & 14 & -7 \\ 0 & 0 & -7 & -14 & 7 \end{bmatrix}$$

$\begin{bmatrix} \text{Step 4':} & \text{Use type (c) row operations} \\ \text{on rows } 2, 3, \ldots, m, \text{ when necessary,} \\ \text{to clear out column } J_1. \text{ This com-} \\ \text{pletes Stage 1.} \end{bmatrix}$

Here, leave row 2 alone ($r_2' = r_2$) and use $r_3' = r_3 - 3r_1$ on row 3 to complete Stage 1.

Note that for Stage 1, Step 4' is identical to Step 4 of Method 2.1.

$$\begin{bmatrix} \underline{1} & -4 & 2 & 9 & 0 \\ 0 & 0 & 7 & 14 & -7 \\ 0 & 0 & -7 & -14 & 7 \end{bmatrix}$$

$\uparrow$

Left-most nonzero column in submatrix.

$$\begin{bmatrix} \underline{1} & -4 & 2 & 9 & 0 \\ 0 & 0 & 7 & 14 & -7 \\ 0 & 0 & -7 & -14 & 7 \end{bmatrix}$$

$$\begin{bmatrix} \underline{1} & -4 & 2 & 9 & 0 \\ 0 & 0 & \underline{1} & 2 & -1 \\ 0 & 0 & -7 & -14 & 7 \end{bmatrix}$$

$$\begin{bmatrix} \underline{1} & -4 & 2 & 9 & 0 \\ 0 & 0 & \underline{1} & 2 & -1 \\ 0 & 0 & 0 & 0 & 0 \end{bmatrix}$$

Stage 2 (for Row 2)

Step 1: Focus on the submatrix consisting of rows 2, 3, etc. For this submatrix, locate the left-most column that does not consist of all zeros. This locates column J_2.

Here the left-most nonzero column is column 3, and thus, $J_2 = 3$.

Step 2: Repeat Step 2 of Stage 1 for submatrix referred to in Step 1 of Stage 2.

Here *no change* is necessary, since $a_{23} \neq 0$ ($a_{23} = 7$ here).

Step 3: Repeat Step 3 of Stage 1 for submatrix referred to in Step 1 of Stage 2.

Here $r_2' = r_2/7$, since we multiply row 2 by 1/7.

Step 4': Use type (c) row operations on rows $3, 4, \dots, m$, when necessary, to clear out the portion of column J_2 *below* row 2. This completes Stage 2.

Use $r_3' = r_3 + 7r_2$ on row 3. Our matrix is now in row echelon form, since rows 1 and 2 have leading entries and row 3 consists entirely of zeros. Note that column 3 is not completely cleared out; it is not necessary for it to be when we are transforming a matrix to row echelon form.

If the matrix is not in row echelon form after two stages, then Stage 3 is used. In the four steps of Stage 3, we would focus on the submatrix consisting of rows $3, 4, \dots, m$. Thus, we see that we have a similar pattern for all stages. We terminate after a particular stage if the matrix has been transformed into row echelon form.

In Example (b), which follows, we illustrate the *back-substitution* process for finding the solution(s) of the original system by working with the row echelon form just obtained.

Example (b): Use the row echelon form just obtained to determine the solution(s) of the original system of equations.

Solution: First, we write out the equations corresponding to the row echelon form:

$$x_1 - 4x_2 + 2x_3 + 9x_4 = \quad 0 \tag{1}$$

$$x_3 + 2x_4 = -1 \tag{2}$$

Next, we solve for the leading variables (here, x_1 and x_3) in terms of the remaining variables:

$$x_1 = 4x_2 - 2x_3 - 9x_4 \tag{1a}$$

$$x_3 = -1 - 2x_4 \tag{2a}$$

[Now, in general, we would begin with the bottom equation and work *backwards*, successively substituting the results previously found for the leading variables into the preceding equations. This process is known as *back-substitution*.]

Here we substitute x_3 from (2a) into (1a) and get

$$x_1 = 2 + 4x_2 - 5x_4$$

Lastly, we assign arbitrary value symbols r, s, etc., to the nonleading variables (here x_2 and x_4).

Complete the work.

— — — — — — — — — —

We get

$$x_1 = \quad 2 + 4r - 5s$$
$$x_2 = \quad\quad\quad r$$
$$x_3 = -1 \quad\quad - 2s$$
$$x_4 = \quad\quad\quad\quad s$$

This is the same result obtained from the reduced row echelon form in frame 15 of Chapter Two.

In vector form, we have

$$\mathbf{x} = \begin{bmatrix} 2 + 4r - 5s \\ r \\ -1 \quad - 2s \\ s \end{bmatrix} = \begin{bmatrix} 2 \\ 0 \\ -1 \\ 0 \end{bmatrix} + r \begin{bmatrix} 4 \\ 1 \\ 0 \\ 0 \end{bmatrix} + s \begin{bmatrix} -5 \\ 0 \\ -2 \\ 1 \end{bmatrix}$$

2.   Note: For a particular matrix there is only one reduced row echelon form. That is, the reduced row echelon form is unique. However, for a particular matrix, the row echelon form is usually not unique. By changing the sequence of row operations, one can arrive at different row echelon forms. For example, for the matrix $\begin{bmatrix} 1 & 4 \\ 3 & 8 \end{bmatrix}$, two possible row echelon forms are (verify) $\begin{bmatrix} 1 & 4 \\ 0 & 1 \end{bmatrix}$ and $\begin{bmatrix} 1 & \frac{8}{3} \\ 0 & 1 \end{bmatrix}$. To get the second form, first interchange rows 1 and 2 in the given matrix. The unique reduced row echelon form is $\begin{bmatrix} 1 & 0 \\ 0 & 1 \end{bmatrix}$.

For the remainder of this section we will focus on linear systems of n equations in n unknowns. (Also, it will ususally be true that the system will have a unique solution.)

Example: For the Example of frame 18 of Chapter Two, use the Gaussian elimination method (Method 9.1) to find a row echelon form, and then the solution (through back-substitution relative to the row echelon form).

Solution: The starting equations and augmented matrix are as follows:

Starting equations:

$$x_1 - x_2 + x_3 = 1$$
$$2x_1 + x_2 + 3x_3 = 4$$
$$3x_1 - x_2 - x_3 = -5$$

Starting augmented coefficient matrix:

$$\left[\begin{array}{rrr|r} 1 & -1 & 1 & 1 \\ 2 & 1 & 3 & 4 \\ 3 & -1 & -1 & -5 \end{array}\right]$$

The sequence of matrices obtained (verify) are as follows:

$$\left[\begin{array}{rrr|r} 1 & -1 & 1 & 1 \\ 2 & 1 & 3 & 4 \\ 3 & -1 & -1 & -5 \end{array}\right] \xrightarrow[r_3' = r_3 - 3r_1,]{r_2' = r_2 - 2r_1} \left[\begin{array}{rrr|r} 1 & -1 & 1 & 1 \\ 0 & 3 & 1 & 2 \\ 0 & 2 & -4 & -8 \end{array}\right] \xrightarrow[r_3' = r_3 - 2r_2]{r_2' = r_2/3;}$$

[Stage 1 is completed]

$$\left[\begin{array}{rrr|r} 1 & -1 & 1 & 1 \\ 0 & 1 & \frac{1}{3} & \frac{2}{3} \\ 0 & 0 & -\frac{14}{3} & -\frac{28}{3} \end{array}\right] \xrightarrow{r_3' = (-3/14)r_3} \left[\begin{array}{rrr|r} 1 & -1 & 1 & 1 \\ 0 & 1 & \frac{1}{3} & \frac{2}{3} \\ 0 & 0 & 1 & 2 \end{array}\right]$$

[Stage 2 is completed]   [Stage 3 is completed]

Note: The sequence of row operations for a particular transition is indicated adjacent to an associated arrow. For example, in going from the second to the third matrix, the row operations are first $r_2' = r_2/3$, and then $r_3' = r_3 - 2r_2$ applied to the matrix that is obtained (not shown) from applying $r_2' = r_2/3$.

The last matrix shown is a row echelon form. The corresponding equations are

$$x_1 - x_2 + x_3 = 1 \tag{1}$$

$$x_2 + x_3/3 = 2/3 \tag{2}$$

$$x_3 = 2 \tag{3}$$

Now solve by using back-substitution. (Thus, each time an equation is encountered in applying this process, it will have only one unknown left.)

— — — — — — — — — —

From Eq. (3) we have $x_3 = 2$. Substituting into Eq. (2), we get

$$x_1 = 2/3 - x_3/3 = \tfrac{2}{3} - \tfrac{2}{3} = 0.$$

Now substituting the values of x_2 and x_3 into Eq. (1), we get

$$x_1 = 1 + x_2 - x_3 = 1 + 0 - 2 = -1.$$

Thus, the unique solution is $x_1 = -1$, $x_2 = 0$, $x_3 = 2$.

3. Let us digress for a while to discuss accuracy of results. In real calculations on a computer or a hand calculator, numbers are handled in their decimal form (for example, $\tfrac{3}{2}$ is treated as 1.5). Also, results are rounded off to a finite number of digits. Typically, for a hand calculator or large computer, the accuracy is to about eight significant digits (large computers do, however, provide for double precision, which extends accuracy to about 16 significant digits, but at a cost of time and storage space).

Notes: (i) It should be noted that for most of the calculations done thus far in this book, we worked with simple numbers, so simple in fact that we could combine fractions easily by reducing to common denominators (as in $\tfrac{3}{4} - \tfrac{1}{3} = \tfrac{5}{12}$, say). In real-world calculations involving matrices, things are quite different, and we are forced to do our calculations on hand calculators or computers. The numbers we would work with are most often expressed in decimal form ($\tfrac{4}{3}$ would be 1.33333 to six significant digits in decimal form), and we sometimes have to pay attention to matters such as significant digits

and accuracy of results.   (ii)   The topics of this and the next frame are treated more extensively in Anton (1977), Grossman (1980a), and Kolman (1980).

To give an indication of the meaning of significant digits, suppose we have
a hypothetical computer or calculator that has accuracy to three significant
digits. Some typical numbers that have three significant digits are:

4.67,   25.2,   8650.,   .0137,   4.50, $-74200.$,   .00370

The count is from the left-most nonzero digit. In the number 8650., there
is no accuracy in the fourth digit, which is zero. In .00370, the trailing zero is
the third digit carried, and it is accurate.

As examples of divisions and multiplications in which we round the result
to three significant digits, consider the following:

$$2.3/.15 = 15.333 \ldots \cong 15.3, \qquad .10/.15 = .6666 \ldots \cong .667,$$

$$156./1.67 = 93.413 \ldots \cong 93.4, \qquad (141.)(9.) = 1269. \cong 1270.$$

Here the "approximately equals" symbol $\cong$ is used to indicate the roundoff
to three significant digits. (You should check the initial parts of the calculations
with a hand calculator.)

Our convention will usually be to drop trailing zeros in our three-significant-
digit numbers. Thus, for the preceding initial list of numbers, we could write
4.50 as 4.5 and .00370 as .0037. Also, for .000, which is how we could write
zero to three significant digits, we will usually write this as 0, and 1.00 will
usually be written as 1.

Suppose in the following that we are working with decimal numbers and
that results are rounded off to a prescribed number of significant digits after
doing calculations. A problem with the Gaussian elimination method (or Gauss–
Jordan elimination method) is that in certain calculations, most notably in
division steps [which occurs in type (b) elementary row operations], we will
occasionally introduce roundoff errors. Thereafter, in subsequent calculations,
such errors can propagate, and finally result in very large errors in x_i values
in the final solution.

In particular, this can happen if initially in a sequence of calculations, we
divide one number by another number that is much smaller in absolute value
and then round our results to the prescribed number of significant digits. [For
an example of roundoff error in dividing one number by a smaller number,
$65.8/.00152 = 43289.47 \ldots$, which is rounded to 43300. to three significant
digits.]

In using the Gaussian elimination method (or the Gauss–Jordan elimination
method), we can remedy this somewhat by dividing by the entry in a column
that has the largest absolute value, during Step 2 of a stage.

The following variation to Method 9.1, will be called the *Gaussian elimination method with pivoting* (also called *Gaussian elimination with partial
pivoting* or *pivotal condensation*).

As with Method 9.1, if the starting matrix is the augmented coefficient matrix for a linear system of equations, the final back-substitition process is applied relative to the resulting row echelon form to find the solution or solutions for the system, if such exist.

We illustrate the method with respect to the following sample system of equations:

$$.15x_1 + .1x_2 + 2.4x_3 = 7.4 \tag{1}$$

$$10x_1 + 5x_2 - 3x_3 = 6 \tag{2}$$

$$3x_1 - 7x_2 - 2x_3 = 7 \tag{3}$$

The exact solution to the sample system is $x_1 = 2$, $x_2 = -1$, $x_3 = 3$; this can be checked by substitution into Eqs. (1), (2), and (3).

In the Gaussian elimination method with pivoting below, the general steps, which appear to the right of the central dividing line, are set off with brackets, as in [ ]. On the left, we start with the augmented coefficient matrix for the sample system given by Eqs. (1), (2), and (3).

After each calculation, results will be rounded to three significant digits.

Method 9.2 (Gaussian Elimination Method with Pivoting)

Stage 1 (for Row 1)

$$\begin{bmatrix} .15 & .1 & 2.4 & 7.4 \\ \boxed{10} & 5 & -3 & 6 \\ 3 & -7 & -2 & 7 \end{bmatrix}$$

[*Step 1:* In the left-most nonzero column, find an entry with the largest absolute value. This entry is called the pivot entry for the column. Circle it.]

The left-most nonzero column is column 1. The pivot entry in column 1 is $a_{21} = 10$.

$$\begin{bmatrix} 10 & 5 & -3 & 6 \\ .15 & .10 & 2.4 & 7.4 \\ 3 & -7 & -2 & 7 \end{bmatrix}$$

[*Step 2:* Perform a row interchange, if necessary, to bring the pivot entry to the top of the column.]

Interchange rows 1 and 2; in symbols, $r_1 \leftrightarrow r_2$.

$$\begin{bmatrix} \underline{1} & .5 & -.3 & .6 \\ .15 & .10 & 2.4 & 7.4 \\ 3 & -7 & -2 & 7 \end{bmatrix}$$

[*Step 3:* Let the entry now at the top of the column found in Step 1 be a. If $a \neq 1$, multiply the first row by $1/a$ (divide by a). If $a = 1$, no multiplication is needed.]

Multiply the first row by $\frac{1}{10}$. In symbols $r_1' = r_1/(10)$.

$$\begin{bmatrix} 1 & .5 & -.3 & | & .6 \\ 0 & .025 & 2.45 & | & 7.31 \\ 0 & -8.5 & -1.1 & | & 5.2 \end{bmatrix}$$

Step 4': Use type (c) row operations on rows below top row, when necessary, to clear out the column located in Step 1.

Here, in symbols, $r_2' = r_2 - .15r_1$ and $r_3' = r_3 - 3r_1$ are applied to previous matrix. Stage 1 is completed.

Stage 2 (for Row 2)

Step 1: Focus on the submatrix consisting of rows 2, 3, etc. For this submatrix, locate the left-most nonzero column, and in this column, locate the entry with the largest absolute value. This entry is the pivot entry for the column. Circle it.

$$\begin{bmatrix} 1 & .5 & -.3 & | & .6 \\ 0 & .025 & 2.45 & | & 7.31 \\ 0 & \boxed{-8.5} & -1.1 & | & 5.2 \end{bmatrix}$$

The left-most nonzero column in the submatrix is column 2. The pivot entry in the submatrix is -8.5.

Step 2: Perform a row interchange, if necessary, to bring the pivot entry to the top of the column of the *submatrix*.

$$\begin{bmatrix} 1 & .5 & -.3 & | & .6 \\ 0 & -8.5 & -1.1 & | & 5.2 \\ 0 & .025 & 2.45 & | & 7.31 \end{bmatrix}$$

Interchange rows 2 and 3. In symbols, $r_2 \leftrightarrow r_3$.

Henceforth, we will *abbreviate* the work. We will list only the calculations for the sample system.

$$\begin{bmatrix} 1 & .5 & -.3 & | & .6 \\ 0 & 1 & .129 & | & -.612 \\ 0 & .025 & 2.45 & | & 7.31 \end{bmatrix}$$

Step 3: Multiply first row of the submatrix by $1/(-8.5)$. In symbols, $r_2' = r_2/(-8.5)$.

$$\begin{bmatrix} 1 & .5 & -.3 & | & .6 \\ 0 & 1 & .129 & | & -.612 \\ 0 & 0 & 2.45 & | & 7.33 \end{bmatrix}$$

Step 4': Clear out column 2 of the *submatrix*. Transform the .025 to zero by using $r_3' = r_3 - .025r_2$.

Stage 2 is completed.

$$\begin{array}{ccc|c} 1 & .5 & -.3 & .6 \\ 0 & 1 & .129 & -.612 \\ [\,0 & 0 & \boxed{2.45} & 7.33\,] \end{array}$$

Stage 3 (for Row 3)

We go right to Step 3 for our particular matrix.

Step 3: Focus on first row of submatrix. The entire submatrix here consists only of row 3. Multiply this row by 1/2.45. In symbols, $r_3' = r_3/(2.45)$.

$$\begin{bmatrix} 1 & .5 & -.3 & .6 \\ 0 & 1 & .129 & -.612 \\ 0 & 0 & 1 & 2.99 \end{bmatrix}$$

Our resulting matrix is a row echelon form so Stage 3, and all of Method 9.2 are completed.

Next, we will apply the back-substitution process to the row echelon form just found and thereby obtain the solution of the original system. The solution will, of course, be approximate, since we have rounded off to three significant digits.

Example: Solve the sample system of equations that corresponds to this row echelon form by back-substitution.

— — — — — — — — —

Solution:

$$\begin{aligned} x_3 &= 2.99 & \text{(1a)} \\ x_2 &= -.612 - .129(2.99) = -1.00 & \text{(2a)} \\ x_1 &= .6 - .5(-1.) + .3(2.99) = 2.00 & \text{(3a)} \end{aligned}$$

Thus, our answers are very accurate. Recall that the exact solution is $x_1 = 2$, $x_2 = -1$, $x_3 = 3$.

4. Example: Consider the sample system handled in frame 3 by the pivoting approach. Solve it by using Gaussian elimination without pivoting. Again, round calculations to three significant digits.

Solution: Here we use our r, r' (old row, new row) symbolism and list only a few of the intermediate matrices.

$$\begin{bmatrix} .15 & .1 & 2.4 & 7.4 \\ 10 & 5 & -3 & 6 \\ 3 & -7 & -2 & 7 \end{bmatrix} \xrightarrow[\substack{r_2' = r_2 - 10r_1, \\ r_3' = r_3 - 3r_1.}]{r_1' = r_1/(.15);} \begin{bmatrix} 1 & .667 & 16. & 49.3 \\ 0 & -1.67 & -163. & -487. \\ 0 & -9.00 & -50.0 & -141. \end{bmatrix}$$

The symbolism means that initially the first row is divided by .15 [$r_1' = r_1/(.15)$], and then the resulting matrix is transformed by the row operations $r_2' = r_2 - 10r_1$ and $r_3' = r_3 - 3r_1$.

The second matrix listed above is that which exists at the end of Stage 1. Now complete the calculations.

--- -- --- --- --- --- -- ---

$$r_2' = r_2/(-1.67); \qquad \begin{bmatrix} 1 & .667 & 16. & 49.3 \\ 0 & 1 & 97.6 & 292. \\ 0 & 0 & 828. & 2490. \end{bmatrix} \qquad r_3' = r_3/(828.)$$
$$r_3' = r_3 + 9r_2.$$

[*End of Stage 2*]

$$\begin{bmatrix} 1 & .667 & 16. & 49.3 \\ 0 & 1 & 97.6 & 292. \\ 0 & 0 & 1 & 3.01 \end{bmatrix}$$

[*End of Stage 3*]

The last matrix shown is also a row echelon form for the sample system. Solving by back-substitution, we get

$$x_3 = 3.01 \tag{1b}$$

$$x_2 = 292. - 97.6(3.01) = -1.776 \cong -1.78 \tag{2b}$$

$$x_1 = 49.3 - .667(-1.78) - 16.(3.01) \cong 2.33 \tag{3b}$$

The values for x_1 and x_2 are far off from the exact values of $x_1 = 2$ and $x_2 = -1$. The large overall errors can be traced to an accumulation of roundoff errors. Actually, if we do the Gaussian elimination calculations with rounding to six significant digits (again with no pivoting), we would obtain the very excellent results $x_3 = 2.99999$, $x_2 = -1.00002$, and $x_1 = 2.00014$. The equation for x_2 would be

$$x_2 = 292.399 - 97.8(2.99999) \cong -1.00002 \tag{2c}$$

Note that the *exact* calculation for x_2 would be

$$x_2 = 292.4 - 97.8(3) = 292.4 - 293.4 = -1 \tag{2d}$$

Now look at Eq. (2b), which led to the poor approximation $x_2 \cong -1.78$. The first term there, 292., is in error by only .137% as compared to the exact value 292.4 [here, $(292.4 - 292)/(292.4) = .4/(292.4) = .00137 = .137\%$], and the second term $97.6(3.01) = 293.776$ is in error by only .128%. The difference of the two terms yields the result -1.78 to three significant digits, and this result is in error by $.78/1 = .78$, or 78%!

At any rate, in summarizing the comparison of the calculations of the last two frames, we see that pivoting appears to be helpful.

Final Comments for Section A: Our work in this section on numerical methods barely scratches the surface of this vast subject. There are several indirect methods, *iterative* in nature, that are used extensively in practice. (In an iterative method, one develops a sequence of approximations to the solution, where, hopefully, the successive approximations get closer and closer to the actual solution.) Two iterative methods used for solving a system of n equations in n unknowns are the *Jacobi iteration* method and the *Gauss–Seidel iteration* method. These are particularly useful if the number of equations is very large (say, n greater than 50), and if the coefficient matrix has many entries that equal zero.

There are also available numerical approximation methods for computing eigenvalues and eigenvectors of an $n \times n$ matrix A. Several of these are the *power method,* the *deflation method,* and *Jacobi's method* (for a symmetric matrix).

For a further discussion of the numerical methods cited above, the reader is referred to the following books cited in the References: Anton (1977), Fadeev and Fadeeva (1963), Fox (1965), Grossman (1980a), Kolman (1980), Noble and Daniel (1977), and Scheid (1968).

B.   COMPUTER METHODS

Preliminary Comments: Henceforth, we shall dispense with our usual format for the final portion of a frame. Thus, our frames will no longer end with a question to be answered or a problem to be solved.

In frame 5, we shall discuss a computer topic that involves an elementary knowledge of FORTRAN programming. In frame 6 we shall discuss a computer topic dealing with BASIC programming. The Sigma 7 computer at Queens College in New York City was used for the computer runs discussed in this section.

5.   FORTRAN is a scientific computer programming language that was developed in the mid-1950s and that revolutionized the computer programming concept because of its similarity to the English language and its relative simplicity (as compared to the programming languages that preceded it). One of the great strengths of FORTRAN lies in the fact that a program may be broken down into several components: often, one *main* (or *calling*) *program* and other program portions, which are known as *subroutines*. In typical usage, the subroutines are linked to the main program. At present, there are available large collections of subroutines distributed by the following company:

International Mathematical and Statistical Libraries, Inc.
7500 Bellaire Boulevard
Houston, Texas 77036
Telephone: (713) 772-1927

The totality of subroutines (written in FORTRAN and to be used in conjunction wth a main FORTRAN program written by the user) is known collectively as the *IMSL Library*. A summary of information dealing with the IMSL Library of subroutines is available in the *IMSL Library Contents Document,* Edition 8, 1980. Contact the preceding address or phone number to acquire this brochure. Many of the subroutines of the library are available at the large computer centers situated at colleges and companies throughout the country. (If so, the subroutines are already stored within the memory section of the computer, ready to spring into action when called upon.)

The Library collection consists of subroutines from many areas of mathematics and statistics, and these are further sorted into 17 smaller subdivisions (or chapters). Of these, three subdivisions are specifically concerned with linear algebra: Eigensystem Analysis; Linear Algebraic Equations; Vector, Matrix Arithmetic. Taken together, there are over 100 subroutines on linear algebra in these three subdivisions. Also, there is a subdivision dealing with the related topic, linear programming, which contains about 20 subroutines.

An important point about using a subroutine is that the user does not have to know much at all about the structure of the subroutine program. All the user has to know is how input data and information are fed into the subroutine and how output data and information come out of the subroutine.

Here we shall illustrate the usage of two IMSL subroutines, namely, VMULFF, whose function is the multiplying of two matrices, and LINV1F, whose function is the computation of the inverse of a square matrix. First, we shall explain how each of these subroutines functions, and then we shall link both of them to a sample main (or calling) program. We shall show how the sample main program works by listing input data, and the output data that results from an actual run of the sample main program. The input data was typed in through a time-sharing terminal.

Subroutine VMULFF is activated by the following general "CALL" statement placed in the main program:

CALL VMULFF (A, B, L, M, N, IA, IB, C, IC, IER)

The arguments (items enclosed in parentheses) are described as follows. Some apply to input and the rest apply to output of data when subroutine VMULFF is activated during the actual run of a program.

- A—an L by M matrix (input).
- B—an M by N matrix (input).
- L—the number of rows in A (input).
- M—the number of columns in A; it is also the number of rows in B (input).
- N—the number of columns in B (input).
- IA—Row dimension of matrix A exactly as specified in the DIMENSION statement in the calling program (input).
- IB—Row dimension of matrix B exactly as specified in the DIMENISON statement in the calling program (input).
- C—an L by N matrix, which is the matrix product C = AB. In programming symbols, C = A*B (output).

- IC—Row dimension of matrix C exactly as specified in the DIMENSION statement in the calling program (input).
- IER—Error variable (output). The output value IER = 129 indicates A, B, or C was dimensioned incorrectly.

Subroutine LINV1F is activated by the following general CALL statement placed in the main program:

CALL LINV1F (A, N, IA, AINV, IDGT, WKAREA, IER)

The arguments are described as follows:

- A—the N by N matrix to be inverted (input).
- N—the number of rows/columns of A (input).
- IA—Row dimension of matrices A and AINV (AINV is the inverse of A—see below) exactly as specified in the DIMENSION statement of the calling program (input).
- AINV—the inverse of matrix A, and hence an N by N matrix (output).
- IDGT—an input option variable that pertains to a test for accuracy of data. If IDGT equals zero, the accuracy test is bypassed (input).
- WKAREA—Work area array variable of dimension which is at least N.
- IER—Error parameter. If IER = 129, this indicates that inverse of A does not exist (determinant of A is zero) (output).

For elementary usage, it is not necessary to be too concerned with the IER, IDGT, and WKAREA variables. In fact, in our sample program this is our attitude.

In Figure 9.1 we list the sample main program. It is written in standard FORTRAN IV (with some minor exceptions to be explained later). A good reference for the FORTRAN IV programming language (this is the language that people usually refer to as "FORTRAN") is McCracken (1972). It is assumed here that the reader has an elementary understanding of FORTRAN. To summarize, the structure of the program calls for the reading in of matrices A, B, and D from the terminal. (Preceding that, the dimensions M, L, N, and N1 are read in. A is M by L, B is L by N, and D is N1 by N1.)

Refer to Figure 9.1. Subroutine VMULFF is called upon (line 15), and the matrix product AB is formed, and labeled as matrix C. Its dimensions are M by N. Then subroutine LINV1F is called (line 23), and matrix D is inverted. The output matrix D^{-1} is symbolized by DINV; it is N1 by N1 in size. During the the course of running the program, matrices A, B, C, D, and DINV will be printed out on the terminal (such printouts will be displayed shortly).

```
C     MATRIX MULTIPLICATION AND INVERSE
      DIMENSION A(10,10),B(10,10),C(10,10),
    1   D(10,10),DINV(10,10),WKAREA(10)
      READ(105,10)M,L,N,N1
   10 FORMAT(4I3)
      READ(105,20)  ((A(I,J),J=1,L),I=1,M),
    1 ((B(I,J),J=1,N),I=1,L)
   20 FORMAT(4F10.0)
      WRITE(108,30) ((A(I,J),J=1,L),I=1,M)
   30 FORMAT(1H1,T11,'MATRIX A,  ROW-WISE',//,
    1(1P4E13.4))
      WRITE(108,40) ((B(I,J),J=1,N),I=1,L)
   40 FORMAT(1H0,T11,'MATRIX B,  ROW-WISE',//,
    1(1P4E13.4))
      CALL VMULFF (A,B,M,L,N,10,10,C,10,IER)
      WRITE(108,60)((C(I,J),J=1,N),I=1,M)
   60 FORMAT(1H0,T11,'MATRIX C = AB,  ROW-WISE',
    1//,(1P4E13.4))
      READ(105,20) ((D(I,J),J=1,N1),I=1,N1)
      WRITE(108,70) ((D(I,J),J=1,N1),I=1,N1)
   70 FORMAT(1H0,T11,'MATRIX D,  ROW-WISE',//,
    1//,(1P4E13.4))
      CALL LINV1F (D,N1,10,DINV,0,WKAREA,IER1)
      WRITE(108,80) ((DINV(I,J),J=1,N1),I=1,N1)
   80 FORMAT(1H0,T11,'MATRIX DINV,  ROW-WISE',
    1//,(1P4E13.4))
      STOP
      END
```

FIGURE 9.1   A FORTRAN program that uses two IMSL subroutines.

Note: It should be noted that the input unit referred to in the READ statements of the computer program is referenced as 105; in most systems the reference number is 5. Likewise, the output print unit is referenced as 108 in the WRITE statements; in most systems the number is 6. In other respects, the preceding program (Figure 9.1) is written in the standard FORTRAN IV language presented in McCracken (1972) and other leading references.

Example: The program was run with the following matrices as inputs:

$$A = \begin{bmatrix} 1. & 2. \\ 3. & 4. \\ 5. & 6. \end{bmatrix}; \qquad B = \begin{bmatrix} 7. & 8. & 9. & 1. \\ 2. & 3. & 4. & 5. \end{bmatrix}; \qquad D = \begin{bmatrix} 2. & 3. & 4. \\ 1. & 2. & 1. \\ 1. & 2. & 3. \end{bmatrix}$$

Thus, M = 3, L = 2, N = 4, and N1 = 3. The multiplication of matrix A times B occurs in frame 12 of Chapter One. The inverting of matrix D occurs in Example (b) of frame 30 in Chapter Two. Note that the DIMENSION statements in the computer program indicate that the maximum allowable value for each of M, L, N, and N1 is 10.

On program lines 4 and 5, the program calls for the reading in of M, L, N, and N1, where each will appear in a field of three columns width (see 4I3 on line 5). If M = 3 (which is the value of M for matrix A above), then the M field will consist of two spaces followed by the digit 3 in the third column. The program then calls for the entries of the matrices A and B to be entered on input data lines by rows (i.e., row-wise). The program determines that each of these input lines will contain at most four entries, each in F10.0 format (see lines 6, 7, 8 of computer program—Figure 9.1). This format indicates that each entry occupies a field whose width is 10 columns, where the decimal point (if used) can be typed anywhere in the 10 columns.

When the program was run with the preceding matrices and size variables (M, L, etc.), the initial input of data appeared as follows. The question mark (?) is the *prompt* that is typed by the computer when it asks for input data. The data typed after the question mark is entered by the *user* (in this case, the author).

```
?   3  2  4  3
?   1.          2.          3.          4.
?   5.          6.          7.          8.
?   9.          1.          2.          3.
?   4.          5.
```

On input line 1, the values for M, L, N, and N1 appear in columns 3, 6, 9, and 12, respectively. The values for a_{11}, a_{12}, a_{21}, and a_{22} are entered (by the user) after the question mark on the second input line. On input line 3, a_{31}, a_{32}, b_{11}, and b_{12} are entered. The remaining six b_{ij} entries are entered row-wise on input lines 4 and 5.

During execution, the program then calls for the row-wise printing out of matrices A and B, with four entries per line, where each matrix listing is preceded by a heading (for example, "MATRIX A, ROW-WISE"). See lines 9–14 of the FORTRAN program, for the corresponding instructions. Then subroutine VMULFF is called and the product of matrices A and B is produced (labeled as C in the program), and printed out in row-wise fashion, four entries per line. See line 15 and lines 16–18 of the program. The computer printout follows.

```
MATRIX A, ROW-WISE

1.0000E 00    2.0000E 00   3.0000E 00   4.0000E 00
5.0000E 00    6.0000E 00

MATRIX B, ROW-WISE

7.0000E 00    8.0000E 00   9.0000E 00   1.0000E 00
2.0000E 00    3.0000E 00   4.0000E 00   5.0000E 00

MATRIX C = AB, ROW-WISE

1.1000E 01    1.4000E 01   1.7000E 01   1.1000E 01
2.9000E 01    3.6000E 01   4.3000E 01   2.3000E 01
4.7000E 01    5.8000E 01   6.9000E 01   3.5000E 01
```

On output each entry is in 1PE13.4 format. This format means that an output value will occupy a field whose width is 13 columns, where coded scientific notation is used. Five digits of each entry are given, one nonzero digit before the decimal point and four digits after (see 1PE13.4). Thus, the number 2437.6 would be transmitted as 2.4376E 03. The E 03 means that 2.4376 is multiplied by 10^3. (Put differently, 2.4376E 03 means 2.4376 × 10^3, which means 2,437.6. Thus, the E 03 indicates the decimal point should be shifted three places to the right. The "." symbol indicates the prior position of the decimal point before the shift.) From the computer printout we see that 3 × 4 matrix C is given by

$$C = \begin{bmatrix} 11. & 14. & 17. & 11. \\ 29. & 36. & 43. & 23. \\ 47. & 58. & 69. & 35. \end{bmatrix}$$

The printout for matrix C is headed by "MATRIX C = AB, ROW-WISE" in the listing.

Next, the computer prompts (?) for matrix D to be read in (inputted). The computer is carrying out the READ instruction of line 19 of the computer program. After this, the execution of the program calls for matrix D to be printed out row-wise, with four entries per line, each in 1PE13.4 format (see computer program lines 20–22). The listings for matrix D follow:

```
?       2.          3.          4.          1.
?       2.          1.          1.          2.
?       3.
```

```
            MATRIX D, ROW-WISE

    2.0000E 00      3.0000E 00      4.0000E 00      1.0000E 00
    2.0000E 00      1.0000E 00      1.0000E 00      2.0000E 00
    3.0000E 00
```

Next, subroutine LINV1F is called (program line 23), and matrix D^{-1} is calculated. Then the program causes D^{-1} to be printed out (program lines 24–26) row-wise, four entries per line, each entry in 1PE13.4 format. The printout listing is as follows:

```
            MATRIX DINV, ROW-WISE

    2.0000E 00     -5.0000E-01     -2.5000E 00     -1.0000E 00
    1.0000E 00      1.0000E 00       .0000E 00     -5.0000E-01
    5.0000E-01
```

The printout indicates that

$$D^{-1} = \begin{bmatrix} 2. & -.5 & -2.5 \\ -1. & 1. & 1. \\ 0. & -.5 & .5 \end{bmatrix},$$

and this checks with the result in frame 30 of Chapter Two. (In the computer listing, 5.0000E-01 is equal to 5. $\times$ 10^{-1}, or .5⌃. That is, we shift the decimal point one place to the left.)

6. The BASIC programming language was developed in the 1960s at Dartmouth College by Professors J. Kemeny and T. Kurtz. It is probably the easiest programming language to learn, and it has become extremely popular in the last 10 years. This last fact is probably related to the spreading influence of personal computers (microcomputers for the home), for which BASIC is the language of choice of most users. BASIC is a fairly powerful and flexible language that is well suited for problems that do not involve excessive amounts of data.

Because matrix calculations occur so often in practice, most BASIC languages at large computer centers (colleges, industrial companies) have been designed to handle some of the usual matrix operations. This has been done by means of a collection of special matrix statements, all of which contain the term MAT. [Note, however, that the BASIC languages for most personal computers do *not* as of this writing (1982) contain these special matrix statements.]

The following list contains examples of several MAT statements, together with notes of explanation of the effects. In the following, the symbols A, B, and C denote matrices. Also, it is understood that A and B have the proper sizes so that the operation under question (addition, multiplication, etc.) is well defined.

Statement	Explanation
MAT B = INV (A)	Inverse of matrix A is found, and the result is stored in matrix B.
MAT C = A + B	Matrices A and B are added, and the result is stored in matrix C.
MAT C = A − B	Matrix B is subtracted from A, and the result is stored in matrix C.
MAT C = A*B	The product of matrices A and B is formed (in the order shown), and the result is stored in matrix C.
MAT C = TRN (A)	The transpose of matrix A is formed and stored in matrix C.
MAT C = (K)*A	The scalar K (K is an ordinary variable) is multiplied by matrix A, and the result is stored in matrix C.
MAT READ A, B	The entries of matrices A and B are read in row by row (i.e., row-wise) from the contents of DATA statements.
MAT PRINT A, B	The matrices A and B will be printed out row-wise with a maximum of five entries per line. Typically, each entry will occupy a field width of 14 columns.

Example: The BASIC program listed in Figure 9.2 will accomplish the same tasks as the FORTRAN program shown in Figure 9.1, when the latter is run with the input data cited in the Example of frame 5.

The actual results of running the BASIC program are shown in Figure 9.3 (output printed on time-sharing terminal). Let us describe the functioning of the BASIC computer program of Figure 9.2. In the DIM statement (statement 10), the matrices A, B, C, D, and E are dimensioned in exactly the way in which they occur in the running of the program (this differs from what we had in the FORTRAN program of frame 5). Here E is to be the inverse of matrix D, once the latter is calculated. Next, the program calls for matrices A, B, and D to be read in from the DATA statements (statements 20, 150, 160). In this program, the entries of A and B are successively listed, row-wise, in DATA statement 150, and the entries of D are listed row-wise in DATA statement 160. Next, the program calls for the printing out of matrices A and B, where each is preceded by a heading (statements 30, 40, 50, 60). Both A and B are listed at the top part of Figure 9.3. Since A is given as 3×2 in the DIM statement, it is printed out on three lines, with two entries per line. That is, matrix A appears in the same way as when it is written in matrix form. See the beginning of the Example of frame 5. Similarly, for the other matrices cited in the BASIC program.

Next, the program calls for the matrix product C = AB to be computed (statement 70). Then the program calls for the printing out of matrix C. Refer again to Figure 9.3 for the actual printout of matrix C when the BASIC program is run.

Next, the program calls for printing out matrix D, preceded by a heading (statements 100, 110). In statement 120, the program calls for matrix E to be computed, where $E = D^{-1}$. Then the program calls for matrix E to be printed out, preceded by the appropriate heading (statements 130, 140). Refer again to Figure 9.3 for the printout of matrix D^{-1} when the program is run.

It is constructive to compare the results from the running of the FORTRAN program (the Example in frame 5) and the running of the BASIC program.

```
10 DIM A(3,2),B(2,4),C(3,4),D(3,3),E(3,3)
20 MAT READ A,B,D
30 PRINT "MATRIX A"
40 MAT PRINT A
50 PRINT "MATRIX B"
60 MAT PRINT B
70 MAT C = A*B
80 PRINT "MATRIX C"
90 MAT PRINT C
100 PRINT "MATRIX D"
110 MAT PRINT D
120 MAT E = INV (D)
130 PRINT "MATRIX D INVERSE"
140 MAT PRINT E
150 DATA 1,2,3,4,5,6,7,8,9,1,2,3,4,5
160 DATA 2,3,4,1,2,1,1,2,3
170 END
```

FIGURE 9.2 A BASIC program containing MAT statements.

```
MATRIX A

   1              2

   3              4

   5              6
MATRIX B

   7              8              9              1

   2              3              4              5
MATRIX C

  11             14             17             11

  29             36             43             23

  47             58             69             35
MATRIX D

   2              3              4

   1              2              1

   1              2              3
MATRIX D INVERSE

   2          -.500000       -2.50000

  -1              1              1

   0          -.500000        .500000
```

FIGURE 9.3   Printouts that result when the BASIC program of Figure 9.2 is run.

7.   Consider again a person who wishes to use a computer to do matrix calculations. The question naturally arises as to what a computer user should do if his computer does not have available either the IMSL FORTRAN subroutines, or provisions for the MAT statements in the BASIC version available on the computer. These are certainly realistic questions, particularly for a person with access to a popular home computer (such as a Radio Shack TRS-80, Apple, Commodore, Sinclair, Texas Instrument, or Atari brand). One way of solving the problem is by writing programs that will accomplish matrix operations such as matrix addition, multiplication, inversion, and so on. Each of these operations can be programmed by making use of various looping procedures, such as "DO" loops in the FORTRAN language or "FOR–NEXT" loops in the BASIC language.

Actually, since matrix manipulations are so important and common in science, engineering, and business, many books are available in which programs for such manipulations appear. Thus, the computer user does not even have to expend effort to write a program if he knows where to look.

For FORTRAN programs dealing with matrix and linear algebra, the reader is referred to Carnahan, Luther, and Wilks (1969), Conte and DeBoor (1972), Gerald (1978), Johnson and Riess (1981), and Lipschutz and Poe (1978).

The following books contain BASIC programs relating to matrix and linear algebra: Agnew and Knapp (1978), Caton and Grossman (1980), Gottfried (1982), and Williams (1978).

Other FORTRAN subroutines (besides those cited in frame 5) are available for the person interested in solving linear algebra problems on the computer. The collection of FORTRAN subroutines known as LINPACK was made available in 1979, and it is useful for solving many problems of numerical linear algebra. A magnetic tape containing the FORTRAN subroutines cost $75 as of 1979, and it is available from the International Mathematical and Statistical Libraries, Inc. (IMSL, Inc. for short—their address and phone number are listed at the beginning of frame 5). A *User's Guide* for LINPACK, prepared by Dongarra, Bunch, Moler, and Stewart is available from

Society for Industrial and Applied Mathematics (SIAM)
33 South 17th Street
Philadelphia, Pa. 19103
Telephone: (215) 564-2929

A collection of subroutines dealing with eigenvalue problems is entitled EISPACK. It also cost about $75 in 1979 and is available from IMSL, Inc. The EISPACK collection is documented in the second edition of Volume 6 of *Lecture Notes in Computer Science,* available from the Springer-Verlag publishing company (175 Fifth Ave.; New York NY 10010).

8. Personal computer software items, which deal with linear algebra, are available from several companies that supply products for personal computers. By the way, the term *software* means any of the following:

(a) Program or programs stored in coded form on a cassette tape.
(b) Program or programs stored in coded form on a diskette (also known as a floppy disk). A diskette is a type of disk (there are also disks known as hard disks). For a disk, information is stored in a magnetic coating.
(c) A listing of a program in BASIC, FORTRAN, or some other language. This is a so-called *source* listing.
(d) A program stored on a module containing one or more ROM chips. (Such modules are featured by the Texas Instrument TI-99/4 home computer. ROM stands for read only memory, which means a computer can read from such memory but cannot write new information into it.)

There are also other forms of software, but the preceding list includes the most typical forms (ca. 1982, anyway).

If one wants to learn about available personal computer software for linear algebra or other fields, one should investigate the various magazines and catalogs that deal with personal computers. Such sources provide extensive amounts of such information. Several of the more popular magazines are *Popular Computing* (formerly *On Computing*), *BYTE, Personal Computing, Recreational Computing, Creative Computing,* and *Soft Side Magazine.*

To be specific, we shall list and describe several definite software packages, which are useful for linear and matrix algebra calculations.

A specific software package that consists of 100 programs dealing with mathematics, statistics, business, etc., is

Master Pac 100 (2nd Ed.)
developed by
H & E Computronics Inc.
50 N. Pascack Road
Spring Valley, NY 10977

Programs 50–54 and 56 of Master Pac 100 deal specifically with matrix calculations. The Master Pac 100 package is available on cassette or diskette for the TRS-80 computer and on diskette for the Apple computer.

A software package entitled *Matrix Routines,* which includes routines to multiply and invert matrices, find determinants, solve a system of linear equations, and compute the dot product is available from

Benchmark Computing Services
P.O. Box 385R
Providence, Utah 84322

It is available either on a cassette tape or as a source listing for the TRS-80 computer.

The package entitled *Matrix Algebra,* which is available from the CONDUIT organization (address listed below), has catalog number MTH267P. It consists of several CAI (Computer Assisted Instruction) units designed for the Commodore Pet personal computer. The units deal with matrix addition, subtraction, scalar and matrix multiplication, matrix inversion, and solution of a linear system of equations. The package consists of either a cassette or diskette, plus an instructor's manual.

This package and others dealing with mathematics and the natural and social sciences were listed in the fall, 1981 catalog, *Pipeline,* published by

CONDUIT
P.O. Box 388
Iowa City, Iowa 52244

C. USING A PROGRAMMABLE CALCULATOR MODULE FOR MATRIX ALGEBRA CALCULATIONS

9. We shall focus our attention on the capabilities of the Texas Instruments (TI) 58C or 59 programmable calculator with respect to matrix algebra computations. Either of these calculators (I own a TI 58C) comes with a removable *Master Library Module*. This module has 25 programs built into it, and two of the programs are specifically concerned with matrix calculations:

ML-02: Matrix Inversion, Determinants, and Simultaneous Equations.
ML-03: Matrix Addition and Multiplication.

Let us digress to discuss the software and physical aspects of this module. The Master Library Module has 25 programs stored in coded form within a solid-state "chip" of silicon (similar in construction to the silicon integrated circuit that is the major component of a calculator). If the module is plugged into the calculator, then these library programs are easily accessed by pressing a few keys on the keyboard in a prescribed order. Even though the library programs can be incorporated into other programs that the user might wish to write, they can also be used on their own to carry out certain calculations. Our emphasis here will deal with this second aspect.

It should be noted that Texas Instruments sells other modules that can be used with the TI Programmable 58C (or 59) calculator. Some of the subjects the other modules deal with are applied statistics, business decisions, mathematics/utilities, electrical engineering, etc. One can learn more about the features of the programmable calculators and available modules by phoning the following toll-free telephone number: 1-800-527-3570.

Note: It is a very easy matter to install and remove a module from the calculator. The manual entitled *TI programmable 58C/59 Master Library* (1979), available from Texas Instruments, Inc. describes the functioning of the 25 programs of the Master Library Module.

We shall explain shortly how to use the ML-02 program of the Master Library Module to find the determinant and inverse of a matrix, and for solving a system of n linear equations in n unknowns.

The keyboard of the TI 58C calculator is shown in Figure 9.4 (with respect to module usage, the keyboard of the TI 59 calculator is identical).

Some comments on the basic operation of the calculator are in order. The ON/OFF switch is on the upper left part of the calculator. The calculator has 45 keys arranged in nine rows, each row having five keys. The "2nd" key (first key in second row) is similar to a "Shift" key on a typewriter or terminal. Each key has a symbolic expression written on it, and each key (except for the "2nd", "INV", "CLR", and "R/S" keys) has a symbolic expression written above it. See Figure 9.4. For example, in the first row, the key labeled "A" has "A'" written above it. In the third row, the key labeled "LRN" has "Pgm" written above it.

FIGURE 9.4　The Keyboard of the TI 58C Programmable Calculator. (Courtesy of Texas Instruments, Incorporated.)

If we first press the key labeled "2nd" and then press another key, then this in effect causes the latter key to have the function described by the symbolic expression that is written *above* the key. For example, if we press the "2nd" key and "LRN" key in succession, this *converts* the latter key to be the "Pgm" key. If we press the "2nd" key and "A" key in succession, this converts the latter key to be the "A'" key.

We will indicate that the pressing of a key has occurred by enclosing the key symbol with a box. Thus, $\boxed{\text{R/S}}$ means that the "R/S" key has been pressed (by the way, "R/S" is the code symbol for "Run/Stop"). If we write $\boxed{\text{2nd}}\ \boxed{\text{Pgm}}$, this means that we have pressed the "2nd" and "LRN" keys in succession. (Remember, pressing "2nd" and then "LRN" will *activate* the "Pgm" expression written above the "LRN" key.) Thus, whenever you see $\boxed{\text{2nd}}$, this means that next boxed symbol will appear on the keyboard *above* the key that is actually pressed.

Thus, 2nd B′ means that the "2nd" and "B" keys are pressed in succession, thereby activating the "B′" expression.

There are 10 keys for the digits 0, 1, 2, 3, . . ., 8, 9 and these appear in the bottom four rows. We will *not* use a box (i.e., □) to indicate the pressing of a digit key but will simply write the digit.

In Figure 9.4 the *display* is at the top of the calculator, under the words *Texas Instruments*. The number 2.7182818 58 is written in the display in Figure 9.4. If the last two digits in the display are shown separated from the preceding digits by a blank space (as in Figure 9.4) or a minus sign ($-$), this means that scientific notation is in effect. (Scientific notation can be activated by pressing the "EE" key.) Thus, in Figure 9.4, the number shown in the display is equal to 2.7182818×10^{58}. If the display shows $5.439 - 02$, this means that the indicated number is equal to 5.439×10^{-2}, or .05439.

Example: We will demonstrate how the Master Library Module can be used to find the determinant and inverse of the matrix

$$D = \begin{bmatrix} 2 & 3 & 4 \\ 1 & 2 & 1 \\ 1 & 2 & 3 \end{bmatrix}.$$

This is the matrix whose inverse was found in frames 5 and 6 by using FORTRAN and BASIC programs, respectively. We originally encountered this matrix in Example (b) of frame 30 of Chapter 2. As we know, the inverse is given by

$$D^{-1} = \begin{bmatrix} 2 & -.5 & -2.5 \\ -1 & 1 & 1 \\ 0 & -.5 & .5 \end{bmatrix}.$$

The calculation of D^{-1} by using the TI 58C (or 59) calculator in conjunction with the Master Library Module is indicated by the listing in Table 9.1. On the first line, 2nd Pgm 0 2 indicates that the program ML-02 is activated. As noted before, this program allows one to calculate the determinant and inverse of a matrix and to find the solution of a system of n linear equations in n unknowns. (When one writes 2nd Pgm 0 2, this means that first the "2nd" key is pressed, then the "LRN" key is pressed, and finally the digit keys "0" and "2" are pressed in succession.)

To *enter* a negative number, the "$+/-$" key (see bottom row) is pressed *after* the keys for the digits (and decimal point, if any) are pressed. Thus, to enter -6.25, we would press keys for "6", ".", "2", "5", and "$+/-$", in that order.

TABLE 9.1 Computing the Determinant and Inverse of a Matrix with the TI 58C Master Library Module

Enter	Press	Display	Comments
	2nd Pgm 0 2		Select program ML-02.
3	A	3.	This is n ($n = 3$ here).
1	B	1.	Start with column 1.
2	R/S	2.	d_{11}
1	R/S	1.	d_{21}
1	R/S	1.	d_{31}
3	R/S	3.	d_{12} The entries of matrix D are
2	R/S	2.	d_{22} entered by *columns* (i.e., col-
2	R/S	2.	d_{32} umn-wise).
4	R/S	4.	d_{13}
1	R/S	1.	d_{23}
3	R/S	3.	d_{33}
	2nd E'	2.	Determinant $|D|$. Here $|D| = 2$.
1	2nd C'	1.	Calculate D^{-1}, starting with column 1.
	R/S	2.	$(D^{-1})_{11}$
	R/S	−1.	$(D^{-1})_{21}$
	R/S	0.	$(D^{-1})_{31}$
	R/S	−0.5	$(D^{-1})_{12}$
	R/S	1.	$(D^{-1})_{22}$ The entries of matrix D^{-1} are
	R/S	−0.5	$(D^{-1})_{32}$ fed out by columns.
	R/S	−2.5	$(D^{-1})_{13}$
	R/S	1.	$(D^{-1})_{23}$
	R/S	0.5	$(D^{-1})_{33}$

Notes: (a) We use the symbol $(D^{-1})_{ij}$ to denote the entry of D^{-1} in row i and column j. Thus, $(D^{-1})_{13} = -2.5$ in Table 9.1. (b) The accuracy of the TI 58C or 59 calculator is effectively eight significant digits if scientific notation is used to represent numbers.*

10. The Master Library Module program ML-02 can also be used to find the solution of a system of n linear equations in n unknowns. In vector terms, the program will solve $A\mathbf{x} = \mathbf{b}$ for $\mathbf{x}$ if $\mathbf{b}$ is given (A symbolizes an $n \times n$ matrix).

---

* Permission to list results generated by using the TI 58C calculator in conjunction with Master Library Module program ML-02 was granted by Texas Instruments, Inc.

TABLE 9.2 Solving a System of n Linear Equations in n Unknowns with the TI 58C Master Library Module

Enter	Press	Display	Comments
	2nd Pgm 0 2		Select program ML-02
3	A	3.	This is n ($n = 3$ here).
1	B	1.	Start with column 1.
2	R/S	2.	d_{11}
1	R/S	1.	d_{21}
1	R/S	1.	d_{31}
3	R/S	3.	d_{12}
2	R/S	2.	d_{22} The entries of matrix D are entered by columns.
2	R/S	2.	d_{32}
4	R/S	4.	d_{13}
1	R/S	1.	d_{23}
3	R/S	3.	d_{33}
	C	2.	Determinant $|D|$. Here $|D| = 2$.
1	D	1.	Next, enter components of **b**, starting with b_1.
− 3.75	R/S	− 3.75	b_1
− 3.875	R/S	− 3.875	b_2 The components of **b** are entered.
− 2.375	R/S	− 2.375	b_3
	CLR E	1.	Calculate **x**.
1	2nd A'	1.	Display components of **x**, starting with x_1.
	R/S	0.375	x_1
	R/S	− 2.5	x_2 The components of **x** are fed out.
	R/S	0.75	x_3

Example: We list the results of using program ML-02 to solve $D\mathbf{x} = \mathbf{b}$ for **x** if D is the 3×3 matrix in the Example of frame 9, and $\mathbf{b} = [-3.75, -3.875, -2.375]^t$. The reader is referred to Table 9.2 for the displays of input data and output results. The answer is revealed to be $\mathbf{x} = [.375, -2.5, .75]^t$.

The beauty of the TI 58C programmable calculator (together with its Master Library Module) is that extensive calculating power is available in this very compact machine, and at a relatively low price (under $80.00 at some stores as of late 1982).

A programmable calculator can properly be categorized as a tiny computer, smaller in capacity than so-called microcomputers such as the TRS-80 or Apple, but nevertheless having the characteristics of a computer.

Another company that builds programmable calculators is Hewlett-Packard. One of the most popular Hewlett-Packard programmable calculators is the HP-41C (or the updated HP-41CV) model, and this very powerful and useful machine also has the capability of being used in conjunction with modules. Some of the plug-in Application Modules that Hewlett-Packard makes are in the following categories: mathematics, statistics, financial decisions, securities, surveying, and so on. The plug-in Application Module in mathematics is especially of interest to us, since it contains routines dealing with matrix operations.

To acquire more information on the Hewlett-Packard programmable calculators (or the HP-85 personal computer), one should write to

Hewlett-Packard
Corvallis Division
1000 N.E. Circle Blvd.
Corvallis, Oregon 97330

SELF-TEST

This Self-Test will help you determine whether or not you have mastered the chapter objectives. Correct answers are given at the end of the test.

For questions 1, 2, and 3, find solutions, if they exist, for the system of equations given. Use Gaussian elimination (Method 9.1). List the row echelon form obtained by applying Method 9.1 to the augmented coefficient matrix of the original system.

1.

$$4x_1 - 8x_2 + x_3 = 18$$
$$x_1 - 2x_2 - x_3 = 2$$
$$3x_1 - 6x_2 - 2x_3 = 8$$

This same system occurs in question 5 of the Self-Test of Chapter Two.

2.

$$2x_2 - 2x_3 = -8$$
$$x_1 + x_2 + x_3 = 2$$
$$x_1 + 2x_2 = 4$$

This system occurs in question 4 of the Self-Test of Chapter Two.

3.

$$2x_1 - 4x_2 + 3x_3 = -8$$

$$3x_1 + 5x_2 - 4x_3 = 4$$

$$4x_1 - 6x_2 + 5x_3 = -12$$

This same system occurs in questions 1 and 3 of the Self-Test of Chapter Two.

4.   For the system of question 3, apply Gaussian elimination with pivoting (Method 9.2) to find the solution. List the row echelon form obtained.

5.   Find the solution to the system below by using ordinary Gaussian elimination (Method 9.1), with rounding to three significant digits. List the row echelon form obtained.

$$.075x_1 - .05x_2 + x_3 = 3.7$$

$$5x_1 - 2.5x_2 - 1.75x_3 = 1.2$$

$$1.5x_1 + 3.5x_2 - .75x_3 = 3.8$$

6.   Repeat question 5 by using Gaussian elimination with pivoting (Method 9.2).

ANSWERS TO SELF-TEST

If your answers to the test questions do not agree with the following answers, review the frames indicated in parentheses after each answer.

1.   The row echelon form obtained is

$$\begin{bmatrix} 1 & -2 & \frac{1}{4} & \Big| & \frac{18}{4} \\ 0 & 0 & 1 & \Big| & 2 \\ 0 & 0 & 0 & \Big| & 0 \end{bmatrix}$$

Solution is given by $x_1 = 2r + 4$, $x_2 = r$, $x_3 = 2$, where r is arbitrary.

(frames 1, 2)

2.   A partial application of Method 9.1 leads to a third row of the form $0\ \ 0\ \ 0\,|\,6$. This indicates that the system has no solution. The row echelon form obtained is

$$\begin{bmatrix} 1 & 1 & 1 & \Big| & 2 \\ 0 & 1 & -1 & \Big| & -4 \\ 0 & 0 & 0 & \Big| & 1 \end{bmatrix}$$

(frames 1, 2)

3. The row echelon form obtained is

$$\begin{bmatrix} 1 & -2 & \frac{3}{2} & -4 \\ 0 & 1 & -\frac{17}{22} & \frac{16}{11} \\ 0 & 0 & 1 & 2 \end{bmatrix}$$

Solution is given by $x_1 = -1$, $x_2 = 3$, $x_3 = 2$.   (frames 1, 2)

4. The row echelon form obtained is

$$\begin{bmatrix} 1 & -\frac{3}{2} & \frac{5}{4} & -3 \\ 0 & 1 & -\frac{31}{38} & \frac{26}{19} \\ 0 & 0 & 1 & 2 \end{bmatrix},$$

and this is different from the form obtained in question 3. Of course, the solution is the same: $x_1 = -1$, $x_2 = 3$, $x_3 = 2$.   (frame 3)

5. The row echelon form is

$$\begin{bmatrix} 1 & -.667 & 13.3 & 49.3 \\ 0 & 1 & -81.8 & -293. \\ 0 & 0 & 1 & 3.60 \end{bmatrix}$$

The approximate solution is $x_1 = 2.41$, $x_2 = 1.48$, $x_3 = 3.60$, and this is far from the exact solution $x_1 = 2$, $x_2 = 1$, $x_3 = 3.6$.   (frames 1, 2, 4)

6. The row echelon form is

$$\begin{bmatrix} 1 & -.5 & -.35 & .24 \\ 0 & 1 & -.0529 & .809 \\ 0 & 0 & 1 & 3.58 \end{bmatrix},$$

and the approximate solution here is $x_1 = 1.99$, $x_2 = .998$, $x_3 = 3.58$. This is fairly close to the exact solution $x_1 = 2$, $x_2 = 1$, $x_3 = 3.6$.   (frames 1–3)

REFERENCES

Agnew, J. and Knapp, R. C. *Linear Algebra with Applications* (Belmont, Calif.: Wadsworth, 1978).

Anton, H. *Elementary Linear Algebra,* 2nd Ed. (New York: Wiley, 1977).

Anton, H., and Kolman, B. *Applied Finite Mathematics* (New York: Academic Press, 1978).

Bloch, N. J., and Michaels, J. G. *Linear Algebra* (New York: McGraw-Hill, 1977).

Carnahan, B., Luther H. A., and Wilks, J. O. *Applied Numerical Methods* (New York: Wiley, 1969).

Caton, G., and Grossman, S. I. *A Computer Supplement to Linear Algebra: BASIC* (Belmont, Calif.: Wadsworth, 1980).

Conte, S. D., and DeBoor, C. W. *Elementary Numerical Analysis: An Algorithmic Approach* (New York: McGraw-Hill, 1972).

Cooper, L., and Steinberg, D. *Methods and Applications of Linear Programming* (Philadelphia: Saunders, 1974).

Fadeev, D. K., and Fadeeva, V. N. *Computational Methods of Linear Algebra* (San Francisco: Freeman, 1963).

Fox, L. *An Introduction to Numerical Linear Algebra* (New York: Oxford University Press, 1965).

Gerald, C. *Applied Numerical Analysis,* 2nd Ed. (Reading, Mass.: Addison-Wesley, 1978).

Gottfried, B. S. *Programming with BASIC,* 2nd Ed. (New York: McGraw-Hill, Schaum's Outline Series, 1982).

Grossman, S. I. *Elementary Linear Algebra* (Belmont, Calif.: Wadsworth, 1980a).

Grossman, S. I. *Applications for Elementary Linear Algebra* (Belmont, Calif.: Wadsworth, 1980b).

Hadley, G. *Linear Algebra* (Reading, Mass.: Addison-Wesley, 1961).

Hoffman, K., and Kunze, R. *Linear Algebra,* 2nd Ed. (Englewood Cliffs, N.J.: Prentice-Hall, 1971).

Johnson, L. W., and Riess, R. D. *Introduction to Linear Algebra* (Reading, Mass.: Addison-Wesley, 1981).

Kemeny, J. G., and Snell, J. L. *Finite Markov Chains* (New York: Springer Verlag, 1976).

Kolman, B. *Introductory Linear Algebra with Applications,* 2nd Ed. (New York: Macmillan, 1980).

Lipschutz, S. *Linear Algebra* (New York: McGraw-Hill, Schaum's Outline Series, 1968).

Lipschutz, S., and Poe, A. *Programming with FORTRAN* (New York: McGraw-Hill, Schaum's Outline Series, 1978).

McCracken, D. D. *A Guide to FORTRAN IV Programming,* 2nd Ed. (New York: Wiley, 1972).

Noble, B., and Daniel, J. W. *Applied Linear Algebra,* 2nd Ed. (Englewood Cliffs, N.J.: Prentice-Hall, 1977).

O'Nan, M. *Linear Algebra,* 2nd Ed. (New York: Harcourt Brace Jovanovich, 1976).

Rorres, C., and Anton, H. *Applications of Linear Algebra*, 2nd Ed. (New York: Wiley, 1979).

Rothenberg, R. I. *Finite Mathematics* (New York: Wiley, Self-Teaching Guides, 1980).

Rothenberg, R. I. *Linear Programming* (New York: Elsevier North Holland, 1979).

Scheid, F. *Numerical Analysis* (New York: McGraw-Hill, Schaum's Outline Series, 1968).

Searle, S. R. *Matrix Algebra for the Biological Sciences* (New York: Wiley, 1966).

Searle, S. R., and Hausman, W. H., et al. *Matrix Algebra for Business and Economics* (New York: Wiley, 1970).

Shields, P. C. *Elementary Linear Algebra,* 3rd Ed. (New York: Worth, 1980).

Strang, G. *Linear Algebra and its Applications,* 2nd Ed. (New York: Academic Press, 1980).

Williams, G. *Computational Linear Algebra with Models,* 2nd Ed. (Boston: Allyn and Bacon, 1978).

Wylie, C. R. *Advanced Engineering Mathematics,* 4th Ed. (New York: McGraw-Hill, 1975).

Yaqub, A., and Moore, H. G. *Elementary Linear Algebra with Applications* (Reading, Mass.: Addison-Wesley, 1980).

CROSS-REFERENCE CHART TO SOME POPULAR LINEAR ALGEBRA TEXTBOOKS

The Textbooks

Agnew, J., and Knapp, R. C. *Linear Algebra with Applications* (Monterey, Calif.: Brooks/Cole, 1978).

Anton, H. *Elementary Linear Algebra,* 2nd Ed. (New York: Wiley, 1977).

Bloch, N. J., and Michaels, J. G. *Linear Algebra* (New York: McGraw-Hill, 1977).

Grossman, S. I. *Elementary Linear Algebra* (Belmont, Calif.: Wadsworth, 1980).

Kolman, B. *Introductory Linear Algebra with Applications,* 2nd Ed. (New York: Macmillan, 1980).

O'Nan, M. *Linear Algebra,* 2nd Ed. (New York: Harcourt Brace Jovanovich, 1976).

Shields, P. C. *Elementary Linear Algebra,* 3rd Ed. (New York: Worth, 1980).

Yaqub, A., and Moore, H. G. *Elementary Linear Algebra with Applications* (Reading, Mass.: Addison-Wesley, 1980).

Chapter in this book	Agnew & Knapp	Anton	Bloch & Michaels	Grossman	Kolman	O'Nan	Shields	Yaqub & Moore
1. Elementary Matrix Algebra	1	1	1	2	1	2	2	1
2. Linear Systems	2	1	1	1	1	1	1	1
3. Determinants	3	2	4	3	2	3	2	2
4. Geometrical Properties of the Vector Space R^n	4	3	2, 3, & 6	4	3	2, 6	3, 4	3, 4
5. Vector Spaces	4, 6	4	3	5	3	4	3	3, 4
6. Linear Transformations	5	5	4	6	4	5	6	5
7. Eigenvalues and Eigenvectors	3, 7, & 8	6	5	7	5	7	5	6
8. Selected Applications	ⓐ	7	ⓐ	ⓐ	6, 7	ⓐ	7 & ⓐ	7 & ⓐ
9. Numerical, Computer, and Calculator Methods	ⓑ	8	—	8	8	—	1, 2, & 5	—

Notes for Chart:
ⓐ Applications are scattered throughout the book.
ⓑ Numerical methods and BASIC programs are scattered throughout the book.

Index